Saas-Fee Advanced Course 45

More information about this series at http://www.springer.com/series/4284

Philip J. Armitage · Wilhelm Kley

From Protoplanetary Disks to Planet Formation

Saas-Fee Advanced Course 45

Swiss Society for Astrophysics and Astronomy
Edited by Marc Audard, Michael R. Meyer
and Yann Alibert

Springer

Authors
Philip J. Armitage
JILA
University of Colorado and NIST
Boulder, CO, USA

Wilhelm Kley
Institute of Astronomy and Astrophysics
University of Tübingen
Tübingen, Germany

Volume Editors
Marc Audard
Department of Astronomy
University of Geneva
Geneve, Switzerland

Michael R. Meyer
Department of Astronomy
University of Michigan
Ann Arbor, MI, USA

Yann Alibert
Physics Institute
University of Bern
Bern, Switzerland

This Series is edited on behalf of the Swiss Society for Astrophysics and Astronomy: Société Suisse d'Astrophysique et d'Astronomie, Observatoire de Genève, ch. des Maillettes 51, CH-1290 Versoix, Switzerland.

ISSN 1861-7980 ISSN 1861-8227 (electronic)
Saas-Fee Advanced Course
ISBN 978-3-662-58686-0 ISBN 978-3-662-58687-7 (eBook)
https://doi.org/10.1007/978-3-662-58687-7

Library of Congress Control Number: 2018966857

Cover credit: NASA/JPL-Caltech/T. Pyle (SSC) Caption: Artist's concept of a protoplanetary disk, or planet-forming, around a young star with a gap caused by the presence of forming gas giant planets

This Springer imprint is published by the registered company Springer-Verlag GmbH, DE part of Springer Nature
The registered company address is: Heidelberger Platz 3, 14197 Berlin, Germany

Preface

Is the Sun and its planetary system special? How did the solar system form? Are there similar systems in the Galaxy? How common are habitable planets? What processes take place in the early life of stars and in their surrounding circumstellar disks that could impact whether life emerges?

The fields of exoplanet research and planet formation have exploded in the past 20 years, thanks largely to the discovery of the first exoplanet by Swiss astronomers, Michel Mayor and Didier Queloz, in 1995. Furthermore, the technological advancements in ground and space observatories have allowed us to study in detail protoplanetary disks where pebbles, rocks, and protoplanets are created. While exoplanet research initially focussed on the detection of planets, going from large Jupiter-mass objects down to Earth-size planets, revealing a wide variety of masses, eccentricities, semi-major axes, etc., and the need to understand their formation became a necessity, fostering interdisciplinary research, linking diverse research communities.

The 45th Saas-Fee Advanced Course "From Protoplanetary Disks to Planet Formation" of the Swiss Society for Astrophysics and Astronomy (SSAA) took place in the alpine resort of Les Diablerets, in the canton of Vaud, from March 15 to 20, 2015, gathering 87 participants from different countries, albeit with a significant Swiss contingent thanks to the Swiss National Center of Competence in Research Planets, established in 2014 to provide an interdisciplinary research program dedicated to the study of the origin, evolution, and characterization of planets. Three lecturers presented three different aspects of planet formation: Prof. Phil Armitage (University of Colorado) focussed on the processes in protoplanetary disks, while Prof. Willy Kley (University of Tübingen) focussed on planet formation and disk-planet interactions. The observational properties of protoplanetary disks were presented by Dr. Leonardo Testi (ESO), at a time when the first ALMA results, such as the exquisite image of the disk of HL Tau, were coming out. The lecturers prepared each eight presentations describing different aspects pertaining to their chapter. The presentations, and the video records of the lectures, are available on the course Web site, which can be found from the Web site of the Swiss Society for Astrophysics and Astronomy, `http://www.ssaa.ch/`.

In a happy coincidence, a partial solar eclipse took place during the course, with the maximum coverage occurring during the coffee break. Organizers and participants gladly brought observing material, from mylar sheets to small telescopes with Hα filters, sharing them with fellow astronomers and passers-by going to ski. In the traditional spirit of the Saas-Fee courses, participants had their afternoon free to go skiing, snowshoeing, shopping and interact with their fellow participants, either informally or scientifically. We further organized activities for the participants, allowing them to discover curling and snowshoeing. The conference banquet closed the course with a Swiss cheese fondue in the chalet close to Eurotel Victoria Hotel, the official hotel of the conference (while the lectures were taking place at the nearby Salle des Conférences), providing a cozy and Swiss ambiance.

In the present book, Armitage and Kley provide chapters based on their lectures given during the course. Unfortunately, the observational aspects could eventually not be included in this book. However, clearly, the fast-changing field of observations of protoplanetary disks would have made this chapter out of date even before its publication.

Group photograph in front of the Massif des Diablerets. *Credit* M. Audard

Group photograph of the organizing committee and the speakers. From left to right: M. Audard, Y. Wang, P. Armitage, M. Meyer, Y. Alibert, M. Logossou, L. Testi, W. Kley, A. Baleisis. *Credit* M. Audard

Participants looking at the partial solar eclipse on March 20, 2015, the maximum coverage conveniently occurring during the coffee break. *Credit* M. Audard

List of participants of the 45th Saas-Fee Advanced Course

AKIMKIN	Vitaly	MATTHEWS	Elisabeth
AL-MAOTAHNI	Anas	MEYER	Michael
ALIBERT	Yann	MIOTELLO	Anna
ALI DIB	Mohamad	MOHANDAS	Gopakumar
ALLART	Romain	MOTALEBI	Fatemeh
ANSDELL	Megan	NINAN	Joe
ARMITAGE	Philip	NOH	Hyerim
ATAIEE	Sareh	ONO	Tomohiro
AUDARD	Marc	OTTONI	Gaël
BERLOK	Thomas	PARKER	Richard
BIANCHI	Eleonora	PERETTI	Sébastien
BØGELUND	Eva	PERICAUD	Jessica
BONEBERG	Dominika	PICOGNA	Giovanni
CALVO	Flavio	POHL	Adriana
CERSULLO	Maria Federica	PRIVITERA	Giovanni
COFFINET	Adrien	QUIROGA-NUÑEZ	Luis Henry
CONOD	Uriel	RAB	Christian
DELISLE	Jean-Baptiste	RAIMBAULT	Manuela
ENGLER	Natalia	RAZDOBURDIN	Dmitry
FACCHINI	Stefano	RIBAS	Álvaro
FERRIZ-MAS	Antonio	RICHERT	Alexander
FLETCHER	Mark	RIGLIACO	Elisabetta
FROSTHOLM	Troels	ROLOFF	Victoria
GARUFI	Antonio	SALINAS	Vachail
GEORGAKARAKOS	Nikolaos	SCHREIBER	Andreas
GREENWOOD	Aaron	SENECAL	Luc
GUIDI	Greta	SERPOLLIER	Cléa
HAGELBERG	Janis	SOTIRIADIS	Sotiris
HAKIM	Kaustubh	STEINER	Daniel
HANDS	Thomas	STOLL	Moritz
HENG	Kevin	SWOBODA	David
HONDA	Mitsuhiko	TAMBURELLO	Valentina
KARSKA	Agata	TAZZARI	Marco
KASPER	David	TEAGUE	Richard
KLEY	Willhelm	TESTI	Leonardo
KLOSTER	Dylan	THIABAUD	Amaury
LAVIE	Baptiste	TOCKNELL	James
LI	Jian	TSAI	Shang-Min
LICHTENBERG	Tim	UDRY	Stéphane

(continued)

(continued)

MALIK	Matej	VENTURINI	Julia
MANZO MARTÍNEZ	Ezequiel	WANG	Yuan
MARBOEUF	Ulysse	WEAVER	Ian
MARSCHALL	Raphael	WYTTENBACH	Aurélien
MARTIN	David		

Geneve, Switzerland
Ann Arbor, USA
Bern, Switzerland
November 2018

Marc Audard
Michael R. Meyer
Yann Alibert

Acknowledgements

We thank all three lecturers for their excellent work to prepare, present, and publish their lectures. We are indebted to Martine Logossou for her participation in the organization of the course: Thanks to her experience and skills, the conference went flawlessly. We thank Yuan Wang for providing further help, in particular the recording of the lectures, and Marie-Claude Dunand for additional help to prepare the material for the course. Additional thanks to Françoise Page for her patience in keeping the account of the course open as long as needed and to Audra Baleisis for her pedagogical input which included collecting feedback from the participants helping the lecturers improve their presentations.

We thank Switch for providing the recording material, Dimitri Cosetto at the Commune of the Les Diablerets for his logistic help at the Salle des Conférences, and the team of the Eurotel Victoria for providing a nice and cozy environment for the participants.

Finally, we thank wholeheartedly the Swiss Society for Astronomy and Astrophysics and the Swiss Academy of Sciences (ScNat) for selecting and funding this course, the NCCR PlanetS for providing significant financial help to reduce the registration fees and providing fee waivers to some participants, and we thank the Société Académique de Genève for further financial help to cover costs for the course and its publication.

The three speakers in front of the Massif des Diablerets. *Credit* M. Audard

Contents

List of Figures

Fig. 2.9 Example of the runaway growth process from planetesimals to planetary embryos using the *N*-body method as described by Kokubo [135]. The basic setup and initial conditions are given in the main text in Sect. 2.3.2.1. The left panel shows the eccentricity and distance distribution of the bodies for the initial setup (top panel) and at times 100,000 and 200,000 yrs (middle and bottom). The right panel shows the cumulative mass distribution of the formed bodies where $n_c(m)$ = number of particles with mass larger than m. The results are shown at the same times as in the left panel: $t = 0$ (vertical thin line), $t =$ 100,000 (dashed) and $t =$ 200,000 (solid). The large bullet in both panels denotes the single outstanding large object that has formed through a runaway process.
Taken from Figs. 3 and 4 in [135], © German Astronomical Society (Astronomische Gesellschaft, AG), reproduced with permission 178

Fig. 2.10 The evolution of the mass and radius of Jupiter in the Solar Nebula. During the whole growth the planet was fixed at its present position (in-situ formation). The core mass (made of solids) is given by the red line, the gas mass by the green line, and the total mass by the dashed blue line. The different phases of the accretion process have been marked by the vertical lines and additional roman numbers in the left panel, see text for more details. In the right panel, the red line refers to the core radius, the blue dashed line to the total radius, and the green line to the capture radius of planetesimals. *Courtesy* Chr. Mordasini. Taken from Fig. 2 (top and bottom left panels) in [195], © ESO, reproduced with permission 188

Fig. 2.11 The growth of a planetary core as a function of time at different radii in the disk. Curves are shown for the Hill regime ($m_p > M_t$, with the transition mass M_t, see Eq. 2.41), the drift regime ($m_p < M_t$) and the standard pebble accretion (PA). The result shown are for a headwind parameter $\Delta \equiv \Delta v_K / c_s =$ 0.05 (see Eq. 2.42) and a dust to gas mass ratio $Z \equiv \Sigma_p / \Sigma = 0.01$.Taken from Fig. 11 in [144], © ESO, reproduced with permission 190

Fig. 2.12 The gas flow around a Jupiter type planet embedded in a protoplanetary disk. The planet is at location $x = -1, y = 0$ (in units of 5.2 AU), and the star is located at the origin. Both panels show a density image (in Cartesian coordinates) of the Roche lobe region near the planet. The motion of the planet around the star would be counter clockwise in an inertial frame of reference. (**Left**) The flow field around the planet, displayed in a reference frame corotating with the planet. The solid white

List of Tables

Chapter 1
Physical Processes in Protoplanetary Disks

Philip J. Armitage

Abstract This review, based on lectures given at the 45th Saas-Fee Advanced Course "From Protoplanetary Disks to Planet Formation", introduces physical processes in protoplanetary disks relevant to accretion and the initial stages of planet formation. After a brief overview of the observational context, I introduce the elementary theory of disk structure and evolution, review the gas-phase physics of angular momentum transport through turbulence and disk winds, and discuss possible origins for the episodic accretion observed in Young Stellar Objects. Turning to solids, I review the evolution of single particles under aerodynamic forces, and describe the conditions necessary for the development of collective gas-particle instabilities. Observations show that disks can exhibit pronounced large-scale structure, and I discuss the types of structures that may form from gas and particle interactions at ice lines, vortices and zonal flows, prior to the formation of large planetary bodies. I conclude with disk dispersal.

1.1 Preamble

The objective of this review is to introduce the physical processes in protoplanetary disks that are relevant to protostellar accretion and the initial stages of planet formation. Protoplanetary disks, as well as being interesting objects of study in their own right, are also simultaneously an outcome of star formation and initial conditions for planet formation. As such, we need to understand the evolution of both the dominant gaseous component and the trace of solid material that is critical for planet formation. Much interesting complexity occurs due to interactions between the two.

The review is organized around three motivating questions; how do protoplanetary gas disks evolve over time, how are solids transported and concentrated within the gas, and how do gas-phase and solid processes interact to form structure within disks? I begin in Sect. 1.2 with a brief summary of the observational context. The review proper starts in Sect. 1.3 by outlining the equilibrium structure of disks. Disk

P. J. Armitage (✉)
JILA, University of Colorado & NIST, Boulder, CO 80309-0440, USA
e-mail: pja@jilau1.colorado.edu

M. Audard et al. (eds.), *From Protoplanetary Disks to Planet Formation*,
Saas-Fee Advanced Course 45, https://doi.org/10.1007/978-3-662-58687-7_1

evolution is described in Sects. 1.4 and 1.5, first following the classical approach in which the origin of angular momentum transport is unspecified, and subsequently in a more modern presentation where angular momentum transport and loss processes are ascribed to specific fluid instabilities. Section 1.6 discusses candidate theoretical explanations, some of them directly related to angular momentum transport processes, for episodic accretion outbursts. In Sect. 1.7 the focus switches to solids, and I review how single solid particles settle, drift, diffuse and concentrate relative to the gas. Section 1.8 then describes how these processes can give rise to structure within the disk on various scales, either at transient "particle traps" or at persistent locations such as ice lines where the disk structure varies rapidly. Finally, Sect. 1.9 reviews what is known about the processes, including photoevaporation, that can disperse protoplanetary disks.

The lectures, even expanded as in this review, could not touch upon all of the physics that an aspiring researcher in the field might require. In an effort to cover as much as possible—and to accommodate the diverse backgrounds of participants (and readers)—the material is presented with varying degrees of detail and rigor. For most topics I begin with a self-contained discussion of essential material that would often be assumed as background in papers or talks. I then discuss the underlying physics of more recent results, and give entry points to the relevant literature.

One caution is in order. Quantities are generally defined and labeled following conventions in the literature, to make it easier for the reader who needs to fill in missing details or to explore further. Given the broad scope of the review there is considerable overloading of notation, with the same symbols being used to represent unrelated quantities in different sections. Take care!

1.2 Observational Context

The challenges of observing protoplanetary disks are formidable. Around Solar mass stars the lifetime of primordial disks of gas and dust is a small fraction—10^{-3}–10^{-4}—of the main sequence lifetime. Disks around stars are hence rare, and even the nearest examples are relatively distant and hard to resolve. A small number of disks, including particularly well-studied examples such as TW Hya, are as close as $d \approx 50$ pc, but for larger samples we need to look out at least as far as low-mass star forming regions such as Taurus, at $d \approx 140$ pc [406]. Regions where massive stars are forming alongside low mass stars, such as Orion, are yet further away at $d \approx 400$ pc [222]. Spatially resolved imaging of disks with scales comparable to that of the Solar System, $\sim 10^2$ AU, requires sub-arcsecond resolution that has only become available in relatively recent times from the *Hubble Space Telescope* [306], from the *Atacama Large Millimeter/submillimeter Array (ALMA)* and earlier mm/sub-mm interferometers [8], and from high-contrast imaging instruments on large ground-based telescopes [133, 152, 295]. Today we have exquisite imaging and spatially resolved spectroscopy of a moderate number of protoplanetary disks, along with a larger amount of data that derives from a time when unresolved observations were the norm.

Williams and Cieza [433] review observations of protoplanetary disks up to a time just prior to the advent of *ALMA*. The focus of this review is on theory, but to set the stage and introduce some ubiquitous terminology we begin with a brief discussion of the observational context. What do we know about disks that might constrain theoretical models, and how do we know it?

1.2.1 The Classification of Young Stellar Objects

Young Stellar Objects (YSOs) frequently show more emission in the infra-red than would be expected from a pre-main-sequence star's photosphere. The IR excess is attributed to the presence of dust in the vicinity of the star, and its strength forms the basis for an empirical classification scheme for YSOs that dates back to work by Lada and Wilking [234] in the early 1980s. We define the slope of the spectral energy distribution (SED) between near-IR and mid-IR wavelengths,

$$\alpha_{\rm IR} \equiv \frac{{\rm d}\log \nu F_\nu}{{\rm d}\log \nu} \equiv \frac{{\rm d}\log \lambda F_\lambda}{{\rm d}\log \lambda}. \tag{1.1}$$

The anchor points in the near- and mid-IR for the determination of $\alpha_{\rm IR}$ vary from study to study, but are typically something like 2 and 25 μm. Based on $\alpha_{\rm IR}$ four or five classes of YSO are recognized [433],

- **Class 0**: heavily obscured sources with no optical or near-IR emission ($\alpha_{\rm IR}$ is therefore undefined).
- **Class I**: $\alpha_{\rm IR} > 0.3$.
- **Flat spectrum sources**: $-0.3 < \alpha_{\rm IR} < 0.3$.
- **Class II**: $-1.6 < \alpha_{\rm IR} < -0.3$.
- **Class III**: $\alpha_{\rm IR} < -1.6$. These sources have at most weak IR-excess emission, and have SEDs that resemble isolated pre-main-sequence stellar photospheres.

Figure 1.1 shows how these empirically derived classes match up against the expected evolution of circumstellar material over time [2]. Stars form from the collapse of molecular cloud cores which have vastly more angular momentum than can be accommodated in a star [157]. Some of that angular momentum can be lost via magnetic braking [249], or subsumed into the orbital angular momentum of binaries [349], but it is an observational fact that enough is commonly left over to form a rotationally supported disk. The free-fall time scale of molecular cloud cores is shorter than the lifetime of disks, so the youngest (Class 0 and Class I) YSOs are surrounded by disks which themselves are fed by infall from envelopes. Older YSOs (Class II) have lost their envelopes but retain relatively massive and often actively accreting gas disks. Finally the gas disk dissipates, leaving behind a Class III YSO.

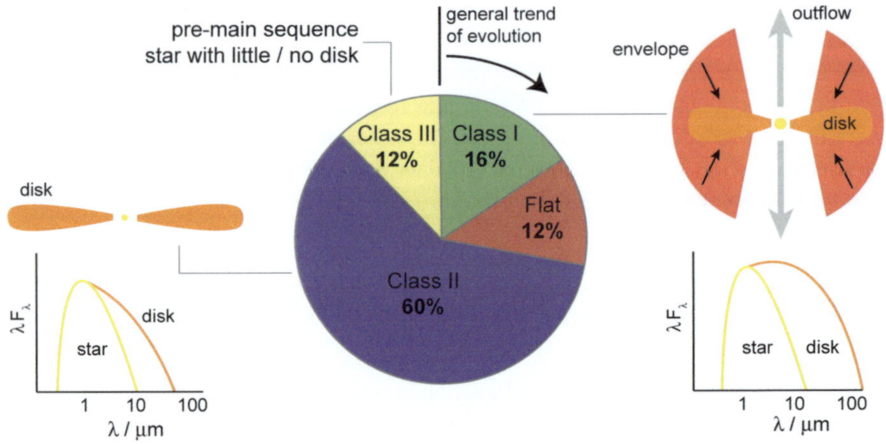

Fig. 1.1 The physical picture underlying the classification of Young Stellar Objects [2, 234] is shown together with statistics from the *Spitzer* c2d Legacy survey for the fraction of sources falling into each class [125]. The classification scheme is based upon the slope of the spectral energy distribution between the near- and mid-infra-red. This maps to a structural (and essentially evolutionary) sequence in which Class 0 and Class I YSOs are accreting from disks, which themselves are being fed by gas falling in from envelopes. Class II YSOs (also called Classical T Tauri stars) are pre-main-sequence stars with surrounding disks, while Class III YSOs (or Weak-lined T Tauri stars) are pre-main-sequence stars with little or no primordial gas remaining in orbit around them

For the subset of YSOs that are optically visible some of the terminology derives from an even older classification scheme that is based on the equivalent width of the Hα line [207]. Hα is a diagnostic of gas accreting on to the star, so Class II sources with disks are largely equivalent to "Classical T Tauri stars" with high Hα equivalent widths."Weak-lined T Tauri stars" are likewise essentially the same objects as Class III sources.

Determining the absolute ages of young stars is a difficult exercise that leads to uncertainty in estimates of the mean duration of the different phases. An easier task is to estimate the *relative* durations. Provided that we survey enough regions of recent and ongoing star formation, we may reasonably assume that the proportions of YSOs in the different classes reflect the amount of time that a typical YSO spends in each class. Figure 1.1 shows the distribution of YSOs among the classes, as determined by the "Cores to Disks" legacy project that used the *Spitzer* space telescope [125]. One sees that Class I sources with envelopes are much less numerous than Flat Spectrum and Class II sources that have disks but little or no envelope. Most of the circumstellar disk lifetime therefore consists of relatively isolated star-disk systems that have completed the primary phase of accretion from the envelope.[1]

There is no uniform definition of the term "protoplanetary disk". In this review, our main focus is on isolated disks where envelope accretion has largely ceased.

[1]Accretion from lower density gas within the star forming region could persist to later times [403].

Observationally this would correspond to Flat Spectrum and Class II YSOs. Most of the action in planet formation, reviewed in Willy Kley's contribution to this volume, is also commonly assumed to occur in an isolated disk environment. One should be aware, however, that scant empirical evidence underlies this assumption, and that the time scales of planet formation processes through to (at least) planetesimal scales can be rapid. Important aspects of planet formation could be already well-advanced during the embedded Class 0 and Class I phases of YSO evolution.

1.2.2 Accretion Rates and Lifetimes

The accretion rate $\dot{M}$ of gas through the disk is one of the most important quantities that we would like to determine. It can only be measured with any confidence at small radii, where the gas from the disk is flowing on to the star. As we will discuss in Sect. 1.4.2.2, for typical Classical T Tauri stars there is abundant evidence that the stellar magnetic field disrupts the inner disk at a magnetospheric radius R_m. Gas interior to R_m follows trajectories that are tied to magnetic field lines, before crashing on to the stellar surface at close to the free fall speed [73]. For a star of mass M_* and radius R_* the accretion luminosity will be,

$$\begin{aligned} L_{\rm acc} &\simeq GM_*\dot{M}\left(\frac{1}{R_*}-\frac{1}{R_m}\right) \\ &\sim 0.2\left(\frac{M_*}{M_\odot}\right)\left(\frac{R_*}{1.5\ R_\odot}\right)^{-1}\left(\frac{\dot{M}}{10^{-8}\ M_\odot\ {\rm yr}^{-1}}\right)L_\odot, \end{aligned} \quad (1.2)$$

where for the numerical estimate we have taken $R_m \gg R_*$ and adopted typical numbers for T Tauri stars. The accretion luminosity will be radiated from shocks or hotspots on the stellar surface. We will be able to estimate $\dot{M}$ provided that we know the basic stellar parameters and can distinguish the emission arising from accretion from the stellar photospheric emission.

In practice the determination of $\dot{M}$ from a stellar spectrum can be attempted in several ways, which differ in observational expense and fidelity. The most direct measurements require access to the entire ultraviolet (UV) spectrum, including short wavelengths that are not accessible from the ground. An example of early work of this type, using data from the *International Ultraviolet Explorer*, are the accretion rate measurements by Gullbring et al. [166]. Next best is observations from the ground, which can cover enough of the UV spectrum (down to about 300 nm) to allow a direct determination of $L_{\rm acc}$ [167, 180]. Recent studies, such as that by Manara et al. [274] take advantage of instruments with a wide spectral coverage to simultaneously determine the accretion rate and the stellar parameters. Finally, the luminosity in a number of different spectral lines has been shown to correlate well with the total accretion luminosity [357], and measurements of line luminosities can be converted to accretion rates at the cost of some additional uncertainty.

The inferred accretion rates in Class II YSOs are found to vary strongly with the stellar mass M_* [299]. In the Chamaeleon I star forming region, for example, Manara et al. determine a stellar mass/accretion rate relation of the form [274],

$$\log\left(\frac{\dot{M}}{M_\odot\ \mathrm{yr}^{-1}}\right) - (1.83 \pm 0.35)\log\left(\frac{M_*}{M_\odot}\right) - (8.13 \pm 0.16). \qquad (1.3)$$

This is a fairly representative result. Mean accretion rates through protoplanetary disks around Solar mass stars are of the order of $10^{-8}\ M_\odot\ \mathrm{yr}^{-1}$, with a stellar mass scaling that is significantly steeper than linear.

By estimating ages from the position of pre-main-sequence stars in the H-R diagram Hartmann et al. [171] showed that the median accretion rate for T Tauri stars in Taurus and Chamaeleon I declined with time. Consistent with this result, and with common sense expectations, the *fraction* of stars that host primordial disks declines with stellar age. This is commonly expressed not on a star-by-star basis, but rather by measuring the fraction of stars in a cluster with near-IR excess, and assigning a representative age to that cluster [170]. Using this method Hernández et al. [181] find that the disk fraction drops to 50% at about 3 Myr. Caution is needed, however, because the ages of young clusters remain hard to determine, with some assessments suggesting that uncertainties may be as large as a factor of two [383].

The astronomically inferred disk lifetime is compatible with the chronology derived independently from the radioactive dating of primitive Solar System materials. The oldest Solar System samples are calcium-aluminum-rich inclusions (CAIs) found within chondritic meteorites. CAIs are dated to 4567.30 ± 0.16 million years ago. Chondritic meteorites also contain (and are named after) *chondrules*—0.1–1 mm spheres of rock that were heated to approximately their melting temperature in short-lived events in the early Solar System (see e.g. the short review by Connolly and Jones [101]). How chondrules formed remains unclear, but it is known that their production (and possibly repeated subsequent heating events) began at the same time as the CAIs and continued for about 4 Myr [71]. If chondrule heating occurred predominantly in a primordial gas disk environment, one would conclude from this that the lifetime of the Solar System's gas disk was fairly typical. At the least, there is no evidence for any inconsistency between astronomical and meteoritic dating results.

1.2.3 Inferences from the Dust Continuum

At disk temperatures $T < 1500$ K the opacity in protoplanetary disks is dominated by the contribution from rocky or icy grains (generically "dust"). Observations of continuum radiation from thermal dust emission provide information about the radial temperature distribution in the disk, the disk mass, and the size of dust particles. To illustrate how this works, consider a toy model of a thin disk in which dust emits thermal radiation with a single temperature at each radius (we defer discussion of realistic complications associated with the vertical structure to Sect. 1.3.2). We assume that the gas surface density Σ and dust temperature T_d are power laws in radius,

$$\Sigma \propto r^{-p},$$
$$T_d \propto r^{-q}, \tag{1.4}$$

and adopt a frequency dependent opacity (defined per unit mass of gas),

$$\kappa_\nu = \kappa_0 \nu^\beta. \tag{1.5}$$

The vertical optical depth through the disk is then $\tau_\nu = \Sigma\kappa_\nu$. For a face-on disk the formal solution of the radiative transfer equation (e.g. [364]) gives the flux density F_ν as,

$$F_\nu = \frac{1}{D^2}\int_{r_{\rm in}}^{r_{\rm out}} B_\nu(T_d)(1-e^{-\tau_\nu})2\pi r \mathrm{d}r, \tag{1.6}$$

where D is the distance to the source, $r_{\rm in}$ and $r_{\rm out}$ are the inner and outer radii of the disk, and B_ν is the Planck function,

$$B_\nu = \frac{2h\nu^3}{c^2}\frac{1}{\exp[h\nu/k_B T_d]-1}. \tag{1.7}$$

Let us consider what we can learn from F_ν in the limit where the disk is either optically thick ($\tau_\nu > 1$) or optically thin.

Taking the $\tau_\nu \gg 1$ limit first, let us assume that we observe the disk at wavelengths where the entire disk is optically thick. This will be true in the near- and mid-IR. Setting $e^{-\tau_\nu} = 0$ in Eq. (1.6) we obtain,[2]

$$\nu F_\nu \propto \nu^{4-2/q}. \tag{1.8}$$

The slope of the infra-red spectral energy distribution thus provides a constraint on the **radial variation of the dust temperature** within the disk.

In the mm/sub-mm region of the spectrum (i.e. $\lambda \sim 1$ mm) the bulk of the emission comes instead from optically thin regions of the disk. Taking the limit where the entire disk is optically thin at the frequencies of interest, we have that $(1-e^{-\tau_\nu}) \approx \tau_\nu = \Sigma\kappa_\nu$. Equation (1.6) then takes the form,

$$F_\nu = \frac{B_\nu(\bar{T_d})\kappa_\nu}{D^2}\int_{r_{\rm in}}^{r_{\rm out}} 2\pi r\Sigma \mathrm{d}r, \tag{1.9}$$

where $\bar{T}$ is an weighted average of the temperature of the emitting material that we can determine from the infra-red part of the SED. From this we can infer two disk properties. First, we note that the integral in the above expression is just the disk mass,

[2]To see this, substitute $x \equiv h\nu/k_B T$ and approximate the limits as $r_{\rm in} = 0$ and $r_{\rm out} = \infty$.

$M_{\rm disk}$. If we know the distance to the source, the opacity, and the disk temperature, a measurement of the flux density at optically thin wavelengths **determines the disk mass**. Second, at sufficiently long wavelengths we will be on the Rayleigh-Jeans tail of the Planck function. In this limit $B_\nu \propto \nu^2$ and we find,

$$\nu F_\nu \propto \nu^{\beta+3}. \tag{1.10}$$

A measurement of the spectral slope at optically thin wavelengths in the mm/sub-mm thus determines the **frequency dependence of the opacity**. The opacity, in turn, is determined by the size distribution, structure, and composition of the solid particles within the disk [114]. For many disks, the integrated emission implies a value for $\beta \approx 1 \pm 0.5$ that is significantly smaller than the value ($\beta \approx 2$) found for dust in either the interstellar medium or in molecular clouds [56]. This result provides evidence for the growth of solid particles up to sizes of at least mm within protoplanetary disks [350, 359].

It is noteworthy that we can extract constraints on the mass, temperature distribution, and particle properties from the SED of even an unresolved disk. Continuum observations become even more powerful if the disk can be spatially resolved at multiple wavelengths. Tazzari et al. [399], for example, perform a multi-wavelength analysis of three disks (DR Tau, AS 209 and FT Tau) using mm and radio data covering the range between 0.88 mm and 1 cm. They find evidence for a radial dependence of the maximum particle size, that is both broadly consistent with and a critical test of theoretical models for particle evolution in disks [65].

With great power comes great responsibility, and several caveats should be borne in mind when interpreting dust continuum observations. There are well-known sources of uncertainty in disk mass estimates derived from mm/sub-mm data:

- The most directly inferred quantity is the mass of *optically thin dust* with particle sizes roughly the same as the observing wavelengths. Additional mass could be hidden, even in similar sized particles, at radii where the emission is optically thick. Moreover, the conversion between the observed flux and the mass depends on the opacity, which (in detail) is a function of the unknown structural and compositional properties of the particles. (Going beyond the simplest continuum observations, some mineralogical information is available by modeling the broad silicate feature seen in disk spectra near 10 μm [312].)
- The total mass of *solid material* could be much larger than the mass inferred from the optically thin emission if (a) particles have grown to radii $s \gg \lambda_{\rm max}$ (where $\lambda_{\rm max}$, the longest wavelength for which thermal emission is securely measurable, is usually a few cm) and, (b) the size distribution places most of the mass in the largest particles. Objects of meter size and above are essentially "dark matter" and unobservable in protoplanetary disks.
- Conversion to a *gas mass* involves educated guesswork, because several physical processes affect the solid and gas components in different ways. For example, radial drift (see Sect. 1.7.1) is expected to change the gas to dust ratio as a function of

both radius and time [436], while photoevaporation (see Sect. 1.9.1) preferentially removes gas from the disk leaving all but the smallest solid particles behind.

Notwithstanding all these uncertainties, the results of sub-mm disk surveys provide highly suggestive hints of the typical environment for planet formation. In Taurus-Auriga, Andrews and Williams [9] deduce a lognormal distribution of disk masses, with a mean $M_{\rm disk} \approx 5 \times 10^{-3}\ M_\odot$ and a median mass ratio between disk and star of 5×10^{-3}. Larger masses (for the disk *plus* envelope) are inferred from modeling of Class I sources in the same region [121]. In Ophiuchus, resolved observations of a sample of relatively more massive disks suggest a disk surface density profile $\Sigma \propto r^{-1}$ [11].

1.2.4 Molecular Line Observations

In addition to the continuum emission in the IR and mm/sub-mm, a number of molecules and molecular ions have been detected in line radiation. In the mm/sub-mm the workhorse molecule is CO (and its isotopologues such as ^{13}CO and $C^{18}O$), but several other fairly simple and abundant species including CS, HCN, N_2H^+ and DCO^+ are observable [118]. Some of the same molecules (albeit at different disk radii) are also observable in the near-IR, including CO. The far-IR provides access to additional molecules, including water [185, 334] and ammonia [365].

Given the difficulty in reliably determining disk masses from dust continuum data, one might hope to do better using molecular emission. This is also hard. H_2 is a homonuclear diatomic molecule with no electric dipole moment, and as a result cannot produce a rotational or vibrational spectrum in the dipole approximation [364]. The bulk of the mass in cold H_2 within protoplanetary disks is thus observationally inaccessible. There are some loopholes. Hot H_2 in the atmosphere of the disk (with $T > 2000$ K) can be excited by stellar Lyα radiation, and detected via electronic transitions that lie in the far-UV [138]. These observations, however, do not furnish any simple path toward measuring disk masses. More promisingly, emission from HD (specifically the $J = 1 \rightarrow 0$ rotational transition at $\lambda \approx 112\ \mu$m) was detected using *Herschel* data for the TW Hya system [62]. The original analysis of this data yielded a disk mass estimate, $M_{\rm disk} \geq 0.05\ M_\odot$, that was surprisingly large given the 3–10 Myr age typically assumed for the star. More recent analyses, using different disk models, confirm that the disk mass is high but give somewhat lower numbers (Kama et al. find $2.3 \times 10^{-2}\ M_\odot$ [208]; Trapman et al. find $7.7 \times 10^{-3}\ M_\odot \leq M_{\rm disk} \leq 2.3 \times 10^{-2}\ M_\odot$ [408]). Two other disks, GM Aur and DM Tau, have HD disk mass estimates [287]. The *Herschel* observatory stopped taking data in 2013, leaving us with no ongoing capability for further measurements. Determining disk masses using HD is, however, a strong science driver for proposed missions sensitive to these far-IR wavelengths.

Other molecules can be used to estimate gas masses. CO itself is not useful as the emission is optically thick, but a combination of CO isotopologues can be used instead [432]. Gas phase CO in disks can be photo-dissociated, freeze-out in cold dense regions, and be processed into other species. A subset of these processes can be observationally constrained. Qi et al. [342], for example, used imaging in N_2H^+ to infer that the CO snow line in TW Hya lies at a radius of about 30 AU. In general, however, a chemical model is needed to estimate the fraction of carbon (typically of the order of 10%) that resides in observable gas-phase CO [294], followed by a final step of converting the CO mass to a total gas mass.

Molecular line observations also provide a wealth of kinematic information. The instrumental resolution of spectra in the mm/sub-mm is typically better than the expected thermal width of molecular lines, so for bright disks that are well-resolved spatially an observation yields a "data cube" expressing the line emission as a function of sky position (x, y) and line of sight velocity $v_{\rm los}$. At the lowest order, such data can be used to measure the rotation profile of the disk gas and hence the mass of the central star. Beyond that we may try to measure the contribution that thermal and turbulent broadening make to the line profile. The strength of disk turbulence, in particular, is a key quantity for both disk evolution (see Sect. 1.5) and planet formation (see Willy Kley's contribution in this volume), that desperately needs empirical constraint. If the turbulence can be represented as a small-scale fluid process[3] the summed contribution to the line width is,

$$\Delta v = \frac{\nu}{c}\sqrt{\frac{2k_B T}{\mu m_{\rm H}} + v_{\rm turb}^2}, \tag{1.11}$$

where μ is the molecular weight of the observed species in units of the mass of a hydrogen atom m_H, T is the temperature, and $v_{\rm turb}$ is a root-mean-square estimator of the turbulence. We generally expect turbulence in disks to be subsonic, but by observing relatively heavy molecules such as CO or CS it is possible to attain sensitivity to $v_{\rm turb}$ values that are significantly below the sound speed. Using molecules with differing optical depths (e.g. isotopologues of CO) opens up the possibility of mapping $v_{\rm turb}$ as a function of height above the disk mid-plane.

Recent attempts to measure the turbulent velocity have focused on large disks with simple kinematics that can be modeled precisely. Flaherty et al. [129, 130] have analyzed *ALMA* data for the disk around the nearby A star HD 163296. Using a combination of molecular species and transitions (CO, ^{13}CO, C^{18}O and DCO$^+$), the observed data cubes were found to be consistent with models that included only orbital motion and thermal broadening. No evidence was found for a turbulent contribution. The derived upper limits are below a tenth of the sound speed throughout the vertical extent of the disk. For TW Hya, Teague et al. [400] analyzed *ALMA* data that included transitions of CO, CN and CS. Turbulent velocities in the range of 0.2–0.4 c_s were inferred for this disk on scales of around 50–100 AU. These measurements, while important and (for HD 163296) provocative to theorists, remain

---

[3]Representing fluid motions as microturbulence is a standard approximation for stellar atmospheres, but whether it is generally valid for disk turbulence is not obvious. Simon et al. [380] showed that it works reasonably well in the case where turbulence is driven by the magnetohydrodynamic instabilities discussed in Sect. 1.5.4.4.

in their infancy. It will be valuable to obtain a larger sample of disks, to acquire data in additional molecular lines, and to compare different analysis techniques.

1.2.5 *Large-Scale-Structure in Disks*

The reliable determination of quantities such as disk lifetimes, masses, and accretion rates has been the focus of observational effort for several decades. Although there remain uncertainties the strengths and weaknesses of the different methods have been exhaustively litigated, and we think we have a decent physical picture for what is going on. The same cannot be said of more recent observations that show a variety of large-scale-structure in disks. Very basic questions—including whether the observed structure is an intrinsic property of the fluid dynamics of disks or rather a sign of planet-disk interactions—remain open. In advance of the theoretical discussion in Sect. 1.8, we summarize here the main families of observed structures.

1.2.5.1 Transition Disks

The term transition disk [124] is an umbrella for a subset of disks that do not fit neatly into the SED-based classification scheme described in Sect. 1.2.1. Strom et al. [390] identified a number of YSOs that had little-to-no near IR excess (resembling Class III sources) but robust mid- and far-IR emission (resembling Class II sources). Observations with the *Spitzer* space telescope showed that this type of SED is by no means uncommon. The exact numbers depend upon the adopted definitions but in Taurus, for example, one study found that the fraction of disks classified as "transitional" or "evolved" (meaning that they are becoming optically thin at both near- and mid-IR wavelengths) is about 15% [265]. The transitional disk class includes several well-known systems, including TW Hya, GM Aur, IRS 48 and LkCa15, for which a wealth of observational data is available.

The geometric interpretation of transition disk SEDs is straightforward. Near-IR excesses originate in the inner disk, which in a normal Class II source is expected to be the region with the highest optical depth. Seeing little or no near-IR emission, while mid-IR emission persists, implies a disk with a hole or cavity in the dust distribution. This inference is supported by imaging in the sub-mm, which directly reveals the presence of dust cavities within transition disks. Andrews et al. [10] used the Submillimeter Array (SMA) to image a sample of 12 transition disks in nearby star-forming regions at 880 μm. The found cavities with radii between 15 and 73 AU, and estimated that 20–25% of the brightest disks (in mm emission) exhibited such inner clearings.

The cavities in transition disks are for the most part not empty. Most transitional sources are accreting, at rates which can be as high as normal Classical T Tauri stars [275] (though the median accretion rate is probably significantly reduced—Kim et al. [214] find a suppression of approximately an order of magnitude for Class II YSOs in Orion A). Moreover, although the defining feature of transitional disks is a cavity in the dust distribution, some transition disks' spectra indicate the presence of very low levels of dust close to the star (for example GM Aur [78]). The overall picture is thus one in which a relatively unexceptional outer disk is truncated at some cavity radius r_{cavity}. Inside the cavity the column density of (observable) dust is severely depleted, but gas persists and continues to accrete on to the star.

1.2.5.2 Rings

Long baseline observations with *ALMA* have enabled imaging of a number of protoplanetary disks with an angular resolution as good as 0.025 arcseconds. A major surprise has been the discovery of multiple rings of emission in several sources. In HL Tau, continuum imaging at wavelengths between 0.87 and 2.9 mm detects seven pairs of bright/dark ring-like structures at orbital radii between 20 and 100 AU [8]. Qualitatively similar rings, though with lesser contrast, are seen between 2.4 and 40 AU in TW Hya [13]. TW Hya's rings are also visible at near-IR wavelengths that probe sub-micron-sized dust particles near the disk surface [419]. Other examples of systems with rings of dust emission imaged by *ALMA* are HD 163296 [193], HD 169142 [127] and AA Tau [258].

1.2.5.3 Azimuthal Asymmetries

Non-axisymmetric structure is also evident in high resolution disk images. One class of non-axisymmetric structure is horseshoe-shaped emission in the mm/sub-mm. A prototypical example is the transition disk system IRS 48. Van der Marel et al. [420] found that the mm-sized particles surrounding the cavity at $r \approx 60$ AU, traced by 0.44 mm continuum emission, were strongly concentrated in a crescent-shaped feature on one side of the star. No such asymmetry was evident in either the gas or in micron-sized dust traced by mid-IR imaging. A broadly similar morphology is observed in HD 142527 [83], in LkHα 330 [194], and in a number of other transition disk sources.

Spiral structure represents a separate class of non-axisymmetric features. Spirals are seen in the very young protostellar source L1448 IRS3B [404], in the disk around the young star Elias 2–27 [327], and in the transitional disk source MWC 758 (in high resolution near-IR imaging [161]).

1.2.5.4 Interpretation

The physical origin of the large-scale-structure seen in protoplanetary disks remains unclear. Planets orbiting within the gas disk can produce cavities and spirals, and can form local pressure maxima and vortices that would trap dust in ring-like or horseshoe-shaped configurations. There is no doubt that planets can form or migrate to the large radii where most disk structure is seen—the HR 8799 system has four super-Jovian mass planets with projected orbital separations between 15 and 70 AU [279, 280]. Some observable disk structure is thus surely of planetary origin. Whether all or even most of the currently observed structure is caused by planets is less clear. Theoretically, as we will discuss in Sect. 1.8, there are a number of processes that may be able to form structures in disks without fully-formed planets. Observationally, direct imaging surveys suggest that wide separation planetary systems with planets of roughly Jupiter mass and above are not common [87, 146, 418]. Empirical constraints on the abundance of lower mass planets, however, remain weak.

- The lifetime of disks around Young Stellar Objects is a few Myr. The bulk of this time is spent in the relatively isolated Class II (or Classical T Tauri star) phase, though critical events for subsequent disk evolution (such as magnetic field loss) and planet formation (such as planetesimal formation) may start earlier while the disk itself is still accreting from an envelope.
- Key observational diagnostics include the optical/UV spectrum (used to measure the accretion rate on to the star), thermal emission from dust in the IR out to mm wavelengths (used to measure disk masses and particle properties, albeit with significant uncertainties), and molecular line emission (used to measure kinematics, constrain chemical models, and infer the strength of disk turbulence).
- High resolution imaging of protoplanetary disks in both thermal emission and scattered light shows that disks can sustain a variety of large-scale-structure, which may be related to intrinsic disk processes or to the interaction of planets with the gas and dust within the disk.

1.3 Disk Structure

The few Myr lifetime of protoplanetary disks equates to millions of dynamical times in the inner disk and thousands of dynamical times in the outer disk at 100 AU. To a first approximation we can treat the disk as evolving slowly through a sequence of axisymmetric static structures as mass accretes on to the star and is lost through disk winds, and our first task is to discuss the physics that determines those structures. Quantities that we are interested in include the density $\rho(r, z)$, the gas and dust temperatures $T(r, z)$ and $T_d(r, z)$, the chemical composition, and the ionization fraction. The density of solid particles ("dust") ρ_d is also important, but we will defer saying much about that until we have discussed turbulence, radial drift, and the aerodynamic coupling of solids and gas.

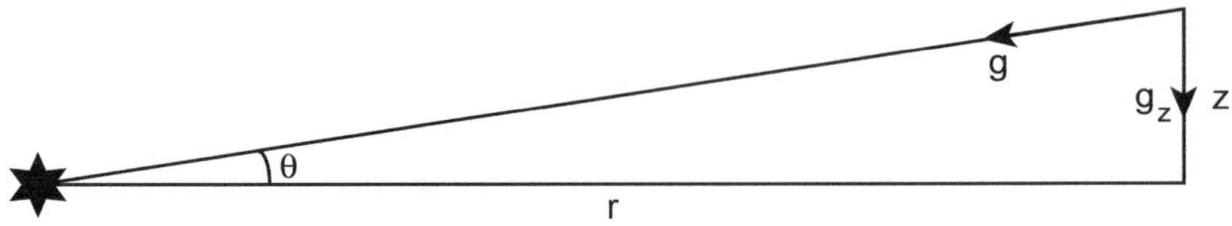

Fig. 1.2 The geometry for calculating the vertical hydrostatic equilibrium of a non-self-gravitating protoplanetary disk. The balancing forces are the vertical component of stellar gravity and the vertical pressure gradient. Taken from Fig. 2.2 in [18], © Cambridge University Press

1.3.1 Vertical and Radial Structure

1.3.1.1 Vertical Structure

The vertical profile of gas density in protoplanetary disks is determined by the condition of hydrostatic equilibrium. The simplest case to consider is an optically thick disk that is heated by stellar irradiation, has negligible mass compared to the mass of the star, and is supported by gas pressure. We can then approximate the optically thick interior of the disk as isothermal, with constant sound speed c_s and pressure $P = \rho c_s^2$. The sound speed is related to the temperature via $c_s^2 = k_B T/\mu m_H$, where k_B is Boltzmann's constant, m_H is the mass of a hydrogen atom, and where under normal disk conditions the mean molecular weight $\mu \simeq 2.3$. In cylindrical co-ordinates, the condition for vertical hydrostatic equilibrium (Fig. 1.2) is,

$$\frac{\mathrm{d}P}{\mathrm{d}z} = -\rho g_z = -\frac{GM_*}{r^2 + z^2} \sin\theta\rho, \tag{1.12}$$

where M_* is the stellar mass. For $z \ll r$,

$$g_z = \frac{GM_*}{(r^2 + z^2)^{3/2}} z \simeq \Omega_{\mathrm{K}}^2 z, \tag{1.13}$$

where $\Omega_{\mathrm{K}} \equiv \sqrt{GM_*/r^3}$ is the Keplerian orbital velocity (here defined at the mid-plane, later in Sect. 1.3.1.2 we will need to distinguish between the mid-plane and other locations). Equation (1.12) then becomes,

$$c_s^2 \frac{\mathrm{d}\rho}{\mathrm{d}z} = -\Omega_{\mathrm{K}}^2 \rho z, \tag{1.14}$$

which integrates to give,

$$\rho(z) = \rho_0 \exp\left[-z^2/2h^2\right], \tag{1.15}$$

where ρ_0 is the mid-plane density and we have defined the *vertical scale height* $h \equiv c_s/\Omega_K$. Because the effective gravity increases with height (and vanishes at the mid-plane) this standard disk profile is gaussian, rather than exponential as in a thin isothermal planetary atmosphere. A consequence is that the scale over which the density drops by a factor of e gets smaller with z; loosely speaking disks become more "two dimensional" away from the mid-plane. Defining the surface density $\Sigma = \int \rho dz$, the central density is,

$$\rho_0 = \frac{1}{\sqrt{2\pi}} \frac{\Sigma}{h}. \tag{1.16}$$

Up to straightforward variations due to differing conventions (e.g. some authors define $h = \sqrt{2}c_s/\Omega_K$) these formulae define the vertical structure of the most basic disk model (isothermal, with a gaussian density profile). For many purposes it is an adequate description, especially if one is mostly worried about conditions within a few h of the mid-plane.

The most obvious cause of gross departures from a gaussian density profile is a non-isothermal temperature profile. If the disk is accreting, gravitational potential energy that is thermalized in the optically thick interior will need a vertical temperature gradient $dT/dz < 0$ in order to be transported to the disk photosphere and radiated. We will return to this effect in Sects. 1.3.2 and 1.4, after assessing the other assumptions inherent in Eq. (1.15). We first note that the simplification to $z \ll r$ is convenient but not necessary, and that we can integrate equation (1.12) without this assumption to give,

$$\rho = \rho_0 \exp\left[\frac{r^2}{h^2}\left(\left(1 + z^2/r^2\right)^{-1/2} - 1\right)\right]. \tag{1.17}$$

Protoplanetary disks *are* geometrically thin, however, with $h/r \approx 0.05$ being fairly typical. In this regime, as is shown in Fig. 1.3, departures from a gaussian are negligibly small. We only really need to worry about the full expression for vertical gravity when considering disk winds, which flow beyond $z \sim r$.

What about the contribution of the disk itself to the vertical component of gravity? Approximating the disk as an infinite sheet with (constant) surface density Σ, Gauss' theorem tells us that the gravitational acceleration above the sheet is independent of height,

$$g_z = 2\pi G \Sigma. \tag{1.18}$$

Comparing this acceleration with the vertical component of the star's gravity at $z = h$, we find that the disk dominates if,

$$\Sigma > \frac{M}{2\pi} \frac{h}{r^3}. \tag{1.19}$$

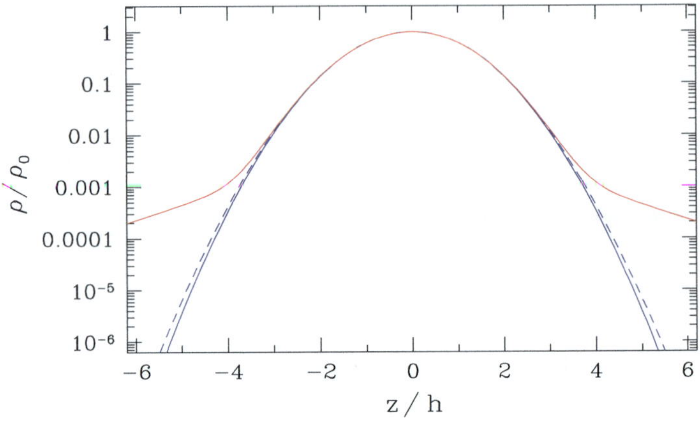

Fig. 1.3 Simple models for the vertical density profile of an isothermal disk, in units of the disk scale height h. The solid blue line shows the gaussian density profile valid for $z \ll r$, the dashed blue line shows the exact solution relaxing this assumption (the two are essentially identical for this disk, with $h/r = 0.05$). The red dashed curve shows a fit to numerical simulations that include a magnetic pressure component [184]

Very roughly we can write the disk mass $M_{\rm disk} \sim \pi r^2 \Sigma$, which allows us to write the condition for the disk's own gravity to matter as,

$$\frac{M_{\rm disk}}{M_*} > \frac{1}{2}\left(\frac{h}{r}\right). \tag{1.20}$$

For $h/r = 0.05$, a disk mass of a few percent of the stellar mass makes a non-negligible change to the vertical structure, and such masses are not unreasonably large. As we will see in Sect. 1.5, however, when disk masses $M_{\rm disk}/M_* \sim h/r$ are encountered we tend to have bigger fish to fry, as this is also the approximate condition for the onset of disk self-gravity, the formation of spiral arms, and substantial departures from axisymmetry.

Magnetic pressure $P_B = B^2/8\pi$ is likely to impact the vertical density profile, at least for $z \gg h$ (here and subsequently, we use units such that B is measured in Gauss). No simple principle for predicting the strength or vertical variation of the magnetic field is known, so we turn to numerical simulations for guidance. Hirose and Turner [184] completed radiation magnetohydrodynamic (MHD) simulations of the protoplanetary disk at 1 AU, adopting fairly typical numbers for the key disk and stellar parameters (a disk surface density $\Sigma = 10^3$ g cm^{-2}, stellar mass $M_* = 0.5\ M_\odot$, stellar effective temperature $T_{\rm eff} = 4000$ K, and stellar radius $R_* = 2\ R_\odot$). Their simulations included Ohmic diffusion but ignored both ambipolar diffusion and the Hall effect (see Sect. 1.5 for further discussion of these processes). They found that the model disk was gas pressure dominated near the mid-plane, but that the atmosphere (or corona) was magnetically dominated. An empirical fit to their density profile is [411],

$$\rho = \frac{\rho_0}{1+\varepsilon}\left[\exp\left(-z^2/2h^2\right) + \varepsilon \exp\left(-|z|/kh\right)\right], \tag{1.21}$$

with $\varepsilon \simeq 1.25 \times 10^{-2}$ and $k \simeq 1.5$. This fit is shown in Fig. 1.3. Magnetic pressure beats out gas pressure for $|z| > 4h$, leading to a low density exponential atmosphere that is much more extended than a standard isothermal disk. Even if the atmosphere itself gives way to a disk wind at still higher altitudes (as suggested by other simulations), these results suggest that observational probes of conditions near the disk surface may be sampling regions where the magnetic field dominates. Moreover, simulations of the inner disk ($r \sim 1$ AU) that include the Hall effect [240] show that it may be possible to generate strong azimuthal magnetic fields whose magnetic pressure may exceed that of the gas within one scale height, or even at the mid-plane. Such fields would lead to larger departures from the standard purely thermal definition of the scale height.

1.3.1.2 Radial Structure

The radial run of the surface density cannot be predicted from considerations of static disk structure, because two dimensional equilibrium density distributions $\rho(r, z)$ can be constructed for a broad class of surface density profiles. Not all such distributions would be stable against rapid hydrodynamic instabilities, but even if we enforce stability as an additional requirement we have no way to discriminate between commonly adopted surface density profiles (e.g. $\Sigma \propto r^{-1}$ versus $\Sigma \propto r^{-3/2}$). Instead, the surface density profile must either be measured observationally [11] or studied using time-dependent models (Sect. 1.4).[4] Given an assumed surface density profile, however, we can derive useful results for the azimuthal velocity v_ϕ. If the disk is static (and even if it is slowly evolving) the azimuthal component of the momentum equation,

$$\frac{\partial \mathbf{v}}{\partial t} + (\mathbf{v}\cdot\nabla)\,\mathbf{v} = -\frac{1}{\rho}\nabla P - \nabla\Phi \tag{1.22}$$

can be written in the mid-plane as,

$$\frac{v_\phi^2}{r} = \frac{GM_*}{r^2} + \frac{1}{\rho}\frac{\mathrm{d}P}{\mathrm{d}r}. \tag{1.23}$$

Here P is the pressure and all quantities are mid-plane values. Let's start with an explicit example of the consequences of this force balance in protoplanetary disks. Consider a disk with $\Sigma \propto r^{-1}$ and central temperature $T_c \propto r^{-1/2}$. We then have $c_s \propto r^{-1/4}$, $\rho \propto r^{-9/4}$ and $P \propto r^{-11/4}$. Substituting into Eq. (1.23) yields,

[4]The *minimum mass Solar Nebula* (MMSN) [175, 429], an approximate lower bound for the amount of disk gas needed to form the planets in the Solar System, can be useful as a reference model despite its tenuous connection to actual conditions in the disk at the time of planet formation. The MMSN has a gas surface density profile $\Sigma(r) = 1.7 \times 10^3 (r/\mathrm{AU})^{-3/2}$ g cm^{-2}.

$$v_\phi = v_\mathrm{K} \left[1 - \frac{11}{4} \left(\frac{h}{r} \right)^2 \right]^{1/2}. \tag{1.24}$$

From this we deduce,

- The deviation from strict Keplerian rotation, $v_\mathrm{K} = \sqrt{GM_*/r}$, is of the order of $(h/r)^2$.
- Its magnitude is *small*. For a disk with $h/r = 0.03$ at 1 AU, the difference between the disk azimuthal velocity and the Keplerian value is about 0.25%, or, in absolute terms $|v_\phi - v_\mathrm{K}| \simeq 70$ m s^{-1}.

When we come to discuss the evolution of particles within disks (Sect. 1.7), it will turn out that this seemingly small effect is of paramount importance. Particles do not experience the radial pressure gradient that is the cause of the mismatch in speeds, and as a result develop a differential velocity with respect to the gas that leads to aerodynamic drag and (usually) inspiral. Because this process is so important it is worth studying not just the magnitude of the effect but also its vertical dependence. To do so, we follow Takeuchi and Lin [395] and consider an axisymmetric vertically isothermal disk supported against gravity by gas pressure. The vertical density profile is then gaussian (Eq. 1.15) and in equilibrium (Eq. 1.22) we have,

$$r\Omega_\mathrm{g}^2 = \frac{GM_*}{\left(r^2 + z^2\right)^{3/2}} r + \frac{1}{\rho} \frac{\partial P}{\partial r}. \tag{1.25}$$

We distinguish between the gas angular velocity, $\Omega_\mathrm{g}(r, z)$, the Keplerian angular velocity $\Omega_\mathrm{K}(r, z) = GM_*/(r^2 + z^2)^{3/2}$, and its mid-plane value $\Omega_\mathrm{K,mid}$. The disk is fully specified by the local power-law profiles of surface density and temperature,

$$\Sigma \propto r^{-\gamma} \tag{1.26}$$

$$T_c \propto r^{-\beta}, \tag{1.27}$$

with $\gamma = 1$ and $\beta = 1/2$ being typically assumed values. Evaluating $\partial P/\partial r$ using Eq. (1.15) with $h = h(r)$ allows us to determine the equilibrium gas angular velocity in terms of the mid-plane Keplerian value,

$$\Omega_\mathrm{g} \simeq \Omega_\mathrm{K,mid} \left[1 - \frac{1}{4} \left(\frac{h}{r} \right)^2 \left(\beta + 2\gamma + 3 + \beta \frac{z^2}{h^2} \right) \right]. \tag{1.28}$$

Provided that the temperature is a locally decreasing function of radius ($\beta > 0$), the sense of the vertical shear is that the gas rotates slower at higher z. Like the sub-Keplerian mid-plane velocities, the magnitude of the shear is only of the order of $(h/r)^2$, but this small effect may nevertheless be detectable with ALMA data [362].

For particle dynamics, what matters is the difference between the gas velocity and the local Keplerian speed. To order z^2/r^2, the vertical dependence of the Keplerian velocity is,

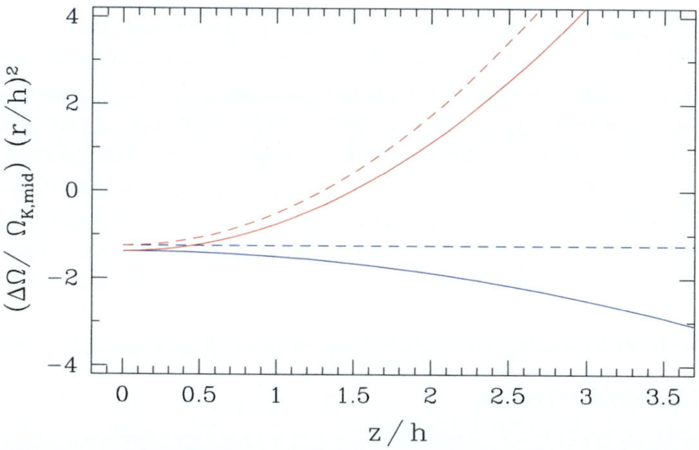

Fig. 1.4 The variation in angular velocity with height in the disk. In blue, the angular velocity of gas relative to the mid-plane Keplerian value, $(\Omega_g - \Omega_{K,mid})/\Omega_{K,mid} \times (r/h)^2$. In red, the difference between the Keplerian angular velocity and the *local* gas angular velocity, $(\Omega_g - \Omega_K)/\Omega_{K,mid} \times (r/h)^2$. The assumed disk has $\Sigma \propto r^{-1}$ and $T_c \propto r^{-1/2}$ (solid curves) or radially constant T_c (dashed curves)

$$\Omega_K \simeq \Omega_{K,mid} \left(1 - \frac{3}{4}\frac{z^2}{r^2}\right). \tag{1.29}$$

This function decreases faster with height than Ω_g, so the difference between them, plotted in Fig. 1.4, switches sign for sufficiently large z,

$$\Omega_g - \Omega_K \simeq -\frac{1}{4}\left(\frac{h}{r}\right)^2 \left[\beta + 2\gamma + 3 + (\beta - 3)\frac{z^2}{h^2}\right] \Omega_{K,mid}. \tag{1.30}$$

Particles that orbit the star at the local Keplerian speed move slower than the gas near the mid-plane (and thus experience a "headwind"), but faster at high altitude. For typical parameters, the changeover occurs at about $z \approx 1.5h$.

The orbital velocity will also deviate from the point mass Keplerian form if the disk mass is sufficiently high. The gravitational potential of a disk is *not* that of a point mass (and does not have a simple form for realistic disk surface density profiles), but for an approximation we assume that it is. Then the modified Keplerian velocity depends only upon the enclosed disk mass,

$$v'_K \simeq v_K \left(1 + \frac{M_{disk}}{M_*}\right)^{1/2}. \tag{1.31}$$

For disk masses $M_{disk} \sim 10^{-2}\, M_*$ the effect on the rotation curve is of comparable magnitude (but opposite sign) to the effect of the radial pressure gradient.

The deviations from Keplerian rotation due to pressure gradients in a planar axisymmetric disk are relatively subtle effects. Larger kinematic departures are possible if the disk is either eccentric or warped. Observationally, there are known examples of strongly warped disks (such as HD 142527 [82]) that can be traced using molecular line observations. Theoretically, we might expect disks formed from the collapse of turbulent molecular cloud cores to start out kinematically disturbed, and the rate of decay such disturbances remains a subject of active research [52]. It's thus worth remembering, especially when interpreting precise kinematic observations, that un-modeled warps or eccentricities as small as $e \sim (h/r)^2$ could be significant.

1.3.2 Thermal Physics

We seek to determine the temperature of gas and dust as $f(r, z)$. Our first task is to calculate the interior temperature of a disk heated solely by starlight. This is straightforward. At most radii of interest the dust opacity is high enough for the disk to be optically thick to both stellar radiation and to its own re-emitted radiation, which hence has a thermal spectrum. It is then a geometric problem to work out how much stellar radiation each annulus of the disk intercepts, and what equilibrium T results. We will then consider the temperatures of gas and dust in the surface layers of disks. These problems are trickier. The surface layers are both optically thin and of low density, so we have to account explicitly for the heating and cooling processes and allow for the possibility that the dust and gas are too weakly coupled to maintain the same temperature. We defer until Sect. 1.4 the question of how accretion heating modifies these solutions.

A disk whose temperature is set by stellar irradiation is described as "passive". The model problem is a flat razor-thin disk that absorbs all incoming stellar radiation and re-emits it locally as a blackbody. We seek the temperature of the blackbody disk emission as $f(r)$. Modeling the star as a sphere of radius R_*, and constant brightness I_*, we define spherical polar coordinates such that the axis of the coordinate system points to the center of the star (Fig. 1.5). The stellar flux passing through a surface at distance r is,

$$F = \int I_* \sin\theta \cos\phi d\Omega, \tag{1.32}$$

where $d\Omega$ represents the element of solid angle. We count the flux coming from the top half of the star only (and later equate that to radiation from only the top surface of the disk), so the integral has limits,

$$\begin{aligned} -\pi/2 < \phi \leq \pi/2 \\ 0 < \theta < \sin^{-1}\left(\frac{R_*}{r}\right). \end{aligned} \tag{1.33}$$

Substituting $d\Omega = \sin\theta d\theta d\phi$, the integral evaluates to,

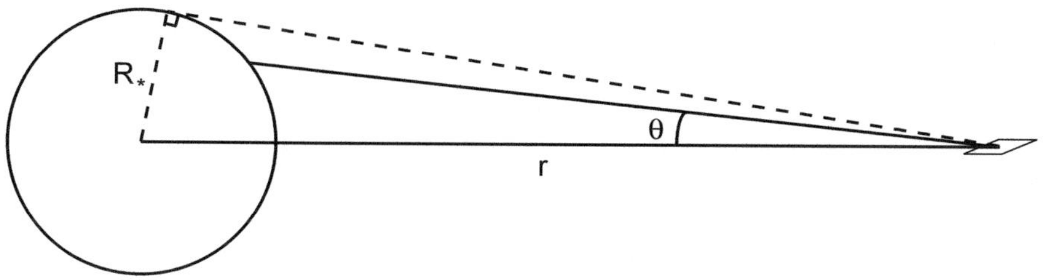

Fig. 1.5 The setup for calculating the radial temperature distribution of an optically thick, razor-thin disk. We consider a ray that makes an angle θ to the line joining the area element to the center of the star. Different rays with the same θ are labeled with the azimuthal angle ϕ; $\phi = 0$ corresponds to the "twelve o'clock" position on the stellar surface. Taken from Fig. 2.3 in [18], © Cambridge University Press

$$F = I_* \left[\sin^{-1}\left(\frac{R_*}{r}\right) - \left(\frac{R_*}{r}\right)\sqrt{1 - \left(\frac{R_*}{r}\right)^2} \right]. \tag{1.34}$$

A star with effective temperature T_* has brightness $I_* = (1/\pi)\sigma T_*^4$, with σ the Stefan-Boltzmann constant. Equating F to the one-sided disk emission $\sigma T_{\rm disk}^4$ the temperature profile is,

$$\left(\frac{T_{\rm disk}}{T_*}\right)^4 = \frac{1}{\pi} \left[\sin^{-1}\left(\frac{R_*}{r}\right) - \left(\frac{R_*}{r}\right)\sqrt{1 - \left(\frac{R_*}{r}\right)^2} \right]. \tag{1.35}$$

The exact result is unnecessarily complicated. To simplify, we expand the right hand side in a Taylor series for $(R_*/r) \ll 1$ (i.e. far from the stellar surface) to obtain,

$$T_{\rm disk} \propto r^{-3/4}, \tag{1.36}$$

as the power-law temperature profile of a thin, flat, passive disk. This implies a sound speed profile, $c_s \propto r^{-3/8}$, and a disk thickness $(h/r) \propto r^{1/8}$. We therefore predict that the disk becomes geometrically thicker ("flares") at larger radii. We can also integrate Eq. (1.35) exactly over r, with the result that the luminosity from both side of the disk sums to $L_{\rm disk}/L_* = 1/4$.

A more detailed calculation of the dust emission from passive disks requires consideration of two additional physical effects [211]. First, as we just noted, the disk thickness as measured by the gas scale height flares to larger radii. If dust is well-mixed with the gas—which may or may not be a reasonable assumption—a flared disk intercepts more stellar radiation in its outer regions than a flat one, which will tend to make it flare even more strongly. We therefore need to solve for the self-consistent *shape* of the disk that simultaneously satisfies hydrostatic and thermal equilibrium at every radius. This is conceptually easy, and the slightly messy geometry required to generalize the flat disk calculation is clearly described by Kenyon and Hartmann [211]. Second, small dust grains that are directly exposed to stellar irradiation (i.e. those where the optical depth to stellar radiation along a line toward the star $\tau < 1$) emit as a dilute blackbody with a temperature higher than if they were true blackbody emitters [211]. The reason for this is that small dust grains, of radius s, have an emissivity $\varepsilon = 1$ only for wavelengths $\lambda \leq 2\pi s$. At longer wavelengths, their emissivity declines. The details depend upon the composition and structure of the dust grains, but roughly the emissivity (and opacity κ) scale inversely with the wavelength. In terms of temperature,

$$\varepsilon = \left(\frac{T}{T_*}\right)^{\beta} \tag{1.37}$$

with $\beta = 1$. A dust particle exposed to the stellar radiation field is then in radiative equilibrium at temperature $T_{\rm s}$ when absorption and emission are in balance,

$$\frac{L_*}{4\pi r^2}\pi s^2 = \sigma T_{\rm s}^4 \varepsilon\left(T_{\rm s}\right) 4\pi s^2. \tag{1.38}$$

The resulting temperature,

$$T_{\rm s} = \frac{1}{\varepsilon^{1/4}}\left(\frac{R_*}{2r}\right)^{1/2} T_*, \tag{1.39}$$

exceeds the expected blackbody temperature by a substantial factor if $\varepsilon \ll 1$.

An illustrative analytic model that incorporates these effects was developed by Chiang and Goldreich [90]. They considered a disk with a surface density profile $\Sigma = 10^3 (r/1\ {\rm AU})^{-3/2}$ g cm^{-2} around a star with $M_* = 0.5\ M_\odot$, $T_* = 4000$ K and $R_* = 2.5\ R_\odot$. Within about 100 AU, their solution has half of the bolometric luminosity of the disk emitted as a blackbody at the interior temperature,

$$T_{\rm i} \approx 150\left(\frac{r}{1\ {\rm AU}}\right)^{-3/7}\ {\rm K}, \tag{1.40}$$

with equal luminosity at each radius emerging from a hot surface dust layer at,

$$T_{\rm s} \approx 550\left(\frac{r}{1\ {\rm AU}}\right)^{-2/5}\ {\rm K}. \tag{1.41}$$

The Chiang and Goldreich solution is a two-layer approximation to dust continuum radiative transfer for a passive, hydrostatic disk. Approximations in the same spirit have been developed that incorporate heating due to accretion [151], but the full problem requires numerical treatment. Several codes are available for the efficient solution of the full radiative transfer problem [67, 122, 358, 386].

Under most circumstances dust dominates both the absorption of starlight and the thermal emission of reprocessed stellar radiation and accretion heating. If the density is high enough, collisions between dust particles and gas molecules will establish a common temperature for both, and there is no need to explicitly consider the thermal physics of the gas. Kamp and Dullemond [209], for example, find that T_g and T_d are within about 10% of each other for an optical extinction $A_V > 0.1$. This criterion will be met in the disk mid-plane within the normal planet forming region (i.e. excluding very large orbital distances where the disk is becoming optically thin). The gas temperature near the surface of the disk is, however, of critical importance for a number of applications,

- Interpretation of sub-mm data, where the observable emission is rotational transitions of molecules such as CO and HCO^+. These observations frequently probe the outer regions of disks, at depths where the molecules are not photo-dissociated but where the gas is warm and not in equilibrium with the dust.
- Interpretation of near-IR and far-IR data, often from the inner disk, where we are seeing ro-vibrational transitions of molecules along with fine-structure cooling lines such as [CII] and [OI].
- Chemistry. It's cold at the disk mid-plane, and chemical reactions are sluggish. Although the densities are much lower in the disk atmosphere, the increased temperatures and exposure to higher energy stellar photons make the upper regions of the disk important for chemistry [177].

The properties of gas near the surface of disks are very closely tied to the incident flux of ultraviolet radiation from the star. Stellar UV radiation ionizes and dissociates atoms and molecules, and heats the gas by ejecting electrons from dust grains (grain photoelectric heating). Depending upon the temperature and density, the heating is balanced by cooling from rotational transitions of molecules (especially CO) and atomic fine structure lines. Also important is energy exchange due to inelastic collisions between gas molecules and dust particles (thermal accommodation)—if this process is too efficient the gas temperature will revert to match the dust which is absorbing and emitting the bulk of the star's bolometric luminosity.

Photoelectric heating [113] is typically the dominant process for dusty gas exposed to an ultraviolet radiation field. The work function of graphite grains (the minimum energy required to free an electron from them) is around 5 eV, so 10 eV FUV photons can eject electrons from uncharged grains with 5 eV of kinetic energy that ultimately heats the gas. Ejection occurs with a probability of the order of 0.1, so the overall efficiency (the fraction of the incident FUV energy that goes into heating the gas rather than the dust) can be rather high, around 5%.

A detailed evaluation of the photoelectric heating rate is involved, and resistant to a fully first-principles calculation. Weingartner and Draine [430] give a detailed description. Here, we sketch the main principles following Kamp and van Zadelhoff [210], who developed models for the gas temperature in A star disks. We consider a stellar radiation spectrum F_ν impinging on grains of graphite (work function $w =$ 4.4 eV [430]) and silicate ($w = 8.0$ eV). For micron-sized grains the work function, which is a property of the bulk material, is equivalent to the ionization potential—the energy difference between infinity and the highest occupied energy level in the solid. Additionally, the probability for absorption when a photon strikes a grain is $Q_{\rm abs} \approx 1$.

Most absorbed photons, however, do *not* eject electrons, rather their energy goes entirely into heating the dust grain. The yield of emitted electrons is some function of photon energy $Y(h\nu)$, and they have some spectrum of kinetic energy E, roughly described by [113] $f(E, h\nu) \propto (h\nu - w)^{-1}$. If the grains are charged (e.g. by prior emission of photoelectrons) then the kinetic energy $(E - eU)$ available to heat the gas is that left over once the electron has escaped the electrostatic potential eU of the grain. The heating rate is then [210],

$$\Gamma_{pe} = 4n_H\sigma \int_{E_{\rm min}}^{E_{\rm max}} \left(\int_{\nu_{\rm th}}^{\nu_{\rm max}} Q_{\rm abs} Y(h\nu) f(E, h\nu) F_\nu {\rm d}\nu \right) (E - eU)\, {\rm d}E, \quad (1.42)$$

where n_H is the number density of hydrogen atoms, σ is the geometric cross-section per hydrogen nucleus, and the lower limits express the minimum frequency $\nu_{\rm th}$ of a photon that can overcome the work function and the minimum energy of a photoelectron that can escape from a charged grain. An assessment of the photoelectric heating rate then requires knowledge of the functions Y and f, specification of the radiation field F_ν, and calculation of the typical charge on grains of different sizes [116, 430]. The physics is conceptually identical but quantitatively distinct when the grains in question are *extremely small* (e.g. Polycyclic Aromatic Hydrocarbons, PAHs) [42, 430].

The rate of energy exchange from inelastic gas-grain collisions can be calculated with a collision rate argument. Consider grains with geometric cross-section $\sigma_d = \pi \langle s^2 \rangle$ and number density n_d, colliding with hydrogen atoms with number density n_H. The thermal speed of the hydrogen atoms is $v_{\rm th} = (8k_B T_g / \pi m_H)^{1/2}$ and the average kinetic energy of the molecules on striking the surface is $2k_B T_g$. The cooling rate per unit volume due to gas-grain collisions can then be written in the form [77],

$$\Lambda_{g-d} = n_d n_g \sigma_d \left(\frac{8k_B T_g}{\pi m_H} \right)^{1/2} \alpha_T \left(2k_B T_g - 2k_B T_d \right), \quad (1.43)$$

where T_g and T_d are the temperatures of the gas and dust respectively. The subtleties of the calculation are reflected in the "accommodation co-efficient" α_T, which is typically $\alpha_T \approx 0.3$ for silicate and carbon grains. For a specified volumetric heating rate (and assumptions as to the gas to dust ratio and properties of the grains), this expression can be used to estimate the density below which the thermal properties of gas and dust decouple.

In addition to cooling that occurs indirectly, as a consequence of gas-grain collisions, gas in the upper layers of disks also cools radiatively. In the molecular layer of the disk, the dominant coolant is typically CO, as this is the most abundant molecule that is not homonuclear (diatomic molecules, such as H_2, have no permanent electric dipole moment and hence radiate inefficiently). At higher temperatures—only attained in the very rarefied uppermost regions of the disk atmosphere—cooling by Lyα emission becomes important. Qualitatively, there are then three distinct layers in the disk:

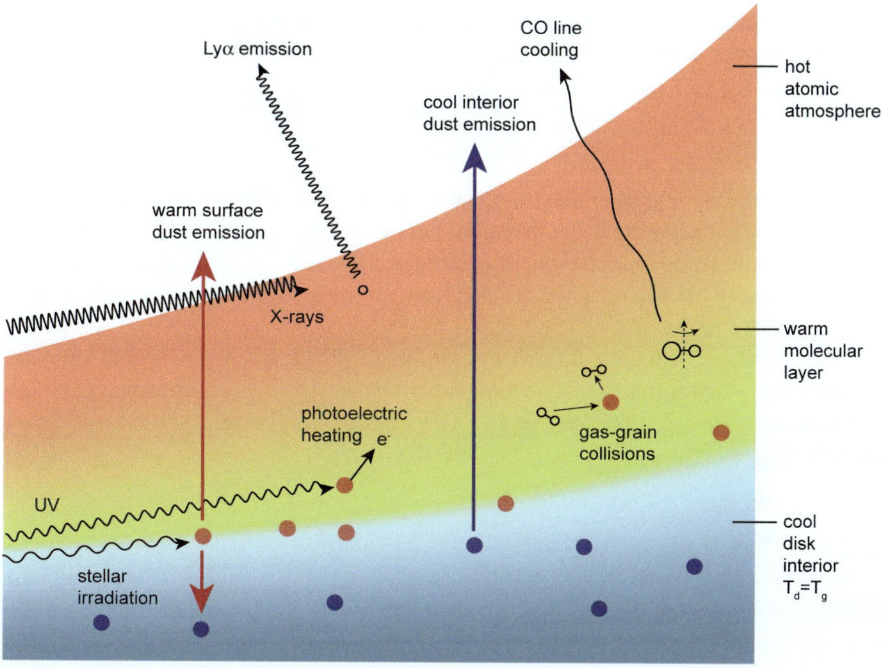

Fig. 1.6 Illustration of some of the physical processes determining the temperature and emission properties of irradiated protoplanetary disks

- A cool mid-plane region, where dust and gas have the same temperature and dust cooling is dominant.
- A warm surface layer in which both dust and gas have temperatures that exceed the mid-plane value. The gas in the warm layer can be substantially hotter than the dust ($T \sim 10^3$ K at 1 AU), and cools both by dust-gas collisions and by CO rotational-vibrational transitions.
- A hot, low-density atmosphere, where Lyα radiation and other atomic lines (e.g. O[I]) cool the gas.

The disk structure that results from these heating and cooling processes is illustrated in Fig. 1.6.

1.3.3 Ionization Structure

The degree of ionization of the gas in protoplanetary disks is important because it is key to understanding how gas couples to magnetic fields, and thence to understanding the role of magnetic fields in the formation of disks, in the sustenance of turbulence within them, and in the generation of jets and magnetohydrodynamic (MHD) winds.

At the most basic level, we care about the ratio of the number density of free electrons n_e to the number density of neutrals,

$$x_e \equiv \frac{n_e}{n_n}, \tag{1.44}$$

though we should remember that dust grains can also bear charges and carry currents. We will consider separately the thermodynamic equilibrium process of *thermal* (or collisional) ionization, which typically dominates above $T \sim 10^3$ K, and *non-thermal* ionization due to photons or particles that have an energy well in excess of the typical thermal energy in the gas.

In anticipation of results that will be derived in Sect. 1.5, we note that very low and seemingly negligible levels of ionization—$x_e \ll 10^{-10}$—often suffice to couple magnetic fields to the fluid. We need to worry about small effects when considering ionization.

1.3.3.1 Thermal Ionization

Thermal ionization of the alkali metals is important in the innermost regions of the disk, usually well inside 1 AU. In thermal equilibrium the ionization state of a single species with ionization potential χ is obeys the Saha equation [364],

$$\frac{n^{\text{ion}} n_e}{n} = \frac{2U^{\text{ion}}}{U} \left(\frac{2\pi m_e k_B T}{h^2} \right)^{3/2} \exp[-\chi / k_B T]. \tag{1.45}$$

Here, n^{ion} and n are the number densities of the ionized and neutral species, and n_e ($=n^{\text{ion}}$) is the electron number density. The partition functions for the ions and neutrals are U^{ion} and U, and the electron mass is m_e. The temperature dependence is not quite just the normal exponential Boltzmann factor, because the ionized state is favored on entropy grounds over the neutral state.

In protoplanetary disks thermal ionization becomes significant when the temperature becomes high enough to start ionizing alkali metals. For potassium, the ionization potential $\chi = 4.34$ eV. We write the abundance of potassium relative to all other neutral species as $f = n_K / n_n$, and define the ionization fraction x,

$$x \equiv \frac{n_e}{n_n}. \tag{1.46}$$

While potassium remains weakly ionized, the Saha equation gives,

$$x \simeq 10^{-12} \left(\frac{f}{10^{-7}} \right)^{1/2} \left(\frac{n_n}{10^{15}\ \text{cm}^{-3}} \right)^{-1/2} \left(\frac{T}{10^3\ \text{K}} \right)^{3/4} \frac{\exp[-2.52 \times 10^4 / T]}{1.14 \times 10^{-11}} \tag{1.47}$$

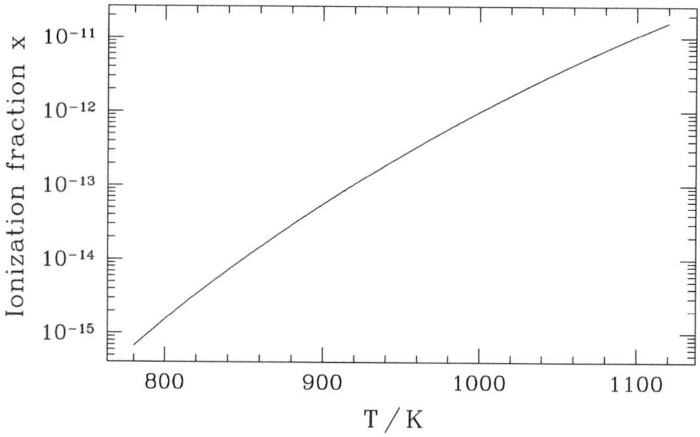

Fig. 1.7 The thermal ionization fraction as a function of temperature predicted by the Saha equation for the inner disk. Here we assume that potassium, with ionization potential $\chi = 4.34$ eV and fractional abundance $f = 10^{-7}$, is the only element of interest for the ionization. The number density of neutrals is taken to be $n_n = 10^{15}$ cm^{-3}. Taken from Fig. 2.8 in [18], © Cambridge University Press

where the final numerical factor in the denominator is the value of the exponent at 10^3 K. The ionization fraction at different temperatures is shown in Fig. 1.7. Ionization fractions large enough to be interesting for studies of magnetic field coupling are reached at temperatures of $T \sim 10^3$ K although the numbers remain extremely small—of the order of $x \sim 10^{-12}$ for these parameters.

1.3.3.2 Non-thermal Ionization

Outside the region close to the star where thermal ionization is possible, any remnant levels of ionization are controlled by non-thermal processes. Considerations of thermodynamic equilibrium are not relevant, and we need to explicitly balance the rate of ionization by high-energy particles or photons against the rate of recombination within the disk gas.

There are several potentially important sources of ionization. Ordering them roughly in order of their penetrating power, ideas that have been suggested include,

- Ultraviolet photons (from the star, or from other stars in a cluster)
- Stellar X-rays
- Cosmic rays
- Energetic protons from a stellar corona [412]
- Particles produced from radioactive decay of nuclides within the disk [387]
- Electric discharges [298].

We will limit our discussion to the first three of these processes. Ionization due to radioactive decay, although undoubtedly present, leads to a very small electron fraction, while the level of ionization due to energetic protons and electric discharges is quite uncertain.[5]

The coronae of T Tauri stars are powerful sources of keV X-rays [336]. Typical luminosities are $L_X \simeq 10^{28}$–10^{31} erg s^{-1}, in X-rays with temperatures $k_B T_X$ of a few keV. The physics of the interaction of these X-rays with the disk gas involves Compton scattering and absorption by photo-ionization, which has a cross-section $\sigma \sim 10^{-22}$ cm^2 for keV energies, decreasing with photon energy roughly as $E_{\rm phot}^{-3}$. Given an input stellar spectrum and assumptions as to where the X-rays originate, the scattering and absorption physics can be calculated using radiative transfer codes to deduce the ionization rate within the disk [123].

Depending upon the level of detail needed for a particular application, the results of numerical radiative transfer calculations can be approximated analytically. For a relatively hard stellar spectrum ($k_B T_X = 5$ keV), the ionization rate fairly deep within the disk scales with radius r and vertical column from the disk surface Σ as $r^{-2} \exp[-\Sigma/8 \text{ g cm}^{-2}]$ [413]. A more detailed fit is given by Bai and Goodman [35]. For an X-ray luminosity scaled to $L_{X,29} = L_X/10^{29}$ erg s^{-1} they represent the numerical results with two components,

$$\frac{\zeta_X}{L_{X,29}} \left(\frac{r}{1 \text{ AU}} \right)^{-2.2} = \zeta_1 \exp[-(\Sigma/\Sigma_1)^\alpha] + \zeta_2 \exp[-(\Sigma/\Sigma_2)^\beta] + \cdots \quad (1.48)$$

where Σ is the vertical column density from the top of the disk and symmetric terms in the column density from the bottom of the disk are implied. For $k_B T_X =$ 3 keV and Solar composition gas the fit parameters are $\zeta_1 = 6 \times 10^{-12}$ s^{-1}, $\zeta_2 = 10^{-15}$ s^{-1}, $\Sigma_1 = 3.4 \times 10^{-3}$ g cm^{-2}, $\Sigma_2 = 1.59$ g cm^{-2}, $\alpha = 0.4$ and $\beta = 0.65$. For $k_B T_X = 5$ keV the fit parameters are $\zeta_1 = 4 \times 10^{-12}$ s^{-1}, $\zeta_2 = 2 \times 10^{-15}$ s^{-1}, $\Sigma_1 = 6.8 \times 10^{-3}$ g cm^{-2}, $\Sigma_2 = 2.27$ g cm^{-2}, $\alpha = 0.5$ and $\beta = 0.7$.

Figure 1.8 shows the estimates for $\zeta(\Sigma)$ for a stellar X-ray luminosity of $L_X = 10^{30}$ erg s^{-1} incident on the disk at 1 AU. If one is mainly interested in regions of the disk more than $\approx$10 g cm^{-2} away from the surfaces the single exponential fit given by Turner and Sano [413] may suffice. The more complex fitting function given by Eq. (1.48) captures the much higher rates of ionization due to X-rays higher up in the disk atmosphere.

Cosmic rays are another potential source of disk ionization. A standard description of the interstellar cosmic ray flux gives them an unattenuated ionization rate of $\zeta_{CR} \sim 10^{-17}$–10^{-16} s^{-1} and an exponential stopping length of 96 g cm^{-2} (substantially

[5] Note that the amount of *power* involved in any of these non-thermal ionization processes is rather small when compared to that liberated by accretion [192]. Any additional processes that could convert even a small fraction of the accretion energy into non-thermal particles would likely matter for the ionization state.

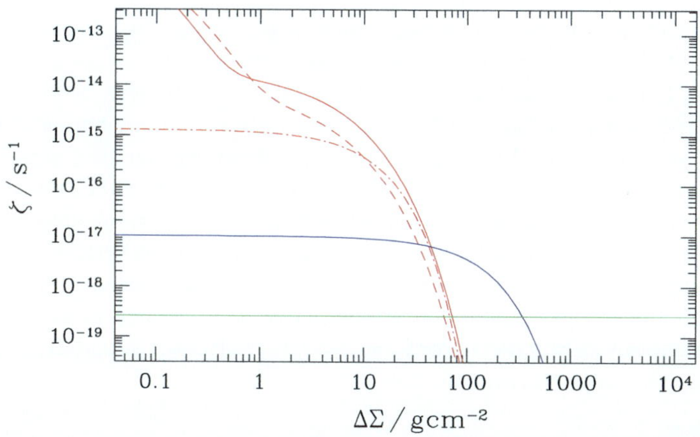

Fig. 1.8 Estimates of the non-thermal ionization rate due to X-rays (red curves), unshielded cosmic rays (blue) and radioactive decay of short-lived nuclides (green), plotted as a function of the vertical column density from the disk surface. The solid red curve shows the Bai and Goodman [35] result for a temperature $k_B T_X = 5$ keV, the dashed curve their result for $k_B T_X = 3$ keV. The dot-dashed red curve shows a simpler formula by Turner and Sano [413]. All of the X-ray results have been normalized to a flux of $L_X = 10^{30}$ erg s^{-1} and a radius of 1 AU

greater than even high energy stellar X-rays).[6] With these parameters, X-rays would remain the primary source of ionization in the upper $\approx$50 g cm^{-2} of the disk, but cosmic rays would dominate in the region between about 50 and 500 g cm^{-2}. It is unclear, however, whether the unattenuated interstellar medium flux of cosmic rays typically reaches the surfaces of protoplanatary disks. The magnetic fields embedded in the *Solar* wind form a partial barrier to incoming cosmic ray particles, whose effect is seen in a modulation of the observed flux with the Solar cycle. T Tauri stars could have much stronger stellar winds that exclude cosmic rays efficiently. Indeed, chemical modeling of molecular line data suggests that cosmic rays are substantially excluded (to a level $\zeta_{CR} \sim 10^{-19}$) from the disk around the nearby star TW Hya [98], though how pervasive this phenomenon is remains unknown. If cosmic rays are not present, the only guaranteed source of ionization at columns more than $\approx$100 g cm^{-2} away from the disk surfaces is radioactive decay.

If our main interest is in conditions at $r \sim 1$ AU the surface density in gas is typically $\Sigma \sim 10^3$ g cm^{-2} and X-rays, which are our main concern, will not reach the mid-plane. The situation is different further out. At 100 AU typical surface densities are much lower—1 g cm^{-2} might be reasonable—and X-rays will sustain a non-zero rate of ionizations throughout that column. On these scales ultraviolet photons can also be important. Stellar FUV radiation will ionize carbon and sulphur atoms near the disk surface, yielding a relatively high electron fraction $x_e \sim 10^{-5}$. The ionized

[6]Umebayashi and Nakano noted that if cosmic rays have an approximately isotropic angular distribution at the disk surface, geometric effects lead to a faster than exponential attenuation deep in the disk [415].

skin that results is shallow, penetrating to a vertical column of just 0.01–0.1 g cm^{-2}, but enough to be significant in the tenuous outer disk [328].

As with ionization, the rate of recombination within the disk can be calculated from complex numerical models that track reactions (often numbering in the thousands) between dozens of different species. The following discussion, which borrows heavily from the description given by Ilgner and Nelson [188] and Fromang [141], is intended only to outline some of the important principles. At the broadest level of discussion we need to consider gas-phase recombination reactions (involving molecular and gas-phase metal ions) along with recombination on the surface of dust grains.

The principles of gas-phase recombination can be illustrated by considering the possible reactions between electrons and generic molecules m and metal atoms M [188, 313]. The basic reactions are then,

- Ionization,
$$\mathrm{m} \rightarrow \mathrm{m}^+ + \mathrm{e}^-, \tag{1.49}$$
with rate ζ. A specific example is $\mathrm{H_2} \rightarrow \mathrm{H_2^+} + \mathrm{e}^-$.
- Recombination with molecular ions,
$$\mathrm{m}^+ + \mathrm{e}^- \rightarrow \mathrm{m}, \tag{1.50}$$
with rate $\alpha = 3 \times 10^{-6} T^{-1/2}$ cm^3 s^{-1}. An example is the dissociative recombination reaction $\mathrm{HCO^+} + \mathrm{e}^- \rightarrow \mathrm{CO} + \mathrm{H}$.
- Recombination with gas-phase metal ions,
$$\mathrm{M}^+ + \mathrm{e}^- \rightarrow \mathrm{M} + h\nu, \tag{1.51}$$
with rate $\gamma = 3 \times 10^{-11} T^{-1/2}$ cm^3 s^{-1}. An example is $\mathrm{Mg^+} + \mathrm{e}^- \rightarrow \mathrm{Mg} + h\nu$.
- Charge exchange reactions,
$$\mathrm{m}^+ + \mathrm{M} \rightarrow \mathrm{m} + \mathrm{M}^+, \tag{1.52}$$
with rate $\beta = 3 \times 10^{-9}$ cm^3 s^{-1}. An example is $\mathrm{HCO^+} + \mathrm{Mg} \rightarrow \mathrm{Mg^+} + \mathrm{HCO}$.

From such a set of reactions we form differential equations describing the time evolution of the number density of species involved. For the molecular abundance n_m, for example, we have,

$$\frac{\mathrm{d}n_\mathrm{m}}{\mathrm{d}t} = -\zeta n_\mathrm{m} + \alpha n_e n_{\mathrm{m}^+} + \beta n_\mathrm{M} n_{\mathrm{m}^+} = 0, \tag{1.53}$$

where the second equality follows from assuming that the system has reached equilibrium. The resulting system of algebraic equations has simple limiting solutions. For example, if there are no significant reactions involving metals then the above equation, together with the condition of charge neutrality ($n_{\mathrm{m}^+} = n_e$), gives an electron fraction $x_e = n_e / n_\mathrm{m}$,

$$x_e = \sqrt{\frac{\zeta}{\alpha n_{\rm m}}}. \tag{1.54}$$

In the more general case, the network yields a cubic equation which can be solved for the electron fraction as a function of the gas-phase metal abundance [188]. Typically the presence of metal atoms and ions is important for the ionization level.

Recombination can also occur on the surfaces of dusty or icy grains. The simplest reactions we might consider are,

$$\begin{aligned} {\rm e}^- + {\rm gr} &\rightarrow {\rm gr}^- \\ {\rm m}^+ + {\rm gr}^- &\rightarrow {\rm gr} + {\rm m}. \end{aligned} \tag{1.55}$$

If the first of these reactions is rate-limiting, then we can write a modified version of the ordinary differential equation (Eq. 1.53) that includes grain processes. Ignoring metals for simplicity,

$$\frac{{\rm d}n_{\rm m}}{{\rm d}t} = -\zeta n_{\rm m} + \alpha n_e n_{\rm m^+} + \sigma v_e n_{\rm gr} n_{\rm e^-}, \tag{1.56}$$

where σ is the cross-section of grains to adhesive collisions with free electrons and v_e is the electron thermal velocity. In the limit where *only* grains contribute to recombination we then find,

$$x_e = \frac{\zeta}{\sigma v_e n_{\rm gr}}. \tag{1.57}$$

If the grains are mono-disperse with radius s, then $x_e \propto 1/\sigma n_{\rm gr} \propto s$, and recombination on grain surfaces will be more important for small grain sizes. We also note that the dependence on the ionization rate ζ is linear, rather than the square root dependence found in the gas-phase case. As for metals, grain populations with commonly assumed size distributions are found to matter for the ionization level.

The above discussion of recombination leaves a great deal unsaid. For grains, an important additional consideration is related to the typical charge state, which needs to be calculated [116]. A good comparison of different networks for the calculation of the ionization state is given by Ilgner and Nelson [188], while Bai and Goodman [35] provide a clear discussion of the important processes. In wading into this literature the reader who encounters a problem involving the ionization level is advised to first evaluate whether a simple analytic approximation is adequate for their application, or whether solution of a full chemical network is required. The accuracy needed from calculations of ionization equilibrium is strongly problem-dependent, and in some cases, such as if we don't know if cosmic rays are present for a particular system, high accuracy may be illusory.

- The vertical density profile of disks is determined by hydrostatic equilibrium. The simplest (vertically isothermal) model yields a gaussian density profile. More complex models need to incorporate vertical thermal gradients and the possible contribution of magnetic pressure. Generally, the optically thin atmospheres of disks have higher temperatures than at the mid-plane, and distinct temperatures for the gas and dust components.
- The radial temperature profile is set by the balance between stellar irradiation and local cooling, supplemented by dissipation of accretion energy if $\dot{M}$ is high enough. These processes yield a mid-plane temperature scaling roughly between $r^{-3/4}$ and $r^{-1/2}$. The radial profile of the surface density cannot be determined by similarly simple physical considerations. However, on average we expect the mid-plane pressure to decline with radius, and this causes the gas to orbit at slightly less than the Keplerian speed.
- The disk close to the star is thermally ionized at levels sufficient to couple magnetic fields to the gas where the temperature exceeds about 10^3 K. At larger radii the ionization is non-thermal, due to stellar X-rays, FUV photons, radioactive decays, and cosmic rays if they are not screened from the disk. The ionization state varies radially and vertically depending on the balance of ionization and recombination reactions, which may occur in the gas phase or on dust grain surfaces.

1.4 Disk Evolution

The population of protoplanetary disks is observed to evolve, but the dominant physical processes responsible for this evolution remain unclear. For a geometrically thin, low-mass disk, the deviation from a point-mass Keplerian rotation curve is small (c.f. Eq. 1.24) and the specific angular momentum,

$$l(r) = r^2 \Omega_{\rm K} = \sqrt{GM_* r} \propto r^{1/2}, \tag{1.58}$$

is an increasing function of orbital radius. To accrete, gas in the disk must lose angular momentum, and the central theoretical problem in disk evolution is to understand this process.

Within any shearing fluid momentum is transported in the cross-stream direction because the random motion of molecules leads to collisions between particles that have different velocities. The classical approach to disk evolution [267, 338] treats the disk as a vertically thin axisymmetric sheet of viscous fluid, and leads to a fairly simple equation for the time evolution of the disk surface density $\Sigma(r, t)$. There appears to be a fatal flaw to this approach, because the molecular viscosity of the gas is much too small to lead to any significant rate of disk evolution. But it's not as bad as it seems. The classical disk evolution equation involves few assumptions beyond the immutable laws of mass and angular momentum conservation, and as we shall see is therefore approximately valid if the "viscosity" is re-interpreted as the outcome of a turbulent process. We will have (much) more to say about the possible origin of disk turbulence in Sect. 1.5.

Redistribution of angular momentum within the gas disk is not the only route to evolution. An almost equally long-studied suggestion [69] is that gas accretes because a magnetohydrodynamic (MHD) wind *removes* angular momentum entirely from the disk. Winds and viscosity have frequently between seen as orthogonal and competing hypotheses for disk evolution, but there is evidence suggesting that both processes are simultaneously important in regions of protoplanetary disks.

1.4.1 The Classical Equations

The evolution of a flat, circular and geometrically thin ($(h/r) \ll 1$) viscous disk follows from the equations of mass and angular momentum conservation [338]. Given a surface density $\Sigma(r,t)$, radial velocity $v_r(r,t)$ and angular velocity $\Omega(r)$, the continuity equation in cylindrical co-ordinates yields,

$$r\frac{\partial \Sigma}{\partial t} + \frac{\partial}{\partial r}(r\Sigma v_r) = 0. \tag{1.59}$$

Angular momentum conservation gives,

$$r\frac{\partial}{\partial t}\left(r^2\Omega\Sigma\right) + \frac{\partial}{\partial r}\left(r^2\Omega \cdot r\Sigma v_r\right) = \frac{1}{2\pi}\frac{\partial G}{\partial r}, \tag{1.60}$$

where $\Omega(r)$ is time-independent but need not be the point mass Keplerian angular velocity. The rate of change of disk angular momentum is given by the change in surface density due to radial flows and by the *difference* in the torque exerted on the annulus by stresses at the inner and outer edges. For a viscous fluid the torque G has the form,

$$G = 2\pi r \cdot \nu\Sigma r\frac{\mathrm{d}\Omega}{\mathrm{d}r} \cdot r, \tag{1.61}$$

where ν is the kinematic viscosity. The torque is the product of the circumference, the viscous force per unit length, and the lever arm r, and scales with the gradient of the angular velocity.

To obtain the surface density evolution equation in its usual form we first eliminate v_r by substituting for $\partial\Sigma/\partial t$ in Eq. (1.60) from Eq. (1.59). This gives an expression for $r\Sigma v_r$, which we substitute back into Eq. (1.59) to yield,

$$\frac{\partial \Sigma}{\partial t} = -\frac{1}{r}\frac{\partial}{\partial r}\left[\frac{1}{\left(r^2\Omega\right)'}\frac{\partial}{\partial r}\left(\nu \Sigma r^3 \Omega'\right)\right], \tag{1.62}$$

where the primes denote differentiation with respect to radius. Specializing to a point mass Keplerian potential ($\Omega \propto r^{-3/2}$) we then find that viscous redistribution of angular momentum within a thin disk obeys an equation,

$$\frac{\partial \Sigma}{\partial t} = \frac{3}{r}\frac{\partial}{\partial r}\left[r^{1/2}\frac{\partial}{\partial r}\left(\nu \Sigma r^{1/2}\right)\right]. \tag{1.63}$$

This equation is a diffusive partial differential equation for the evolution of the gas, which has a radial velocity,

$$v_r = -\frac{3}{\Sigma r^{1/2}}\frac{\partial}{\partial r}\left(\nu \Sigma r^{1/2}\right). \tag{1.64}$$

The equation is linear if the viscosity ν is independent of Σ.

Some useful rules of thumb for the rate of evolution implied by Eq. (1.63) can be deduced with a change of variables. Defining,

$$X \equiv 2r^{1/2}, \tag{1.65}$$

$$f \equiv \frac{3}{2}\Sigma X, \tag{1.66}$$

and taking the viscosity ν to be constant, we get a simpler looking diffusion equation,

$$\frac{\partial f}{\partial t} = D\frac{\partial^2 f}{\partial X^2}, \tag{1.67}$$

with a diffusion coefficient D,

$$D = \frac{12\nu}{X^2}. \tag{1.68}$$

The diffusion time scale across a scale ΔX for an equation of the form of Eq. (1.67) is just $(\Delta X)^2/D$. Converting back to the physical variables, the time scale on which viscosity will smooth surface density gradients on a scale Δr is $\tau_\nu \sim (\Delta r)^2/\nu$. For a disk with characteristic size r, the surface density at all radii will evolve on a time scale,

$$\tau_\nu \approx \frac{r^2}{\nu}. \tag{1.69}$$

This is the *viscous time scale* of the disk.

We can gain some intuition into how Eq. (1.63) works by inspecting time-dependent analytic solutions that can be derived for special forms of the viscosity $\nu(r)$. For $\nu =$ constant a Green's function solution is possible. Suppose that at $t = 0$ the gas lies in a thin ring of mass m at radius r_0,

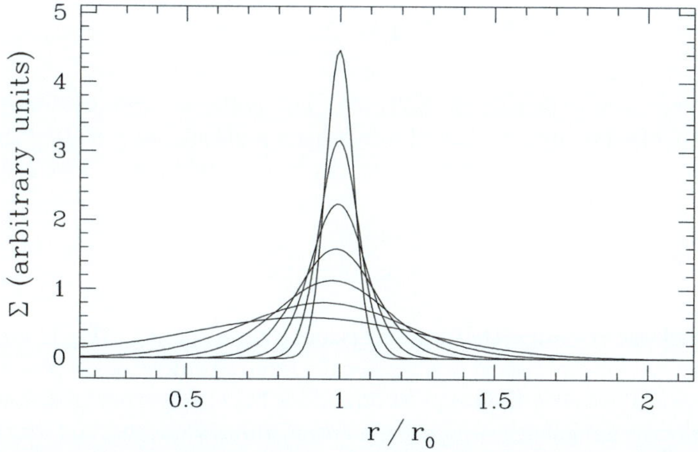

Fig. 1.9 The time-dependent solution to the disk evolution equation with ν = constant, showing the spreading of a ring of gas initially orbiting at $r = r_0$. From top down the curves show the surface density as a function of the scaled time variable $\tau = 12\nu r_0^{-2}t$, for $\tau = 0.004$, $\tau = 0.008$, $\tau = 0.016$, $\tau = 0.032$, $\tau = 0.064$, $\tau = 0.128$, and $\tau = 0.256$. Taken from Fig. 3.3 in [18], © Cambridge University Press

$$\Sigma(r, t = 0) = \frac{m}{2\pi r_0}\delta(r - r_0), \tag{1.70}$$

where $\delta(r - r_0)$ is a Dirac delta function. With boundary conditions that impose zero-torque at $r = 0$ and allow for free expansion toward $r = \infty$ the solution is [267],

$$\Sigma(x, \tau) = \frac{m}{\pi r_0^2}\frac{1}{\tau}x^{-1/4}\exp\left[-\frac{(1 + x^2)}{\tau}\right] I_{1/4}\left(\frac{2x}{\tau}\right), \tag{1.71}$$

in terms of dimensionless variables $x \equiv r/r_0$, $\tau \equiv 12\nu r_0^{-2}t$, and where $I_{1/4}$ is a modified Bessel function of the first kind. This solution is plotted in Fig. 1.9 and illustrates generic features of viscous disk evolution. As t increases the ring spreads diffusively, with the mass flowing toward $r = 0$ while the angular momentum is carried by a negligible fraction of the mass toward $r = \infty$. This segregation of mass and angular momentum is generic to the evolution of a viscous disk, and must occur if accretion is to proceed without overall angular momentum loss (for example in a magnetized disk wind).

1.4.1.1 Limits of Validity

Protoplanetary disks are not viscous fluids in the same way that honey is a viscous fluid (or, for that matter, in the same way as the mantle of the Earth is viscous). To order of magnitude precision, the viscosity of a gas $\nu \sim v_{\text{th}}\lambda$, where v_{th} is the thermal speed of the molecules and the mean-free path λ is,

$$\lambda \sim \frac{1}{n\sigma}. \tag{1.72}$$

Here n is the number density of molecules with collision cross-section σ. Taking σ to be roughly the physical size of a hydrogen molecule, $\sigma \sim \pi(10^{-8}\ \text{cm})^2$, and conditions appropriate to 1 AU ($n \sim 10^{15}\ \text{cm}^{-3}$, $v_{\text{th}} \sim 10^5\ \text{cm s}^{-1}$) we estimate,

$$\begin{aligned} \lambda &\sim 3\ \text{cm} \\ \nu &\sim 3 \times 10^5\ \text{cm}^2\ \text{s}^{-1}. \end{aligned} \tag{1.73}$$

This is not a large viscosity. The implied viscous time according to Eq. (1.69) is of the order of 10^{13} yr, far in excess of observationally inferred time scales of protoplanetary disk evolution. If we nevertheless press on and use Eq. (1.63) to model disk evolution, we are implicitly modeling a system that is not an ordinary viscous fluid with a viscous equation. We need to understand when this is a valid approximation.

The first possibility, introduced by Shakura and Sunyaev [370] in their paper on black hole accretion disks retains the idea that angular momentum is conserved within the disk system, but supposes that turbulence rather than molecular processes is the agent of angular momentum transport. Looking back at the derivation of the disk evolution Eq. (1.63), we note that the fluid properties of molecular viscosity only enter twice, (i) in the specific expression for G (for example in the fact that the torque is hardwired to be linear in the rate of shear) and (ii) in the more basic assumption that angular momentum transport is determined by the *local* fluid properties. The rest of the derivation involves only conservation laws that hold irrespective of the nature of transport. Plausibly then, a disk in which angular momentum is redistributed by the action of turbulence should still be describable by a diffusive equation, provided that the turbulence is a local process. Proceeding rigorously, Balbus and Papaloizou [47] showed that MHD turbulence is in principle local in this sense, whereas angular momentum transport by self-gravity is in principle not. At the level of the basic axisymmetric evolution equation then—before we turn to questions of what determines ν, or how boundary layers behave where the shear is reversed—we have not committed any cardinal sin in starting from the viscous disk equation.

Greater care is needed in situations where the disk flow is no longer axisymmetric. Fluids obey the Navier-Stokes equations, but there is no guarantee that a turbulent disk with a complex geometry will behave in the same way as a viscous Navier-Stokes flow with effective kinematic and bulk viscosities. In eccentric disks, for example, even the most basic properties (such as whether the eccentricity grows or decays) depend upon the nature of the angular momentum transport [308].

The disk evolution equation will also need modification if there are external sources or sinks of mass or angular momentum. If the disk gains or loses mass at a rate $\dot{\Sigma}(r, t)$, and *if that gas has the same specific angular momentum as the disk*, then the modification is trivial,

$$\frac{\partial \Sigma}{\partial t} = \frac{3}{r}\frac{\partial}{\partial r}\left[r^{1/2}\frac{\partial}{\partial r}\left(\nu \Sigma r^{1/2}\right)\right] + \dot{\Sigma}. \tag{1.74}$$

Disk evolution in the presence of thermally driven winds (such as photo-evaporative flows) can be described with this equation. Alternatively, we may consider a disk subject to an external torque that drives a radial flow with velocity $v_{r,ext}$. This adds an advective term,

$$\frac{\partial \Sigma}{\partial t} = \frac{3}{r}\frac{\partial}{\partial r}\left[r^{1/2}\frac{\partial}{\partial r}\left(\nu \Sigma r^{1/2}\right)\right] - \frac{1}{r}\frac{\partial}{\partial r}\left(r \Sigma v_{r,ext}\right). \tag{1.75}$$

The qualitative evolution of the disk—for example the tendency for the outer regions to expand to conserve angular momentum, or the steady-state surface density profile at small radii [394]—can be changed if there is even a modest external torque on the system.

From an observational point of view the relative simplicity of Eq. (1.63) means that it is often used to model the evolution of disk populations and to fit the surface density profile of individual disks. This provides a useful connection to disk theory, but it should be remembered that the general validity of the simple diffusion equation is itself an open question. In the outer regions, especially, it is possible that the initial surface density distribution is modified more by thermal or magnetic winds than by internal redistribution of angular momentum.

1.4.1.2 The α Prescription

Molecular viscosity depends in a calculable way upon the density, temperature and composition of the fluid. Can anything similar be said about the "effective viscosity" present in disks? The standard approach is to write the viscosity as the product of characteristic velocity and spatial scales in the disk,

$$\nu = \alpha c_s h, \tag{1.76}$$

where α is a dimensionless parameter. This ansatz (introduced in a related form in [370]) is known as the Shakura-Sunyaev α prescription.

We can view the α prescription in two ways. The weak version is to regard it as a re-parameterization of the viscosity that describes the leading order scaling expected in disks (so that α is a more slowly varying function of temperature, radius etc than ν). This is useful, and along with convention is the reason why numerical simulations of turbulent transport are invariably reported in terms of an effective α. One can also adopt a strong version of the prescription in which α is assumed to be a constant. This is powerful as it allows for the development of a predictive theory of disk structure that is based on only one free parameter (for a textbook discussion see Frank, King and Raine [139], or for an application to protoplanetary disks see, e.g. [58]). However, its use must be justified on a case by case basis, as there is no reason why α should be a constant. Constant α models probably work better in highly ionized disks around black holes and neutron stars, where angular momentum transport across a broad range of radii occurs via the simplest version of the magnetorotational instability

[45], than in protoplanetary disks where the physical origin of angular momentum transport is more complex [19].

1.4.2 Boundary Conditions

Solving Eq. (1.63) requires the imposition of boundary conditions. The most common, and simplest, is a zero-torque inner boundary condition, which exactly conserves the initial angular momentum content of the disk. If the star has a dynamically significant magnetic field, however, or if the disk is part of a binary system, other boundary conditions may be more appropriate.

1.4.2.1 Zero-Torque Boundary Conditions

A steady-state solution to Eq. (1.63) with a zero-torque inner boundary condition is derived by starting from the angular momentum conservation equation (Eq. 1.60). Setting the time derivative to zero and integrating we have,

$$2\pi r \Sigma v_r \cdot r^2 \Omega = 2\pi r^3 \nu \Sigma \frac{\mathrm{d}\Omega}{\mathrm{d}r} + \text{constant}. \tag{1.77}$$

In terms of the mass accretion rate $\dot{M} = -2\pi r \Sigma v_r$ we can write this in the form,

$$- \dot{M} \cdot r^2 \Omega = 2\pi r^3 \nu \Sigma \frac{\mathrm{d}\Omega}{\mathrm{d}r} + \text{constant}, \tag{1.78}$$

where the constant of integration, which is an angular momentum flux, is as yet undermined. To specify the constant we note that if there is a point where $\mathrm{d}\Omega/\mathrm{d}r = 0$ the viscous stress vanishes, and the constant is just the advective flux of angular momentum,

$$\text{constant} = -\dot{M} \cdot r^2 \Omega. \tag{1.79}$$

The physical situation where $\mathrm{d}\Omega/\mathrm{d}r = 0$ is where the protoplanetary disk extends all the way down to the surface of a slowly rotating star. The disk and the star form a single fluid system, and the angular velocity (shown in Fig. 1.10) must be a continuous function that connects $\Omega \approx 0$ in the star to $\Omega \propto r^{-3/2}$ within the disk. The viscous stress must then vanish at some radius $R_* + r_{\rm bl}$, where $r_{\rm bl}$ is the width of the *boundary layer* that separates the star from the Keplerian part of the disk. Within the boundary layer the angular velocity increases with radius, and gravity is balanced against a combination of rotation and radial pressure support. Elementary arguments [337] show that in many cases the boundary layer is narrow, so that $R_* + r_{\rm bl} \simeq R_*$. We then find that,

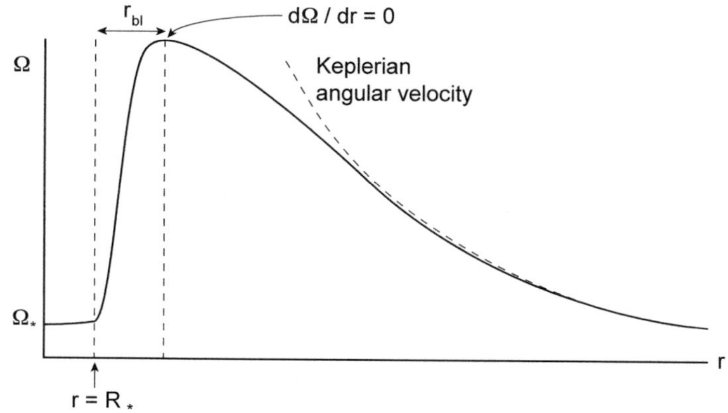

Fig. 1.10 A sketch of what the angular velocity profile $\Omega(r)$ must look like if the disk extends down to the surface of a slowly rotating star. By continuity there must be a point—usually close to the stellar surface—where $\mathrm{d}\Omega/\mathrm{d}r = 0$ and the viscous stress vanishes. Taken from Fig. 3.2 in [18], © Cambridge University Press

$$\text{constant} \simeq -\dot{M} R_*^2 \sqrt{\frac{GM_*}{R_*^3}}, \tag{1.80}$$

and the steady-state solution for the disk simplifies to,

$$\nu\Sigma = \frac{\dot{M}}{3\pi}\left(1 - \sqrt{\frac{R_*}{r}}\right). \tag{1.81}$$

For a specified viscosity this equation gives the steady state surface density profile of a disk with a constant accretion rate $\dot{M}$. Away from the inner boundary $\Sigma(r) \propto \nu^{-1}$, and the radial velocity (Eq. 1.64) is $v_r = -3\nu/2r$.

In obtaining Eq. (1.81) we have derived an expression for a *Keplerian* disk via an argument that relies on the *non-Keplerian* form of $\Omega(r)$ in a boundary layer. The resulting expression for the surface density is valid in the disk at $r \gg R_*$, but would not work well close to the star *even if there is a boundary layer*. To model the boundary layer properly, we would need equations that self-consistently determine the angular velocity along with the surface density [335].

1.4.2.2 Magnetospheric Accretion

For protoplanetary disks the stellar magnetic field can have a dominant influence on the disk close to the star [219]. The simplest magnetic geometry involves a dipolar stellar magnetic field that is aligned with the stellar rotation axis and perpendicular to the disk plane. The unperturbed field then has a vertical component at the disk surface,

$$B_z = B_* \left(\frac{r}{R_*} \right)^{-3}. \qquad (1.82)$$

In the presence of a disk, the vertical field will thread the disk gas and be distorted by differential rotation between the Keplerian disk and the star. The differential rotation twists the field lines that couple the disk to the star, generating an azimuthal field component at the disk surface B_ϕ and a magnetic torque per unit area (counting both upper and lower disk surfaces),

$$T = \frac{B_z B_\phi}{2\pi} r. \qquad (1.83)$$

Computing the perturbed field accurately is hard (for simulation results see, e.g. [360]), but it is easy to identify the qualitative effect that it has on the disk. For a star with rotation period P, we define the co-rotation radius $r_{\rm co}$ as the radius where the field lines have the same angular velocity as that of Keplerian gas in the disk,

$$r_{\rm co} = \left(\frac{GM_* P^2}{4\pi^2} \right)^{1/3}. \qquad (1.84)$$

There are then two regions of star-disk magnetic interaction:

- Interior to co-rotation ($r < r_{\rm co}$) the disk gas has a greater angular velocity than the field lines. Field lines that link the disk and the star here are dragged forward by the disk, and exert a braking torque that removes angular momentum from the disk gas.
- Outside co-rotation ($r > r_{\rm co}$) the disk gas has a smaller angular velocity than the field lines. The field lines are dragged backward by the disk, and there is a positive torque on the disk gas.

Young stars are typically rapid rotators [75], so the co-rotation radius lies in the inner disk. For $P = 7$ days, for example, the co-rotation radius around a Solar mass star is at $r_{\rm co} \simeq 15\ R_\odot$ or 0.07 AU.

The presence of a stellar magnetic torque violates the assumption of a zero-torque boundary condition, though the steady-state solution we derived previously (Eq. 1.81) will generally still apply at sufficiently large radius. The strong radial dependence of the stellar magnetic torque means that there is only a narrow window of parameters where the torque will be significant yet still allow the disk to extend to the stellar surface. More commonly, a dynamically significant stellar field will disrupt the inner disk entirely, yielding a magnetospheric regime of accretion in which the terminal phase of accretion is along stellar magnetic field lines. The disruption (or magnetospheric) radius r_m can be estimated in various ways [219], but all yield the same scaling as the spherical Alfvén radius that is obtained by equating the magnetic pressure of a dipolar field to the ram pressure of spherical infall,

$$r_m \simeq \left(\frac{k B_*^2 R_*^6}{\dot{M} \sqrt{G M_*}} \right)^{2/7}. \tag{1.85}$$

Here B_* is the stellar surface field (defined such that $B_* R_*^3$ is the dipole moment) and k a constant of the order of unity. Taking $k = 1$ for a Solar mass star,

$$r_m \simeq 14 \left(\frac{B_*}{\text{kG}} \right)^{4/7} \left(\frac{R_*}{2\, R_\odot} \right)^{12/7} \left(\frac{\dot{M}}{10^{-8}\, M_\odot\, \text{yr}^{-1}} \right)^{-2/7} R_\odot. \tag{1.86}$$

Often, the magnetospheric radius is comparable to the co-rotation radius. This is to some extent expected, since if r_m is substantially different from $r_{\rm co}$ the magnetic torque acting on the star will tend to modify P in the direction of reducing the difference.

1.4.2.3 Accretion on to and in Binaries

Boundary conditions for disk evolution also need modification in binary systems. For a coplanar disk orbiting *interior* to a prograde stellar binary companion, tidal torques from the companion remove angular momentum from the outer disk and prevent it from expanding too far [323]. The tidal truncation radius roughly corresponds to the largest simple periodic orbit in the binary potential [321], which is at about 40% of the orbital separation for a binary with mass ratio $M_2/M_1 = 0.5$.[7] The tidal torque is a function of radius, but to a first approximation one may assume that tides impose a rigid no-expansion condition at $r = r_{\rm out}$. From Eq. (1.64),

$$\left. \frac{\partial}{\partial r} \left(\nu \Sigma r^{1/2} \right) \right|_{r=r_{\rm out}} = 0. \tag{1.87}$$

This type of boundary condition may also be an appropriate approximation for a circumplanetary disk truncated by stellar tides [283].

An exterior *circumbinary* disk will also experience stellar gravitational torques, which in this case add angular momentum to the disk and slow viscous inflow. How best to model these torques is an open question, particularly in the case of extreme mass ratio binaries composed of a star and a massive planet. Pringle [339] derived an illuminating analytic solution for circumbinary disk evolution under the assumption that tidal torques *completely prevent* inflow past some radius $r = r_{\rm in}$. With this assumption the boundary condition at $r = r_{\rm in}$ is $v_r = 0$, and the task is to find a solution to Eq. (1.63) with this finite radius boundary condition. A simple solution is possible for $\nu = kr$, with k a constant. Defining scaled variables,

---

[7]The size of the disk (and even whether it is tidally truncated at all) will be different if the disk is substantially misaligned with respect to the orbital plane of the binary [261, 293].

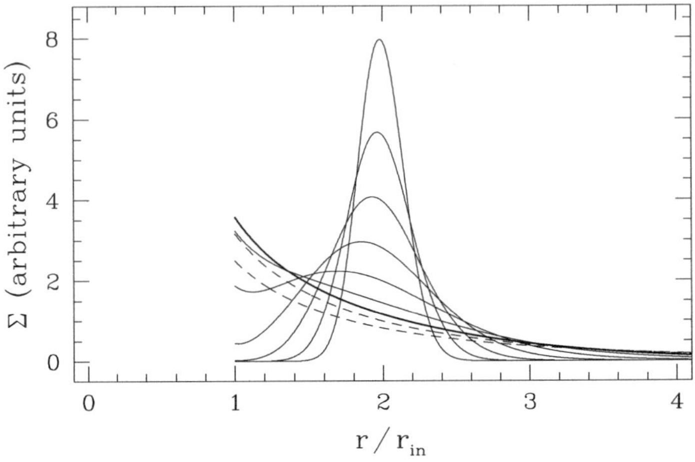

Fig. 1.11 The time-dependent analytic solution (Eq. 1.90) to the disk evolution equation with a $v_r = 0$ boundary condition at $r = 1$ for the case $\nu = r$. The solid curves show the evolution of a ring of gas initially at $r = 2$, at times $t = 0.002$, $t = 0.004$, $t = 0.008$ etc. The bold curve is at $t = 0.128$, and dashed curves show later times. Gas initially accretes, but eventually *decretes* due to the torque being applied at the boundary

$$x = r^{1/2}$$
$$\sigma = \Sigma r^{3/2}, \tag{1.88}$$

the $t > 0$ solution for an initial mass distribution,

$$\sigma(x, t = 0) = \sigma_0 \delta(x - x_1), \tag{1.89}$$

is [339],

$$\sigma = \frac{\sigma_0 t^{-1/2}}{4(3\pi k)^{1/2}} \left\{ \exp\left[-(x - x_1)^2/3kt\right] + \exp\left[-(x + x_1 - 2x_{\rm in})^2/3kt\right] \right\}. \tag{1.90}$$

The solution, plotted in Fig. 1.11, can be compared to the zero-torque solution (Fig. 1.9, though note this is for a constant viscosity). The initial evolution is similar, but at late times the torque that precludes inflow past $r_{\rm in}$ causes qualitatively different behavior. The disk switches from an accretion to a *decretion* disk, with an outward flow of mass driven by the binary torque.

The classical decretion disk solution was developed as a model for disk evolution around binaries. It is not clear, however, whether it is ever realized in the binary context. Numerical simulations of the interaction between a binary and a circumbinary disk show that angular momentum transfer to the disk co-exists with persistent inflow into a low density cavity containing the binary [25, 112]. How best to represent this complexity in a one-dimensional model is not entirely obvious. The decretion disk solution may be a better description of other astrophysical situations, such as disks around rapidly rotating and strongly magnetized stars.

1.4.3 Viscous Heating

Although stellar irradiation is often the dominant source of heat for protoplanetary disks (Sect. 1.3.2), dissipation of gravitational potential energy associated with accretion is also important. Ignoring irradiation for the time being, we can derive the effective temperature profile of a steady-state viscous disk. In the regime where the classical equations are valid, the fluid dissipation per unit area is [338],

$$Q_+ = \frac{9}{4}\nu\Sigma\Omega^2. \tag{1.91}$$

Using the steady-state solution for $\nu\Sigma$ (Eq. 1.81) we equate Q_+ to the rate of local energy loss by radiation. If the disk is optically thick, the disk radiates (from both sides) at a rate $Q_- = 2\sigma T_{\rm disk}^4$, with σ being the Stefan-Boltzmann constant. This yields an effective temperature profile,

$$T_{\rm disk}^4 = \frac{3GM_*\dot{M}}{8\pi\sigma r^3}\left(1 - \sqrt{\frac{R_*}{r}}\right). \tag{1.92}$$

Away from the inner boundary, the steady-state temperature profile for a viscous disk ($T_{\rm disk} \propto r^{-3/4}$) is steeper than for irradiation. For any accretion rate, we then expect viscous heating to be most important in the inner disk, whereas irradiation always wins out at sufficiently large radii.

The viscous disk temperature profile is **not** what we get from considering just the local dissipation of potential energy. The gradient of the potential energy per unit mass ε, is $\mathrm{d}\varepsilon/\mathrm{d}r = GM_*/r^2$. For an accretion rate $\dot{M}$, the luminosity available to be radiated from an annulus of width Δr due to local potential energy release would be,

$$L = \frac{1}{2}\frac{GM_*\dot{M}}{r^2}\Delta r, \tag{1.93}$$

where the factor of a half accounts for the fact that half the energy goes into increased kinetic energy, with only the remainder available to be thermalized and radiated. Equating this luminosity to the black body emission from the annulus, $2\sigma T_{\rm disk}^4 \cdot 2\pi r\Delta r$, would give a profile that is a factor three different from the asymptotic form of Eq. (1.92). The difference arises because the radial transport of angular momentum is accompanied by a radial transport of energy. The local luminosity from the disk surface at any radius then has a contribution from potential energy liberated closer in.

The optically thick regions of irradiated protoplanetary disks will be vertically isothermal. When viscous heating dominates, however, there must be a vertical temperature gradient to allow energy to be transported from the mid-plane toward the photosphere. What this gradient looks like, in detail, depends on the vertical distribution of the heating, which is not well known. However, an approximation to $T(z)$ can be derived assuming that the energy dissipation due to viscosity is strongly concentrated toward the mid-plane. We define the optical depth to the disk mid-plane,

$$\tau = \frac{1}{2}\kappa_R \Sigma, \tag{1.94}$$

where κ_R is the Rosseland mean opacity. The vertical density profile of the disk is $\rho(z)$. If the vertical energy transport occurs via radiative diffusion then for $\tau \gg 1$ the vertical energy flux $F(z)$ is given by the equation of radiative diffusion [364]

$$F_z(z) = -\frac{16\sigma T^3}{3\kappa_R \rho}\frac{\mathrm{d}T}{\mathrm{d}z}. \tag{1.95}$$

Now assume for simplicity that *all* of the dissipation occurs at $z = 0$. In that case $F_z(z) = \sigma T_{\rm disk}^4$ is constant with height. We integrate from the mid-plane to the photosphere at $z_{\rm ph}$ assuming that the opacity is also constant,

$$-\frac{16\sigma}{3\kappa_R}\int_{T_c}^{T_{\rm disk}} T^3 dT = \sigma T_{\rm disk}^4 \int_0^{z_{\rm ph}} \rho(z')dz' \tag{1.96}$$

$$-\frac{16}{3\kappa_R}\left[\frac{T^4}{4}\right]_{T_c}^{T_{\rm disk}} = T_{\rm disk}^4 \frac{\Sigma}{2}, \tag{1.97}$$

where the final equality relies on the fact that for $\tau \gg 1$ almost all of the disk gas lies below the photosphere. For large optical depth $T_c^4 \gg T_{\rm disk}^4$ and the equation simplifies to,

$$\frac{T_c^4}{T_{\rm disk}^4} \simeq \frac{3}{4}\tau. \tag{1.98}$$

Often *both* stellar irradiation and accretional heating contribute significantly to the thermal balance of the disk. If we define $T_{\rm disk,visc}$ to be the effective temperature that would result from viscous heating in the absence of irradiation (i.e. the quantity called $T_{\rm disk}$, with no subscript, above) and $T_{\rm irr}$ to be the irradiation-only effective temperature, then,

$$T_c^4 \simeq \frac{3}{4}\tau T_{\rm disk,visc}^4 + T_{\rm irr}^4 \tag{1.99}$$

is an approximation for the central temperature, again valid for $\tau \gg 1$.

These formulae can be applied to estimate the location of the snow line. In the Solar System meteoritic evidence [296] places the transition between water vapor and water ice, which occurs at a mid-plane temperature of 150–180 K, at around

2.7 AU. This is substantially further from the Sun than would be expected if the only disk heating source was starlight. Including viscous heating, however, an accretion rate of $2 \times 10^{-8}\, M_\odot\, \mathrm{yr}^{-1}$ could sustain a mid-plane temperature of 170 K at 2.7 AU in a disk with $\Sigma = 400\ \mathrm{g\ cm}^{-2}$ and $\kappa_R = 1\ \mathrm{cm}^2\ \mathrm{g}^{-1}$. This estimate (from Eq. 1.98) is consistent with more detailed models for protoplanetary disks [58], though considerable variation in the location of the snow line is introduced by uncertainties in the vertical structure [281].

1.4.4 Warped Disks

The classical equation for surface density evolution needs to be rethought if the disk is non-planar. Disks may be warped for several reasons; the direction of the angular momentum vector of the gas that forms the disk may not be constant, the disk may be perturbed tidally by a companion [236, 305], or warped due to interaction with the stellar magnetosphere [235].

A warp affects disk evolution through physics that is independent of its origin (for a brief review, see [304]). In a warped disk, neighboring annuli have specific angular momenta that differ in direction as well as in magnitude. If we define a unit tilt vector $\mathbf{l}(r, t)$ that is locally normal to the disk plane, the shear then has a vertical as well as a radial component [309],

$$\mathbf{S} = r\frac{\mathrm{d}\Omega}{\mathrm{d}r}\mathbf{l} + r\Omega\frac{\partial \mathbf{l}}{\partial r}. \tag{1.100}$$

The most important consequence of the vertical shear is that it introduces a periodic vertical displacement of radially separated fluid elements. As illustrated in Fig. 1.12 this displacement, in turn, results in a horizontal pressure gradient that changes sign across the mid-plane and is periodic on the orbital frequency. In a Keplerian disk this forcing frequency is resonant with the epicyclic frequency.

How the disk responds to the warp-generated horizontal forcing depends on the strength of dissipation [324]. If the disk is sufficiently viscous, specifically if,

$$\alpha > \frac{h}{r}, \tag{1.101}$$

the additional shear is damped locally. The equation for the surface density and tilt evolution (the key aspects of which are derived in [340], though see [307] for a complete treatment) then includes terms which diffusively damp the warp at a rate that is related to the radial redistribution of angular momentum. Normally, warp damping is substantially faster than the viscous evolution of a planar disk. Even for a Navier-Stokes viscosity—which is fundamentally isotropic—the *effective* viscosity which damps the warp is a factor $\approx 1/2\alpha^2$ larger than its equivalent in a flat disk. Rapid evolution also occurs in the opposite limit of an almost inviscid disk with,

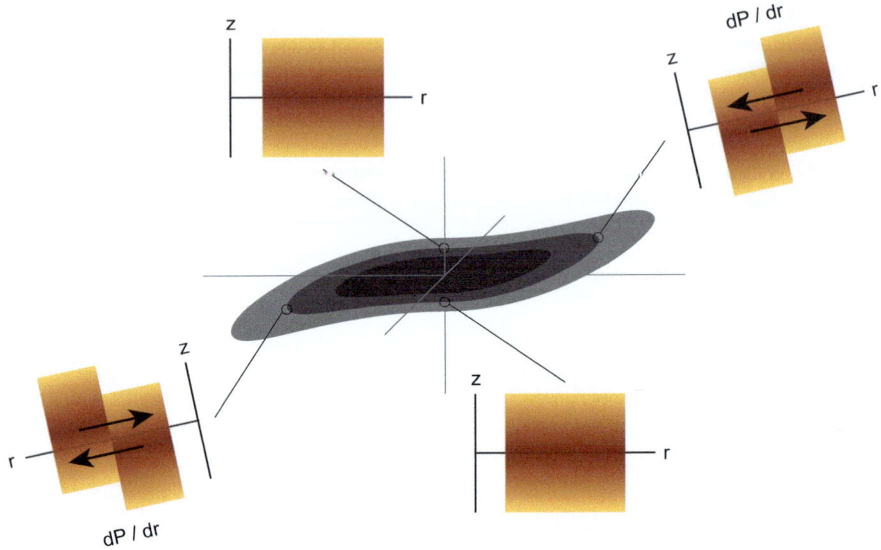

Fig. 1.12 Illustration (after [254, 304]) of how a warp introduces an oscillating *radial* pressure gradient within the disk. As fluid orbits in a warped disk, vertical shear displaces the mid-planes of neighboring annuli. This leads to a time-dependent radial pressure gradient $\mathrm{d}P/\mathrm{d}r(z)$. Much of the physics of warped disks is determined by how the disk responds to this warp-induced forcing

$$\alpha < \frac{h}{r}, \tag{1.102}$$

but in this case the component of angular momentum associated with the warp is communicated radially in the form of a wave. For a strictly inviscid disk the linearized fluid equations for the evolution of the tilt vector take a simple form [262],

$$\frac{\partial^2 \mathbf{l}}{\partial t^2} = \frac{1}{\Sigma r^3 \Omega} \frac{\partial}{\partial r} \left(\Sigma r^3 \Omega \frac{c_s^2}{4} \frac{\partial \mathbf{l}}{\partial r} \right). \tag{1.103}$$

The speed of the warp wave is $v_w \approx c_s/2$.

In most cases we expect protoplanetary disks to have $\alpha < h/r$, and warps will evolve in the wave-like regime. We expect, however, that the details of warp evolution will depend upon the nature of angular momentum transport, and little is known about how warps behave for the transport mechanisms (such as non-ideal MHD) most relevant to protoplanetary disks. The extent of the differences between warp evolution with realistic transport and that with a Navier-Stokes viscosity are undetermined.

1.4.5 Disk Winds

Viscous evolution driven by redistribution of angular momentum is a consequence of turbulent (or possibly laminar) stresses that are internal to the fluid. Evolution can also be driven by external torques, the most important of which is the magnetic torque that an MHD wind exerts on the surface of the disk. An excellent pedagogical introduction to disk winds is the review by Spruit [384], while Königl and Salmeron [221] provide a more recent account that addresses the peculiarities specific to protoplanetary disks.

Winds that are driven solely by pressure gradients ("thermal winds") are of interest as a mechanism for disk dispersal (Sect. 1.9) but do not change the qualitative character of disk evolution. Instead, we consider here MHD winds, with the simplest case being that of a well-ionized disk is threaded by a large-scale ordered poloidal magnetic field. In the ideal MHD limit the fluid is tied to magnetic field lines, which can facilitate acceleration of the wind while exerting a back reaction on the disk that removes angular momentum. This type of MHD wind is known as a Blandford-Payne wind [69]. Other varieties of MHD outflow could also occur. Winds could be launched by a gradient of *toroidal* magnetic field pressure [266], perhaps during or shortly after the collapse of the cloud that forms both the star and the disk. Jets can also originate from the interaction between the stellar magnetosphere and the disk [372].

Returning to the specific case of Blandford-Payne winds, their structure, illustrated in Fig. 1.13, generically has three regions. Within the disk the energy density in the magnetic field, $B^2/8\pi$, is smaller than ρc_s^2, the thermal energy.[8] Due to flux conservation, however, the energy in the vertical field component, $B_z^2/8\pi$, is roughly constant with height for $z < r$, while the gas pressure typically decreases at least exponentially with a scale height $h \ll r$. This leads to a region above the disk surface where magnetic forces dominate. The magnetic force per unit volume is,

$$\frac{\mathbf{J} \times \mathbf{B}}{c} = -\nabla \left(\frac{B^2}{8\pi} \right) + \frac{\mathbf{B} \cdot \nabla \mathbf{B}}{4\pi}, \tag{1.104}$$

where the current,

$$\mathbf{J} = \frac{c}{4\pi} \nabla \times \mathbf{B}. \tag{1.105}$$

The force, in general, can be written as shown above as the sum of a magnetic pressure gradient and a force due to magnetic tension. In the disk wind region where magnetic forces dominate, the requirement that they exert a finite acceleration on the low density gas implies that the force approximately vanishes, i.e. that,

$$\mathbf{J} \times \mathbf{B} \approx 0. \tag{1.106}$$

[8]We can also consider situations where the magnetic pressure in the disk is stronger than the gas pressure, though it must always be weaker than ρv_ϕ^2.

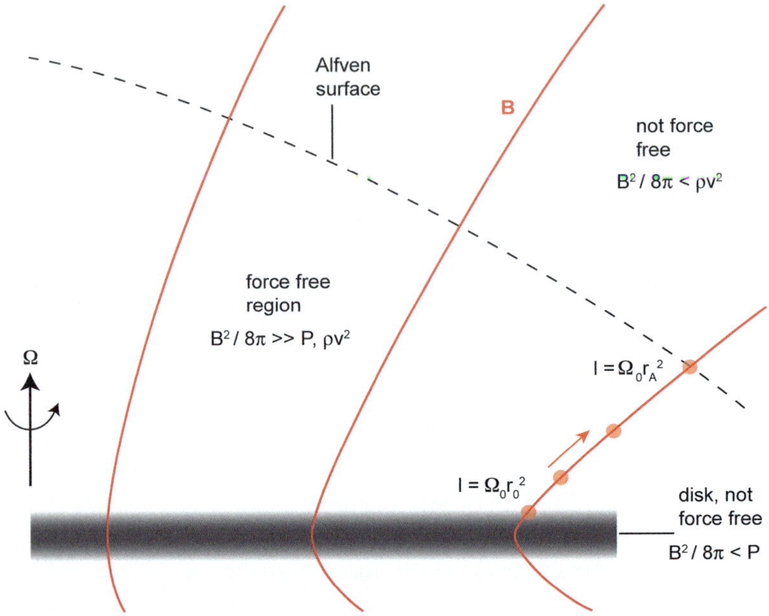

Fig. 1.13 Illustration, after Fig. 1 in [384], of the different regions of a disk wind solution

The structure of the magnetic field in the magnetically dominated region is then described as being "force-free", and in the disk wind case (where B changes slowly with z) the field lines must be approximately straight to ensure that the magnetic tension term is also small. If the field lines support a wind, the force-free structure persists up to where the kinetic energy density in the wind, ρv^2, first exceed the magnetic energy density. This is called the Alfvén surface. Beyond the Alfvén surface, the inertia of the gas in the wind is sufficient to bend the field lines, which tend to wrap up into a spiral structure as the disk below them rotates.

Magneto-centrifugal driving can launch a wind from the surface of a cold gas disk if the magnetic field lines are sufficiently inclined to the disk normal. The critical inclination angle in ideal MHD can be derived via an *exact* mechanical analogy. To proceed, we note that in the force-free region the magnetic field lines are (i) basically straight lines, and (ii) enforce rigid rotation out to the Alfvén surface at an angular velocity equal to that of the disk at the field line's footpoint. The geometry is shown in Fig. 1.14. We consider a field line that intersects the disk at radius r_0, where the angular velocity is $\Omega_0 = \sqrt{GM_*/r_0^3}$, and that makes an angle θ to the disk normal. We define the spherical polar radius r, the cylindrical polar radius ϖ, and measure the distance along the field line from its intersection with the disk at $z = 0$ as s. In the frame co-rotating with Ω_0 there are no magnetic forces along the field line to affect the acceleration of a wind; the sole role of the magnetic field is to constrain the gas to move along a straight line at constant angular velocity. Following this line of

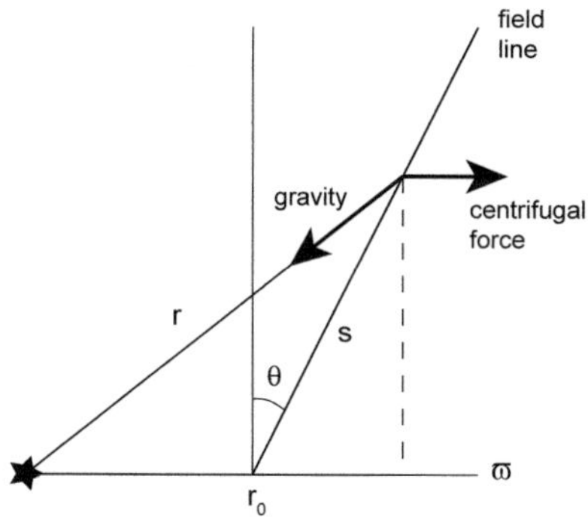

Fig. 1.14 Geometry for the calculation of the critical angle for magneto-centrifugal wind launching. A magnetic field line **s**, inclined at angle θ from the disk normal, enforces rigid rotation at the angular velocity of the foot point, at cylindrical radius $\varpi = r_0$ in the disk. Working in the rotating frame we consider the balance between centrifugal force and gravity

argument, the acceleration of a wind can be fully described in terms of an effective potential,

$$\Phi_{\rm eff}(s) = -\frac{GM_*}{r(s)} - \frac{1}{2}\Omega_0^2\varpi^2(s). \tag{1.107}$$

The first term is the gravitational potential, while the second describes the centrifugal potential in the rotating frame.

Written out explicitly, the effective potential is,

$$\Phi_{\rm eff}(s) = -\frac{GM_*}{(s^2 + 2sr_0 \sin\theta + r_0^2)^{1/2}} - \frac{1}{2}\Omega_0^2 (r_0 + s\sin\theta)^2 . \tag{1.108}$$

This function is plotted in Fig. 1.15 for various values of the angle θ. If we consider first a vertical field line ($\theta = 0$) the effective potential is a monotonically increasing function of distance s. For modest values of θ there is a potential barrier defined by a maximum at some $s = s_{\rm max}$, while for large enough θ the potential decreases from $s = 0$. In this last case purely magneto-centrifugal forces suffice to accelerate a wind off the disk surface, even in the absence of any thermal effects. The critical inclination angle of the field can be found by computing $\theta_{\rm crit}$, specified though the condition,

$$\left.\frac{\partial^2\Phi_{\rm eff}}{\partial s^2}\right|_{s=0} = 0. \tag{1.109}$$

Evaluating this condition, we find,

$$\begin{aligned} 1 - 4\sin^2\theta_{\rm crit} &= 0 \\ \Rightarrow \theta_{\rm crit} &= 30^\circ, \end{aligned} \tag{1.110}$$

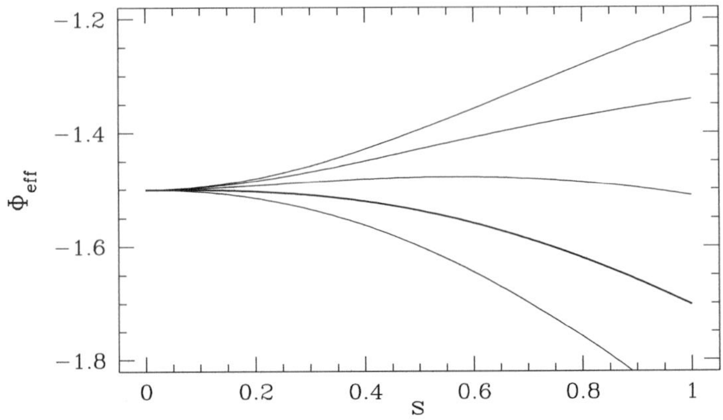

Fig. 1.15 The variation of the disk wind effective potential $\Phi_{\rm eff}$ (in arbitrary units) with distance s along a field line. From top downwards, the curves show field lines inclined at 0°, 10°, 20°, 30° (in bold) and 40° from the normal to the disk surface. For angles of 30° and more from the vertical, there is no potential barrier to launching a cold MHD wind directly from the disk surface

as the minimum inclination angle from the vertical needed for unimpeded wind launching in ideal MHD [69]. Since most of us are more familiar with mechanical rather than magnetic forces, this derivation in the rotating frame offers the easiest route to this result. But it can, of course, be derived just as well by working in the inertial frame of reference [384].

The rigid rotation of the field lines interior to the Alfvén surface means that gas being accelerated along them increases its specific angular momentum. The magnetic field, in turn, applies a torque to the disk that removes a corresponding amount of angular momentum. If a field line, anchored to the disk at radius r_0, crosses the Alfvén surface at (cylindrical) radius r_A, it follows that the angular momentum flux is,

$$\dot{L}_w = \dot{M}_w \Omega_0 r_A^2, \tag{1.111}$$

where $\dot{M}_w$ is the mass loss rate in the wind. Removing angular momentum at this rate from the disk results in a local accretion rate $\dot{M} = 2\dot{L}_w/\Omega_0 r_0^2$. The ratio of the disk accretion rate to the wind loss rate is,

$$\frac{\dot{M}}{\dot{M}_w} = 2\left(\frac{r_A}{r_0}\right)^2. \tag{1.112}$$

If r_A substantially exceeds r_0 (by a factor of a few, which is reasonable for detailed disk wind solutions) a relatively weak wind can carry away enough angular momentum to support a much larger accretion rate.

The behavior of a disk that evolves under wind angular momentum loss depends on how the wind and the poloidal magnetic field respond to the induced accretion. It is not immediately obvious that a steady accretion flow is even possible. The form of the effective potential (Fig. 1.15) suggests that the rate of mass and angular momentum loss in the wind ought to be a strong function of the inclination of the field lines—for $\theta < 30°$ there is a potential barrier to wind launching, while for $\theta \geq 30°$ there is no barrier at all. How θ responds to changes in the inflow rate through the disk is of critical importance [80, 264, 310], and there is no simple analog of the diffusive disk evolution equation. Despite this, viscous and wind-driven disks exhibit some qualitative difference that may enable observational tests. The classical test is the evolution of the outer disk radius, which expands in viscous models (*if* there is no mass loss, even in the form of a thermal wind) but contracts if an MHD wind dominates. Old and almost forgotten observations [382] of disk radius changes in dwarf novae (accreting white dwarfs in mass transfer binary systems) provided empirical support for viscous disk evolution *in those specific systems*. Disk winds also remove energy, and so another potential test is to look for evidence of the dissipation of accretion energy within the disk that is present in viscous models but absent for winds. At fixed accretion rate a wind-driven disk will have a lower effective temperature at small radii than its viscous counterpart, and this will alter the predicted spectral energy distribution (at large radii the temperature of both types of disk is set by irradiation, and no significant differences are expected).

1.4.5.1 Magnetic Field Transport

The strength and radial profile of the vertical magnetic field threading the disk are important quantities for disk winds, and for turbulence driven by MHD processes. Disks form from the collapse of magnetized molecular clouds, and it is inevitable that they will inherit non-zero flux at the time of formation. The poloidal component of that flux can subsequently be advected radially with the disk gas, diffuse relative to the gas, or (if the flux has varying sign across the disk) reconnect.

A theory for the radial transport of poloidal flux within geometrically thin accretion disks was developed by Lubow, Papaloizou and Pringle [263]. They considered a disk within which turbulence generates an effective viscosity ν and an effective magnetic diffusivity η. The disk is threaded by a vertical magnetic field $B_z(r,t)$, which is supported by a combination of currents within the disk and (potentially) a current external to the disk. Above the disk, as in Fig. 1.13 the field is force-free. The field lines bend within the disk, such that the poloidal field has a radial component $B_{rs}(r,t)$ at the disk surface.

To proceed (following the notation in [165]), we first write the poloidal field in terms of a magnetic flux function ψ, such that $\mathbf{B} = \nabla\psi \times \mathbf{e}_\phi$, where $\mathbf{e}_\phi$ is a unit vector in the azimuthal directions. The components of the field are,

$$
\begin{aligned}
B_r &= -\frac{1}{r}\frac{\partial \psi}{\partial z}, \\
B_z &= \frac{1}{r}\frac{\partial \psi}{\partial r}.
\end{aligned}
\tag{1.113}
$$

From the second of these relations we note that ψ is (up to a factor of 2π) just the vertical magnetic flux interior to radius r. We split ψ into two pieces, a piece $\psi_{\rm disk}$ due to currents within the disk, and a piece ψ_∞ due to external currents ("at infinity"),

$$\psi = \psi_{\rm disk} + \psi_\infty. \tag{1.114}$$

The external current generates a magnetic field that is uniform across the disk.

With these definitions, the evolution of the poloidal field in the simplest analysis [165, 263] obeys,

$$\frac{\partial \psi}{\partial t} + r\left(v_{\rm adv} B_z + v_{\rm diff} B_{rs}\right) = 0, \tag{1.115}$$

where $v_{\rm adv}$ is the advective velocity of magnetic flux and $v_{\rm diff}$ its diffusive velocity due to the turbulent resistivity within the disk. The disk component of the flux function is related to the surface radial field via an integral over the disk. Schematically,

$$\psi_{\rm disk}(r) = \int_{r_{\rm in}}^{r_{\rm out}} F(r, r') B_{rs}(r') {\rm d}r', \tag{1.116}$$

where F is a rather complex function that can be found in Guilet and Ogilvie [165]. The appearance of this integral reflects the inherently global nature of the problem—a current at some radius within the disk affects the poloidal magnetic field everywhere, not just locally—and makes analytic or numerical solutions for flux evolution more difficult. Nonetheless, Eq. (1.116) can be inverted to find B_{rs} from ψ, after which the more familiar Eq. (1.115) can be solved for specified transport velocities to determine the flux evolution.

Equation (1.115) expresses a simple competition, the inflow of gas toward the star will tend to drag in poloidal magnetic field, but this will set up a radial gradient and be opposed by diffusion. The physical insight of Lubow et al. [263] was to note that although both of these are processes involving turbulence (and, very roughly, we might guess that $\nu \sim \eta$), the scales are quite distinct. From Fig. 1.13, we note that because the field lines bend *within the disk*, a moderately inclined external field (with $B_{rs} \sim B_z$) above the disk only has to diffuse across a scale $\sim h$ to reconnect with its oppositely directed counterpart below the disk. Dragging in the field with the mean disk flow, however, requires angular momentum transport across a larger scale r. In terms of transport velocities, in a steady-state we have,

$$\begin{aligned} v_{\rm adv} &\sim \frac{\nu}{r}, \\ v_{\rm diff} &\sim \frac{\eta}{h} \frac{B_{\rm rs}}{B_z}, \end{aligned} \tag{1.117}$$

so that for $\nu \sim \eta$ and $B_{rs} \sim B_z$ diffusion beats advection by a factor $\sim(h/r)^{-1} \gg 1$. Defining the magnetic Prandtl number $P_m = \nu/\eta$ as the ratio of the turbulent viscosity to the turbulent resistivity, we would then expect that in steady-state [263],

$$\frac{B_{rs}}{B_z} \sim \frac{h}{r} P_m. \tag{1.118}$$

This argument is the origin of the claim that *thin disks do not drag in external magnetic fields*. It suggests that obtaining enough field line bending to launch a magneto-centrifugal wind ought to be hard, and that whatever primordial flux the disk is born with may be able to escape easily.

The physical arguments given above are robust, but a number of authors have emphasized that the calculation of the transport velocities that enter into Eq. (1.115) involves some subtleties [164, 165, 260, 310, 396]. The key point is that the viscosity and resistivity that enter into the equation for flux transport should not be computed as density-weighted vertical averages, but rather (in the case of the induction equation) as conductivity-weighted averages [164]. This makes a large difference for protoplanetary disks, where the conductivity is both generally low, and highest near the disk surface where the density is small. The derived transport velocities are, moreover, functions of the poloidal field strength, in the sense that diffusion becomes relatively less efficient as the field strength decreases. It should be noted that none of the flux transport calculations fully includes all of the MHD effects expected to be present in protoplanetary disks (see Sect. 1.5.4.3). It seems possible, though, that the variable efficiency of flux diffusion could simultaneously allow,

- For rapid flux loss from the relatively strongly magnetized disks formed from star formation [249], averting overly rapid wind angular momentum loss that would be inconsistent with observed disk lifetimes.
- For convergence toward a weak but non-zero net poloidal flux (possibly with a ratio of thermal to poloidal field magnetic pressure at the mid-plane $\beta \sim 10^4 - 10^7$) later in the disk lifetime [165].

As we will discuss in the next section, poloidal field strengths in roughly this range are of interest for their role in stimulating MHD instabilities within weakly ionized disks, so this is a speculative but interesting scenario.

- Isolated protoplanetary disks are expected to evolve under a combination of (i) internal redistribution of angular momentum, often referred to informally as viscosity, (ii) mass loss due to accretion and thermal winds, and (iii) mass and angular momentum loss in MHD winds. The relative importance of these processes for disk evolution is not clearly established.
- In the limit where local internal redistribution of angular momentum dominates, the disk surface density evolves according to a diffusion equation that matches the one derived for a viscous fluid. The viscosity in this equation must be interpreted as an effective viscosity resulting from a turbulent process.

- The use of the α prescription, and the extension of viscous models to complex non-axisymmetric geometries, are common and sometimes useful approximations. Neither, however, has a clear physical justification.
- The dissipation of accretion energy within the disk modifies its vertical temperature structure, and can lead to much higher mid-plane temperatures within a few AU of the star.
- Magnetic winds can be launched along sufficiently inclined field lines that thread the disk surface. The long-term evolution of such winds depends upon the efficiency of magnetic flux transport relative to gas accretion.

1.5 Turbulence

Turbulence within protoplanetary disks is important for two independent reasons. First, if it is strong enough and has the right properties, it could account for disk evolution by redistributing angular momentum much faster than molecular viscosity. Second, turbulence has its fingers in a plethora of planet formation processes, ranging from the collision velocities of small particles [314] to the formation of planetesimals [197] and the migration rate of low-mass planets [218]. For these reasons we would like to understand disk turbulence, even if (as is possible) it is not always responsible for disk evolution.

The first order of business when considering possibly turbulent fluid systems is usually to estimate the Reynolds number, which is a dimensionless measure of the relative importance of inertial and viscous forces. For a system with characteristic size L, velocity U, and (molecular) viscosity ν, the Reynolds number is defined as,

$$\mathrm{Re} = \frac{UL}{\nu}. \tag{1.119}$$

There is no unique or "best" definition of U and L for protoplanetary disks, but whatever choice we make gives a very large number. For example, taking $L = h$ and $U = c_s$ then our estimate of the viscosity at 1 AU (Eq. 1.73) implies $\mathrm{Re} \sim 10^{11}$. By terrestrial standards this is an enormous Reynolds number. Experiments on flow through pipes, for example—including those of Osborne Reynolds himself—show that turbulence is invariably present once $\mathrm{Re} > 10^4$ [119]. If turbulence is present within protoplanetary disks there is no doubt that viscous forces will be negligible on large scales, and the turbulence will exhibit a broad inertial range.

Figure 1.16 lists some of the possible sources of turbulence within protoplanetary disks. It's a long list! We can categorize the candidates according to various criteria,

- The physics involved in generating the turbulence. The simplest possibility (which appears unlikely) is that turbulence develops spontaneously in an isothermal, purely hydrodynamical shear flow. More complete physical models invoke entropy

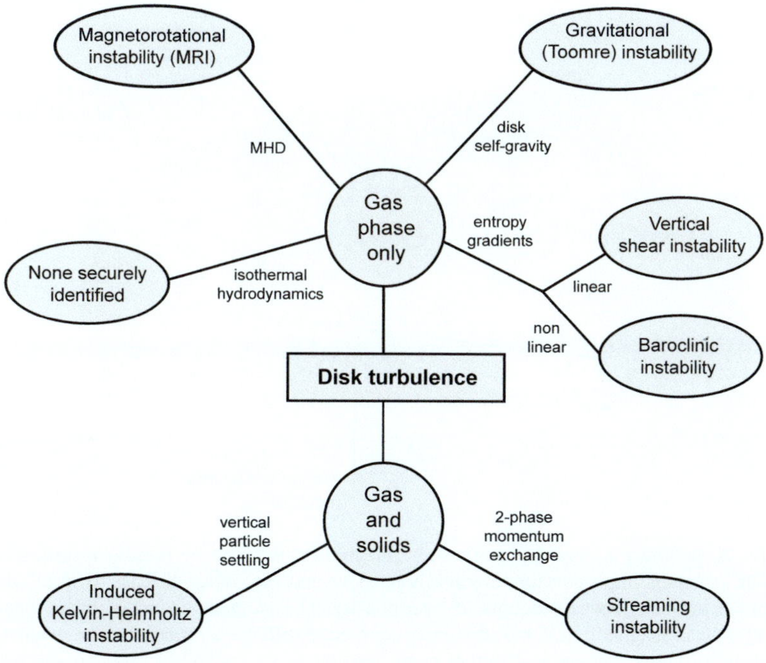

Fig. 1.16 A menu of the leading suspects for creating turbulence within protoplanetary disks

gradients, disk self-gravity or magnetic fields as necessary elements for the origin of turbulence.
- The origin of the free energy that sustains the turbulence, which could be the radial or vertical shear, heating from the star, or velocity differences between gas and solid particles.
- The character of the instabilities proposed to initiate turbulence from an initially non-turbulent flow. *Linear* instabilities grow exponentially from arbitrarily small perturbations, while *non-linear* instabilities require a finite amplitude disturbance. Demonstrating the existence of linear instabilities is relatively easy, whereas proving that a fluid system is non-linearly stable is very hard.
- The species involved. In this section we concentrate on instabilities present in purely gaseous disks; additional instabilities are present once we consider how gas interacts aerodynamically with its embedded solid component (Sect. 1.7.3).

Figure 1.17 illustrates the dominant fluid motions or forces involved in some of the most important disk instabilities.

For each candidate instability we would like to know the disk conditions under which it would be present, its growth rate, and the strength and nature of the turbulence that eventually develops. For disk evolution we are particularly interested in how efficiently the turbulence transports angular momentum (normally characterized by an effective α). In most cases the efficiency of transport can only be determined

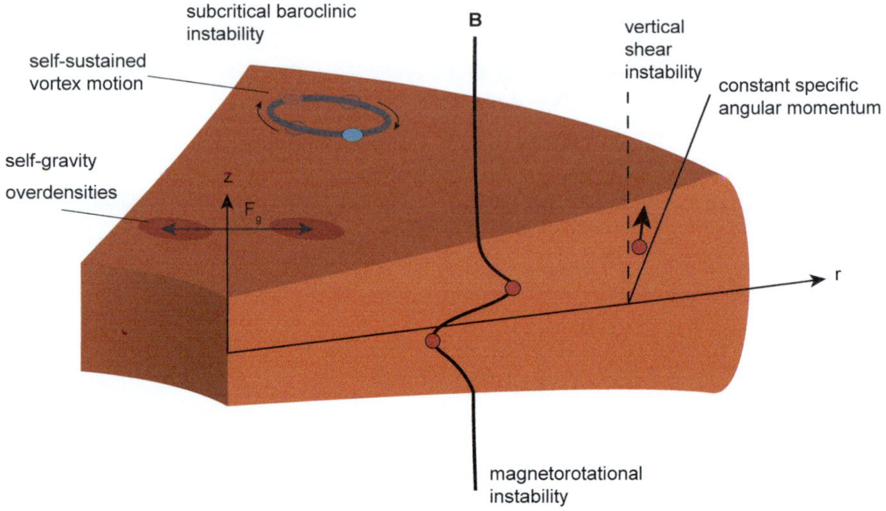

Fig. 1.17 A summary of the most important instabilities that can be present in protoplanetary disks. *Self-gravity* is important for sufficiently massive and cold disks. It leads to spiral arms and gravitational torques between regions of over-density. The *magnetorotational instability* occurs whenever a weak magnetic field is sufficiently coupled to differential rotation. The magnetic field acts to couple fluid elements at different radii, leading to an instability that can sustain MHD turbulence and angular momentum transport. The *vertical shear instability* feeds off the vertical shear that is set up in disks with realistic temperature profiles. It is a linear instability characterized by near-vertical growing modes. The *subcritical baroclinic instability* is a non-linear instability that operates in the presence of a sufficiently steep radial entropy gradient. It resembles radial convection, and leads to self-sustained vortices within the disk

using numerical simulations, whose fluctuating velocity and magnetic fields can be analyzed to determine α via the relation [45],

$$\alpha = \left\langle \frac{\delta v_r \delta v_\phi}{c_s^2} - \frac{B_r B_\phi}{4\pi \rho c_s^2} \right\rangle_\rho , \tag{1.120}$$

where the angle brackets denote a density weighted average over space (and time, in some instances). The first term in this expression is the Reynolds stress from correlated fluctuations in the radial and perturbed azimuthal velocity, the second term is the Maxwell stress from MHD turbulence. We speak of the stress as being "turbulent" if the averages in the above relation are dominated by contributions from small spatial scales. It is also possible for a disk to sustain large scale stresses—for example at some radius we might have non-zero *mean* radial and azimuthal magnetic fields—which are normally described as being "laminar". Note that it is possible for the velocity field to exhibit turbulence on small scales even if the stress is dominated by large scale contributions.

1.5.1 Hydrodynamic Turbulence

The dominant motion in protoplanetary disks is Keplerian orbital motion about a central point mass. Simplifying as much as possible, we first ask whether, in the absence of magnetic fields,[9] the radial shear present in a low-mass disk would be unstable to the development of turbulence. We first consider (rather unrealistically) a radially isothermal disk, where according to Eq. (1.28) there is no *vertical* shear. We then turn to the more general case where the temperature varies with radius, giving rise both to vertical shear and qualitatively distinct possibilities for instability.

1.5.1.1 Linear and Non-linear Stability

The linear stability of a shear flow with a smoothly varying $\Omega(r)$ against axisymmetric perturbations is given by Rayleigh's criterion (this is derived in most fluids textbooks, see e.g. [341]). The flow is stable if the specific angular momentum increases with radius,

$$\frac{\mathrm{d}l}{\mathrm{d}r} = \frac{\mathrm{d}}{\mathrm{d}r}\left(r^2\Omega\right) > 0 \rightarrow \text{stability}. \tag{1.121}$$

A Keplerian disk has $l \propto \sqrt{r}$ and is linearly stable.

There is no mathematical proof of the non-linear stability of Keplerian shear flow, but nor is there any known instability. The apparently analogous cases of pipe flow and Cartesian shear flows—which are linearly stable but undergo non-linear transitions to turbulence—are in fact sufficiently different problems as to offer no guidance [45]. There are analytic and numerical arguments against the existence of non-linear instabilities [46], which although not decisive [348] essentially rule out the hypothesis that a non-linear instability could result in *astrophysically interesting* levels of turbulence [241]. The same conclusion follows from laboratory experiments that have studied the stability of quasi-Keplerian rotation profiles in Taylor-Couette experiments [120]. One caveat is that laboratory experiments, and most theoretical work, consider the stability of unstratified cylindrical shear flows. Marcus and collaborators have identified a new instability (of a distinct character, related to the existence of locations in the flow known as critical layers, for a review of this physics see [285]) that can arise when the vertical stratification present in disks is included [276, 277]. The existence of this instability, which leads to self-replication of vortices, has been reproduced independently [245], and also shown to depend upon the radiative properties of the disk. Conditions in most regions of protoplanetary disks do not appear especially propitious for it to play a major role, but investigations remain in their infancy at the time of writing.

---

[9] Ignoring magnetic fields in astrophysical accretion flows is generally a stupid thing to do, and indeed there is broad consensus that the magnetorotational instability (MRI) [45] is responsible for turbulence and angular momentum transport in most accretion disks. In protoplanetary disks, however, the low ionization fraction means that the dominance of MHD instabilities is much less obvious, and purely hydrodynamic effects could in principle be important.

1.5.1.2 Entropy-Driven Instabilities

A separate class of purely hydrodynamic instabilities (no self-gravity, no magnetic fields) are what might loosely be called "entropy-driven" instabilities, in that they rely on the existence of a non-trivial temperature structure. The prototypical entropy-driven instability is of course convection, which could occur in the vertical direction if dissipation (associated with the physical process behind angular momentum transport) sets up an unstable entropy profile. This is evidently only conceivable in the region where viscous dissipation dominates, as irradiation prefers a nearly isothermal vertical structure. Even there, convective turbulence in disks is less efficient at transporting angular momentum than it is in transporting heat [242], and this disparity creates a formidable barrier to creating consistent models in which convection is the primary source of disk turbulence. Convection may still be present in some regions of disks, perhaps especially at high accretion rates, but as a byproduct of independent angular momentum transport processes (for an example in dwarf novae, see [183]).

A disk that has a radial temperature gradient necessarily has vertical shear (Eq. 1.28). The free energy associated with the vertical shear can be accessed via the vertical shear instability (VSI) analyzed by Nelson, Gressel and Umurhan [302]. The VSI is a disk application of the Goldreich-Schubert-Fricke instability [140, 156] of rotating stars, and was proposed as a source of protoplanetary disk transport by Urpin and Brandenburg [417].[10] The VSI is a linear instability with a maximum growth rate that is of the order of $h\Omega_{\rm K}$ [252], but which is strongly dependent on the radiative properties of the disk. The reason is that to access the free energy in the vertical shear requires vertical fluid displacements, which are easy in the limit that the disk is strictly vertically isothermal but strongly suppressed if it is stably stratified. The local *cooling time* of the fluid is thus a critical parameter, and the VSI will only operate in regions of the disk where radiative cooling and heating processes result in a cooling time that is the same or shorter than the dynamical time $\Omega_{\rm K}^{-1}$. This, in practice, restricts the application of the VSI to specific radii that depend upon the disk structure (Lin and Youdin suggest 5–50 AU [252], Malygin et al. 15–180 AU [273]), and limits its effectiveness if the dust opacity is reduced (due to coagulation into large particles). Under the right conditions, however, numerical simulations suggest that the VSI can generate relatively small but possibly significant levels of transport, with both Nelson et al. [302] and Stoll and Kley [389] finding α of a few $\times 10^{-4}$.

The radial entropy gradient may itself be unstable. The simplest instability would be radial convection (a linear instability). For a disk with pressure profile $P(r)$, density profile $\rho(r)$, and adiabatic index γ, we define the Brunt-Väisälä frequency,

$$N_r^2 = -\frac{1}{\gamma\rho}\frac{\mathrm{d}P}{\mathrm{d}r}\frac{\mathrm{d}}{\mathrm{d}r}\ln\left(\frac{P}{\rho^\gamma}\right). \qquad (1.122)$$

The Solberg-Hoïland criterion indicates that a Keplerian disk is convectively unstable if,

[10] As we shall see, a general rule is that all disk instabilities have long histories and pre-histories.

$$N_r^2 + \Omega_K^2 < 0. \tag{1.123}$$

Protoplanetary disks never (or at least almost never) have a steep enough profile of entropy to meet this condition, so radial convection will not set in. A different instability (the subcritical baroclinic instability, SBI) is possible, however, if the weaker condition $N_r^2 < 0$ (which is just the Schwarzschild condition for non-rotating convection) is satisfied [244, 330, 331]. The SBI, which is likely related to observations of vortex formation in earlier numerical simulations [216], is a non-linear instability that can be excited by finite amplitude perturbations. (Confusingly, it is unrelated to the linear "baroclinic instability" studied in planetary atmospheres.) The SBI relies on radial buoyancy forces to sustain vortical motion via baroclinic driving. This type of effect is possible in disks in which surfaces of constant density are not parallel to surface of constant pressure. Mathematically, for a fluid with vorticity $\omega = \nabla \times \mathbf{v}$, we can take the curl of the momentum equation to get an equation for the vortensity ω/ρ,

$$\frac{D}{Dt}\left(\frac{\omega}{\rho}\right) = \left(\frac{\omega}{\rho}\right) \cdot \nabla \mathbf{v} - \frac{1}{\rho}\nabla\left(\frac{1}{\rho}\right) \times \nabla P. \tag{1.124}$$

The baroclinic term, which for the SBI is responsible for generating and maintaining vorticity in the presence of dissipation, is the second term on the right hand side. The SBI, as with the VSI, is sensitive to the cooling time [244, 344], in this case because the baroclinic driving depends on the disk neither cooling too fast (which would eliminate the buoyancy effect) nor too slow (which would lead to constant temperature around the vortex). In compressible simulations, Lesur and Papaloizou [244] found that under favorable disk conditions the SBI could lead to outward transport of angular momentum with $\alpha \sim 10^{-3}$.

1.5.2 Self-gravity

A disk is described as *self-gravitating* if it is unstable to the growth of surface density perturbations when the gravitational force between different fluid elements in the disk is included along with the force from the central star. For a disk with sound speed c_s, surface density Σ and angular velocity Ω (assumed to be *close* to Keplerian) a linear analysis (for textbook treatments, see e.g. [17, 341]) shows that a disk becomes self-gravitating when the Toomre Q [405],

$$Q \equiv \frac{c_s \Omega}{\pi G \Sigma} < Q_{\rm crit}, \tag{1.125}$$

where $Q_{\rm crit} \sim 1$. We can deduce this result informally using an extension of the time scale argument that gives the thermal Jeans mass. We first note that pressure will prevent the gravitational collapse of a clump of gas, on scale Δr, if the sound-crossing time $\Delta r/c_s$ is shorter than the free-fall time $\sqrt{\Delta r^3/G\Delta r^2\Sigma}$. (We're ignoring factors of 2, π and so on.) Equating these time scales gives the minimum scale that might be vulnerable to collapse as $\Delta r \sim c_s^2/G\Sigma$. On larger scales, collapse can be averted if the free-fall time is longer than the time scale on which radial shear will separate initially neighboring fluid elements. For a Keplerian disk this time scale is $\sim\Omega^{-1}$. If the disk is just on the edge of instability the minimum collapse scale set by pressure support must equal the maximum collapse scale set by shear. Imposing this condition for marginal stability we obtain $c_s\Omega/G\Sigma \sim 1$, in accord with the formal result quoted above.

To glean some qualitative insight into where a disk might be self-gravitating, consider a steady-state disk that is described by an α model in which the transport arises from some process *other* than self-gravity. Collecting some previous results, the steady-state condition implies $\nu\Sigma = \dot{M}/3\pi$, the α prescription is $\nu = \alpha c_s h$, and hydrostatic equilibrium gives $h = c_s/\Omega$. Substituting into Eq. (1.125) we find,

$$Q = \frac{3\alpha c_s^3}{G\dot{M}}. \tag{1.126}$$

Protoplanetary disks generically get colder (and hence have lower c_s) at larger distances from the star, and this is where self-gravity is most likely to be important.

The disk mass required for self-gravity to become important can be estimated. Ignoring radial gradients of all quantities, we write the disk mass $M_{\rm disk} \sim \pi r^2\Sigma$, and again use the hydrostatic equilibrium result $h = c_s/\Omega$. Equation (1.125) then gives,

$$\frac{M_{\rm disk}}{M_*} > \left(\frac{h}{r}\right), \tag{1.127}$$

as the condition for instability. This manipulation of a local stability criterion into some sort of global condition is ugly, and begs the question of *where* in the disk $M_{\rm disk}$ and h/r should be evaluated. We can safely conclude, nonetheless, that for a typical protoplanetary disk with $(h/r) \simeq 0.05$ a disk mass of $10^{-2}\ M_*$ will not be self-gravitating, whereas one with $0.1\ M_*$ may well be.

There are two possibles outcomes of self-gravity in a disk,

- The disk may establish a (quasi) stable state, characterized globally by trailing spiral overdensities. Gravitational torques between different annuli in the disk *transport angular momentum* outward, leading to accretion.
- The pressure and tidal forces, which by definition are unable to prevent the onset of gravitational collapse, may never be able to stop it once it starts. In this case the disk *fragments* into bound objects, which interact with (and possibly accrete) the remaining gas.

Both possibilities are of interest. Angular momentum transport due to self-gravity may be dominant, at least on large scales, at early times while the disk is still massive. Fragmentation, which was once considered a plausible mechanism for forming the Solar System's giant planets [229], remains of interest as a way to form sub-stellar objects and (perhaps) very massive planets. Kratter and Lodato [225] review the physics of disk self-gravity in both the angular momentum transporting and fragmenting regimes. Here we summarize some mostly elementary arguments.

Gravity is a long-range force, and it is not at all obvious that we can deploy the machinery developed for viscous disks to study angular momentum transport in a self-gravitating disk. The transport could be largely non-local, driven for example by large-scale structures in the density field (such as bars) or by waves that transport energy and angular momentum a significant distance before dissipating [47]. There is no precise criterion for when self-gravitating transport can be described using a local theory, but numerical simulations indicate that this is a reasonable approximation for low-mass disks with $M_{\rm disk}/M_* \approx 0.1$ [103, 136, 255, 385]. Transport in more massive disks, such as might be present during the Class 0 and Class I phases of star formation, cannot be described locally (for multiple reasons, e.g. [256, 410]).

In cases where a local description of the transport is valid, we can use a thermal balance argument to relate the efficiency of angular momentum transport to the cooling time. Adopting a one-zone model for the vertical structure, we define the thermal energy of the disk, per unit surface area, as,

$$U = \frac{c_s^2 \Sigma}{\gamma(\gamma - 1)}, \quad (1.128)$$

where c_s is the mid-plane sound speed and γ is the adiabatic index. The cooling time (analogous to the Kelvin-Helmholtz time for a star) is then,

$$t_{\rm cool} = \frac{U}{2\sigma T_{\rm disk}^4}, \quad (1.129)$$

where $T_{\rm disk}$ is the effective temperature. Equating the cooling rate, $Q_- = 2\sigma T_{\rm disk}^4$, to the local viscous heating rate, $Q+ = (9/4)\nu\Sigma\Omega^2$ (Eq. 1.91), and adopting the α-prescription (Eq. 1.76), we find,

$$\alpha = \frac{4}{9\gamma(\gamma - 1)} \frac{1}{\Omega t_{\rm cool}}. \quad (1.130)$$

This relation, which is a general property of α disks quite independent of self-gravity, just says that a rapidly cooling disk needs efficient angular momentum transport if it to generate heat fast enough to remain in thermal equilibrium.

For most sources of angular momentum transport we are no more able to determine $t_{\rm cool}$ from first principles than we are α, so the above relation does not move us forward. Self-gravitating disks, however, have the unusual property that their Toomre Q, measured in the *saturated* (non-linear) state, is roughly constant and similar to the critical value $Q_{\rm crit}$ determined from *linear* theory. This property arises, roughly speaking, because the direct dependence of the linear stability criterion on temperature (via c_s) invites a stabilizing feedback loop—a disk that cools so that $Q < Q_{\rm crit}$ is more strongly self-gravitating, and produces more heating, while one that heats so that $Q > Q_{\rm crit}$ shuts off the instability. It is therefore reasonable to assume that a self-gravitating disk that does not fragment maintains itself close to marginal stability, as conjectured by Paczynski [322].

If we assume that $Q = Q_0$ exactly (where Q_0 is some constant presumably close to $Q_{\rm crit}$) then we have enough constraints to explicitly determine the functional form of α for a self-gravitating disk. Since Q depends on the mid-plane sound speed, $c_s = \sqrt{k_B T_c / \mu m_H}$, the condition of marginal stability directly gives us $T_c(\Sigma, \Omega)$,

$$T_c = \pi^2 Q_0^2 G^2 \left(\frac{\mu m_H}{k_B} \right) \frac{\Sigma^2}{\Omega^2}. \tag{1.131}$$

We can use this to determine $t_{\rm cool}$, and from that α, with the aid of the vertical structure relations developed in Sect. 1.4.3. To keep things simple, we adopt an opacity,

$$\kappa_R = \kappa_0 T_c^2 \tag{1.132}$$

that is appropriate for ice grains, and assume the disk is optically thick. The opacity law, together with the relations for the optical depth, $\tau = (1/2)\Sigma\kappa_R$, and the mid-plane temperature, $T_c^4 / T_{\rm disk}^4 \simeq (3/4)\tau$, then leads to,

$$\alpha = \frac{64\pi^2 Q_0^2 G^2 \sigma}{27\kappa_0} \left(\frac{\mu m_H}{k_B} \right)^2 \Omega^{-3}, \tag{1.133}$$

which coincidentally (for this opacity law) is only a function of Ω. It may look cumbersome—and the numerical factors are *certainly* not to be trusted—but what we have shown is that for a locally self-gravitating disk α is simply a constant times a determined function of Σ and Ω. This result allows for the evolution of low-mass self-gravitating disks to be modeled as a pseudo-viscous process [93, 246, 345].

Self-gravity is typically important in protoplanetary disks at large radii, where irradiation is usually the dominant factor determining the disk's thermal state (except at high accretion rates). The generalization of the self-regulation argument given above is obvious; if irradiation is not so strong as to stabilize the disk on its own then viscous heating from the self-gravitating "turbulence" has to make up the difference. The partially irradiated regime of self-gravitating disks has been studied using local numerical simulations [352], and the analytic generalization for the effective α that results can be found in Rafikov [346].

Ignoring irradiation again (and trusting the numerical factors that we just said were not to be trusted) we can examine the implications of Eq. (1.133) for protoplanetary disks. Taking $Q_0 = 1.5$ and $\kappa_0 = 2 \times 10^{-4}$ cm^2 g^{-1} K^{-2} [59] we find, across the region of the disk where ice grains would be the dominant opacity source, that,

$$\alpha \sim 0.3 \left(\frac{r}{50 \text{ AU}} \right)^{9/2}. \tag{1.134}$$

The steep radial dependence of the estimated self-gravitating α means that, unless all other sources of transport are *extremely* small, it will play no role in the inner disk. In the outer disk, on the other hand, we predict vigorous transport. The physical origin of the transport is density inhomogeneities that are caused by self-gravity, which become increasing large as α grows (explicitly, it is found [103] that the average fractional surface density perturbation $\delta\Sigma/\Sigma \propto \alpha^{1/2}$). Since even the linear threshold for gravitational instability implies that pressure forces can barely resist collapse, we expect that beyond some critical strength of turbulence a self-gravitating disk will be unable to maintain a steady-state. Rather, it will fragment into bound objects that are not (at least not immediately) subsequently sheared out or otherwise disrupted.

Our discussion up to this point might lead one to conjecture that the threshold for fragmentation could be written in terms of a critical dimensionless cooling time,

$$\beta_{\rm crit} \equiv \Omega t_{\rm cool,crit}, \tag{1.135}$$

or via a maximum $\alpha_{\rm crit}$ that a self-gravitating disk can sustain without fragmenting (these are almost equivalent, but defining the threshold in terms of α incorporates the varying compressibility as expressed through γ). Gammie [150], using local two dimensional numerical simulations, obtained $\beta_{\rm crit} \simeq 3$ for a two-dimensional adiabatic index $\gamma = 2$. Early global simulations by Rice et al. [351], which were broadly consistent with Gammie's estimate, implied a maximum effective transport efficiency $\alpha_{\rm crit} \simeq 0.1$ [354].

The idea that the fragmentation threshold is uniquely determined by a single number is too simplistic. Several additional physical effects matter. First, fragmentation requires that collapsing clumps can radiate the heat generated by adiabatic compression. There is therefore a dependence not just on the magnitude of the opacity, but also on how it scales with density and temperature [102, 204]. Second, if we view fragmentation as requiring a critical over-density in a random turbulent field there should be a time scale dependence, with statistically rarer fluctuations that lead to collapse becoming probable the longer we wait [319]. (This introduces an additional implicit dependence on γ, because the statistics of turbulent density fields depend upon how compressible the gas is [128].) Finally, disks can be prompted to fragment not only if they cool too quickly, but also if they accrete mass faster than self-gravity can transport it away. This regime is clearly relevant for Class 0 and Class I disks, where envelope accretion is ongoing. Accretion-induced fragmentation appears inevitable for very massive disks, where it would lead to binary formation [226].

In addition to these physical complexities, work by Meru and Bate [289] initiated a debate as to whether the critical cooling time scale derived from early simulations was robust. A number of simulations—run with both Smooth Particle Hydrodynamics (SPH) and grid-based methods—showed an increase in $\beta_{\rm crit}$ at higher numerical resolution, with little or no evidence for convergence. Recent work has revisited the problem using a Godunov-type mesh-free Lagrangian hydrodynamics scheme as implemented within the GIZMO code [187]. The use of a Riemann solver allows for significantly less numerical dissipation compared to SPH (which is also mesh-free and Lagrangian), which appears to be at the root of the convergence problem. Using GIZMO, Deng et al. [109] obtain a converged estimate $\beta_{\rm crit} \approx 3$ from global simulations of isolated self-gravitating disks with a disk-to-star mass ratio of 0.1. Baehr et al. [30] obtain the same threshold value from local three-dimensional simulations at high resolution. The fact that the most recent work largely agrees with the earliest relatively crude simulations, but not with the assuredly better calculations carried out in the intervening years, appears to be coincidental.

We can combine numerical estimates of $\alpha_{\rm crit}$ with the formula for $\alpha(r)$ (Eq. 1.134) to determine where isolated protoplanetary disks ought to be vulnerable to fragmentation. For a Solar mass star, fragmentation is expected beyond $r \sim 10^2$ AU [93, 286, 345], with an uncertainty in that estimate of perhaps a factor of two. In most (but perhaps not all) cases, it is expected that the disk conditions that allow fragmentation would lead to objects with masses in the brown dwarf regime, or above [227].

1.5.3 Magnetohydrodynamic Turbulence and Transport

The Rayleigh stability criterion (Eq. 1.121) applies to a fluid disk. It does not apply to a disk containing even an arbitrarily weak magnetic field, if that field is perfectly coupled to the gas (the regime of ideal MHD). In ideal MHD a weakly magnetized disk has entirely different stability properties from an unmagnetized one, and is unstable provided only that the angular velocity decreases outward. This is the magnetorotational instability (MRI) [44, 45], which is accepted as the dominant source of turbulence in well-ionized accretion disks (winds could still contribute to or dominate angular momentum loss). In protoplanetary disks the ideal MHD version of the MRI applies only in the thermally ionized region close to the star; across most of the disk we also need to consider both the dissipative (Ohmic diffusion, ambipolar diffusion) and the non-dissipative (the Hall effect) effects of non-ideal MHD. The Ohmic and ambipolar terms can be considered as modifying—albeit very dramatically—the ideal MHD MRI, while the Hall term introduces new effects (in part) via the Hall shear instability [230], which is a different beast unrelated to the ideal MHD MRI. The phenomenology of disk instabilities in non-ideal MHD is rich, and appears to give rise to both turbulent and laminar angular momentum transport as well as phenomena, such as MHD disk winds, that may be observable.

1.5.4 The Magnetorotational Instability

The MRI [44] is an instability of cylindrical shear flows that contain a weak (roughly, if the field is vertical, sub-thermal) magnetic field.[11] In ideal MHD the condition for instability is simply that,

[11] The mathematics of the MRI was worked out by Velikhov [422] and Chandrasekhar [85] around 1960. Thirty years passed before Balbus and Hawley [44] recognized the importance of the instability for accretion flows.

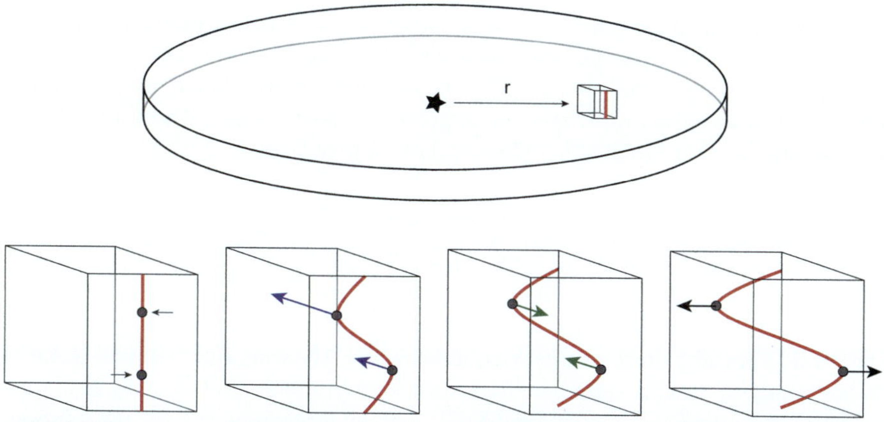

Fig. 1.18 Illustration showing why a weak vertical magnetic field destabilizes a Keplerian disk (the *magnetorotational instability* [45]). An initially uniform vertical field (weak enough that magnetic tension is not dominant) is perturbed radially. Due to the shear in the disk, an inner fluid element coupled to the field advances azimuthally faster than an outer one. Magnetic tension along the field line then acts to *remove* angular momentum from the inner element, and *add* angular momentum to the outer one. This causes further radial displacement, leading to an instability

$$\frac{\mathrm{d}\Omega^2}{\mathrm{d}r} < 0. \tag{1.136}$$

The fact that this condition is always satisfied in disks (though not in star-disk boundary layers) accounts for the MRI's central role in modern accretion theory.

Figure 1.18 illustrates what is going on to destabilize a disk that contains a magnetic field. A basically vertical field is slightly perturbed radially, so that it links fluid elements in the disk at different radii. Because of the shear in the disk, the fluid closer to the star orbits faster than the fluid further out, creating a toroidal field component out of what was initially just vertical and radial field. The tension in the magnetic field linking the two elements (which can be thought of, even mathematically, as being analogous to a stretched spring) imparts azimuthal forces to both the inner fluid (in the direction *opposite* to its orbital motion) and the outer fluid (*along* its orbital motion). The tension force thus reduces the angular momentum of the inner fluid element, and increases that of the outer element. The inner fluid then moves further inward (and the outer fluid further outward) and we have an instability.

We can derive the MRI instability condition in a very similar setup as Fig. 1.18. Consider a disk with a power-law angular velocity profile, $\Omega \propto r^{-q}$, that is threaded by a uniform vertical magnetic field B_0. We ignore any radial or vertical variation in density (and consistent with that, ignore the vertical component of gravity) and adopt an isothermal equation of state, $P = \rho c_s^2$, with c_s a constant. Our task is to determine whether infinitesimal perturbations to this equilibrium state are stable, or whether instead they grow exponentially with time, signaling a linear instability.

To proceed (largely following [141]) we define a locally Cartesian patch of disk that corotates at radius r_0, where the angular frequency is Ω_0. The Cartesian co-ordinates (x, y, z) are related to cylindrical co-ordinates (r, ϕ, z') via,

$$\begin{aligned} x &= r - r_0, \\ y &= r_0 \phi, \\ z &= z'. \end{aligned} \tag{1.137}$$

The local "shearing-sheet" (or in three dimensions,"shearing box") model is useful for both analytic stability studies, and for numerical simulations [173, 238]. In this co-rotating frame, the equations of ideal MHD pick up terms representing the fictitious Coriolis and centrifugal forces,

$$\begin{aligned} \frac{\partial \rho}{\partial t} + \nabla \cdot (\rho \mathbf{v}) &= 0, \\ \frac{\partial \mathbf{v}}{\partial t} + (\mathbf{v} \cdot \nabla)\mathbf{v} &= -\frac{1}{\rho}\nabla P + \frac{1}{4\pi\rho}(\nabla \times \mathbf{B}) \times \mathbf{B} - 2\Omega_0 \times \mathbf{v} + 2q\Omega_0^2 x \hat{\mathbf{x}}, \\ \frac{\partial \mathbf{B}}{\partial t} &= \nabla \times (\mathbf{v} \times \mathbf{B}). \end{aligned} \tag{1.138}$$

Here $\hat{\mathbf{x}}$ is a unit vector in the x-direction. As noted above, the initial equilibrium has uniform density, $\rho = \rho_0$, and a magnetic field $\mathbf{B} = (0, 0, B_0)$. There are no pressure or magnetic forces, so the velocity field is determined by a balance between the Coriolis and centrifugal terms,

$$2\Omega_0 \times \mathbf{v} = 2q\Omega_0^2 x \hat{\mathbf{x}}. \tag{1.139}$$

The equilibrium velocity field that completes the definition of the initial state is,

$$\mathbf{v} = (0, -q\Omega_0 x, 0)\,, \tag{1.140}$$

which has a linear shear (with $q = 3/2$ for a Keplerian disk) around the reference radius r_0.

To assess the stability of the equilibrium, we write the density, velocity and magnetic field as the sum of their equilibrium values plus a perturbation. We can recover the MRI with a particularly simple perturbation which depends on z and t only.[12] For the velocity components, for example, we write,

[12] An analysis that retains the x-dependence can be found in the original Balbus and Hawley paper [44], and follows an essentially identical approach. Studying the stability of non-axisymmetric perturbations (in y), however, requires a different and more involved analysis [104, 311, 329, 401].

$$\begin{aligned} v_x &= v_x'(z,t), \\ v_y &= -q\Omega_0 x + v_y'(z,t), \\ v_z &= v_z'(z,t), \end{aligned} \tag{1.141}$$

and do likewise for the density and magnetic field. We substitute these expressions into the continuity, momentum and induction equations, and discard any terms that are quadratic in the primed variables, assuming them to be *small* perturbations. This would give us seven equations in total (one from the continuity equation, and three each from the other equations), but the x and y components of the momentum and induction equations are all we need to derive the MRI. The relevant linearized equations are,

$$\begin{aligned} \frac{\partial v_x'}{\partial t} &= \frac{B_0}{4\pi\rho_0}\frac{\partial B_x'}{\partial z} + 2\Omega_0 v_y', \\ \frac{\partial v_y'}{\partial t} - q\Omega_0 v_x' &= \frac{B_0}{4\pi\rho_0}\frac{\partial B_y'}{\partial z} - 2\Omega_0 v_x', \\ \frac{\partial B_x'}{\partial t} &= B_0\frac{\partial v_x'}{\partial z}, \\ \frac{\partial B_y'}{\partial t} &= B_0\frac{\partial v_y'}{\partial z} - q\Omega_0 B_x'. \end{aligned} \tag{1.142}$$

We convert these linearized differential equations into algebraic equations by taking the perturbations to have the form, e.g.,

$$B_x' = \bar{B}_x' e^{i(\omega t - kz)}, \tag{1.143}$$

where ω is the frequency of a perturbation with vertical wave-number k. The time derivatives then pull down a factor of $i\omega$, while the spatial derivatives become ik. Our four equations simplify to,

$$\begin{aligned} i\omega v_x' &= -ik\frac{B_0 B_x'}{4\pi\rho_0} + 2\Omega_0 v_y', \\ i\omega v_y' &= -ik\frac{B_0 B_y'}{4\pi\rho_0} + (q-2)\Omega_0 v_x', \\ i\omega B_x' &= -ikB_0 v_x', \\ i\omega B_y' &= -ikB_0 v_y' - q\Omega_0 B_x'. \end{aligned} \tag{1.144}$$

(We've dropped the bars on the variables for clarity.) Eliminating the perturbation variables from these equations, we finally obtain the *MRI dispersion relation*,

$$\omega^4 - \omega^2\left[2k^2 v_A^2 + 2(2-q)\Omega_0^2\right] + k^2 v_A^2\left[k^2 v_A^2 - 2q\Omega_0^2\right] = 0, \tag{1.145}$$

where $v_A^2 = B_0^2/(4\pi\rho_0)$ is the Alfvén speed associated with the net field.

If $\omega^2 > 0$ then ω itself will be real and the perturbation $e^{i\omega t}$ will oscillate in time. Instability requires $\omega^2 < 0$, since in this case ω is imaginary and the perturbation will grow exponentially. Solving the dispersion relation we find the instability criterion is,

$$(kv_A)^2 - 2q\Omega_0^2 < 0. \tag{1.146}$$

Letting the field strength go to zero ($B_z \to 0$, $v_A \to 0$) we find that the condition for instability is simply that $q > 0$, i.e. that the angular velocity decrease outward. Even for an arbitrarily weak field, the result is completely different from Rayleigh's for a strictly hydrodynamic disk.

The growth rate of the instability and what it means for the magnetic field to be "weak" can also be derived from Eq. (1.145). Specializing to a Keplerian rotation law with $q = 3/2$ the dispersion relation takes the form shown in Fig. 1.19. For a fixed magnetic field strength (and hence a fixed Alfvén speed v_A) the flow is unstable for wavenumbers $k < k_{\rm crit}$ (i.e. on large enough spatial scales), where,

$$k_{\rm crit} v_A = \sqrt{3}\Omega_0. \tag{1.147}$$

As the magnetic field becomes stronger, the *smallest* scale $\lambda = 2\pi/k_{\rm crit}$ which is unstable grows, until eventually it exceeds the disk's vertical extent $\approx 2h$. For stronger vertical fields no unstable MRI modes fit within the disk, and the instability is suppressed. Using $h = c_s/\Omega$, the condition that the vertical magnetic field is weak enough to admit the MRI (i.e. that $\lambda < 2h$) becomes

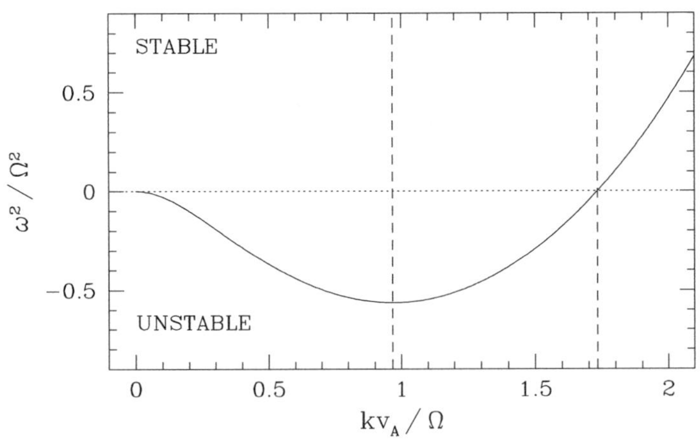

Fig. 1.19 The unstable branch of the MRI dispersion relation is plotted for a Keplerian rotation law. The flow is unstable ($\omega^2 < 0$) for all spatial scales smaller than $kv_A < \sqrt{3}\Omega$ (rightmost dashed vertical line). The most unstable scale (shown as the dashed vertical line at the center of the plot) is close to $kv_A \simeq \Omega$. Taken from Fig. 3.5 in [18], © Cambridge University Press

$$B_0^2 < \frac{12}{\pi} \rho c_s^2 \tag{1.148}$$

If we define the plasma β parameter as the ratio of gas to magnetic pressure,

$$\beta \equiv \frac{8\pi P}{B_0^2} \tag{1.149}$$

this condition can be expressed alternatively as,

$$\beta > \frac{2\pi^2}{3}. \tag{1.150}$$

A magnetic field whose *vertical* component approaches equipartition with the thermal pressure ($\beta \sim 1$) will be too strong to admit the existence of linear MRI modes, but a wide range of weaker fields are acceptable.

The maximum growth rate is determined by setting $d\omega^2/d(kv_A) = 0$ for the unstable branch of the dispersion relation plotted in Fig. 1.19. The most unstable scale for a Keplerian disk is,

$$(kv_A)_{\max} = \frac{\sqrt{15}}{4} \Omega_0, \tag{1.151}$$

where the growth rate is,

$$|\omega_{\max}| = \frac{3}{4} \Omega_0. \tag{1.152}$$

This result implies an *extremely* vigorous growth of the instability, with an exponential growth time scale that is a fraction of an orbital period. This means that if a disk is unstable to the MRI its growth rate will invariably be faster than that of hydrodynamic instabilities that may additionally be present. One cannot strictly be sure that, when several instabilities co-exist, the total angular momentum transport is dominated by the one with the fastest linear growth rate. Simulations, however, show that the MRI in the ideal MHD limit saturates to yield a turbulent state with a moderately high efficiency of angular momentum transport $\alpha \approx 0.02$ [107, 378]. It is likely to overwhelm plausible hydrodynamic sources of transport.

1.5.4.1 Non-ideal MHD

The MRI in its ideal MHD guise is relevant to protoplanetary disks only in the thermally ionized region close to the star (Sect. 1.3.3.1), where $T > 10^3$ K. The very weakly ionized gas further out is imperfectly coupled to the magnetic field, and this both modifies the properties of the MRI and leads to new MHD instabilities. We will begin by sketching the derivation of the non-ideal MHD equations (following Balbus [43], who justifies several of the approximations that we will make), and then estimate the magnitude of the extra terms that arise in protoplanetary disks.

The physics of how magnetic fields affect weakly-ionized fluids is easy to visualize. We consider a gas that is almost entirely neutral, with only a small admixture of ions and electrons (analogous considerations apply if the charge carriers are dust particles, but we will not go there). Magnetic fields exert Lorentz forces on the charged species, but not on the neutrals. Collisions between the neutrals and either the ions or the electrons lead to momentum exchange whenever the neutral fluid has a velocity differential with respect to the charged fluids.

We begin by considering the momentum equation. For the neutrals we have,

$$\rho \frac{\partial \mathbf{v}}{\partial t} + \rho (\mathbf{v} \cdot \nabla) \mathbf{v} = -\nabla P - \rho \nabla \Phi - p_{nI} - p_{ne}. \tag{1.153}$$

Here ρ, $\mathbf{v}$ and P (without subscripts) refer to the neutral fluid, and p_{nI} and p_{ne} are the rate of momentum exchange due to collisions between the neutrals and the ions/electrons respectively. Identical equations apply to the charged species, but for the addition of Lorentz forces,

$$\begin{aligned} \rho_e \frac{\partial \mathbf{v}_e}{\partial t} + \rho_e (\mathbf{v}_e \cdot \nabla) \mathbf{v}_e &= -\nabla P_e - \rho_e \nabla \Phi - e n_e \left(\mathbf{E} + \frac{\mathbf{v}_e \times \mathbf{B}}{c} \right) - p_{en}, \\ \rho_I \frac{\partial \mathbf{v}_I}{\partial t} + \rho_I (\mathbf{v}_I \cdot \nabla) \mathbf{v}_I &= -\nabla P_I - \rho_I \nabla \Phi + Z e n_I \left(\mathbf{E} + \frac{\mathbf{v}_I \times \mathbf{B}}{c} \right) - p_{In}. \end{aligned} \tag{1.154}$$

In these equations $\mathbf{E}$ and $\mathbf{B}$ are the electric and magnetic fields, the ions have charge Ze, where $-e$ is the charge on an electron, and of course $p_{ne} = -p_{en}$ and $p_{nI} = -p_{In}$. Having three momentum equations looks complicated, but we can make a large simplification to the system by noting that the time scale for macroscopic evolution of the fluid is generally much longer than the time scale for collisional or magnetic forces to alter a charged particle's momentum. We can then ignore *everything* in the charged species' momentum equations, except for the Lorentz and collisional terms. For the ions we have,

$$Z e n_I \left(\mathbf{E} + \frac{\mathbf{v}_I \times \mathbf{B}}{c} \right) - p_{In} = 0, \tag{1.155}$$

with a similar equation for the electrons. Imposing charge neutrality, $n_e = Z n_I$, we eliminate the electric field between the ion and electron equations to find an expression for the sum of the momentum transfer terms,

$$p_{In} + p_{en} = \frac{e n_e}{c} (\mathbf{v}_I - \mathbf{v}_e) \times \mathbf{B}. \tag{1.156}$$

The current density $\mathbf{J} = e n_e (\mathbf{v}_I - \mathbf{v}_e)$, so we can write this as,

$$p_{In} + p_{en} = \frac{\mathbf{J} \times \mathbf{B}}{c}. \tag{1.157}$$

Finally, we go to Maxwell's equations, and note that the current can be written as,

$$\frac{4\pi}{c}\mathbf{J} = \nabla \times \mathbf{B} + \frac{1}{c}\frac{\partial \mathbf{E}}{\partial t}. \tag{1.158}$$

The second term in Maxwell's equation is the displacement current, which is $\mathcal{O}(v^2/c^2)$ and consistently ignorable in non-relativistic MHD. Doing so, we substitute Eq. (1.157) in the neutral equation of motion to obtain,

$$\rho\frac{\partial \mathbf{v}}{\partial t} + \rho(\mathbf{v}\cdot\nabla)\mathbf{v} = -\nabla P - \rho\nabla\Phi + \frac{1}{4\pi}(\nabla \times \mathbf{B}) \times \mathbf{B}. \tag{1.159}$$

This is identical to the *ideal* MHD momentum equation (stated without derivation as Eq. 1.138) and pleasingly simple; we have reduced the three momentum equations to an equation for a single (neutral) fluid with a magnetic force term whose dependence on **B** is independent of the make-up of the gas. As one might guess, the consistent simplification of non-ideal MHD to a momentum equation for a single fluid is not always possible. Roughly speaking it works provided that the plasma's inertia is negligible compared to that of the neutral fluid, the coupling between charged and neutral species is strong, and the recombination time is short. Zweibel [449] gives an accessible account of the conditions necessary for a valid single-fluid description. Although some of the early analytic and numerical work on the MRI in weakly-ionized disks utilized a two-fluid approach [68, 174], in many protoplanetary disk situations a single fluid model is both justified [31] and substantially simpler. In the single fluid limit all of the complexities of non-ideal MHD enter only via the induction equation.

Deducing the non-ideal induction equation requires us to specify the form of the momentum coupling terms. Writing these in standard notation (which is different for the two terms, somewhat obscuring the symmetry),

$$\begin{aligned} p_{ne} &= n_e \nu_{ne} m_e (\mathbf{v} - \mathbf{v}_e), \\ p_{nI} &= \rho\rho_I\gamma(\mathbf{v} - \mathbf{v}_I), \end{aligned} \tag{1.160}$$

where ν_{ne} is the collision frequency of an electron with the neutrals, and γ is called the drag-coefficient. The ion-neutral coupling involves longer-range interactions than the electron-neutral coupling, and is accordingly stronger [43].

We now go back to the force balance deduced from the electron momentum equation,

$$-en_e\left(\mathbf{E} + \frac{\mathbf{v}_e \times \mathbf{B}}{c}\right) - p_{en} = 0, \tag{1.161}$$

and attempt to write the terms involving $\mathbf{v}_e$ and p_{en} entirely in terms of the current. We start with the exactly equivalent expression,

$$\mathbf{E}+\frac{1}{c}\left[\mathbf{v}+(\mathbf{v}_e-\mathbf{v}_I)+(\mathbf{v}_I-\mathbf{v})\right]\times\mathbf{B}+\frac{\nu_{ne}m_e}{e}\left[(\mathbf{v}_e-\mathbf{v}_I)+(\mathbf{v}_I-\mathbf{v})\right]=0, \tag{1.162}$$

and deal with the terms in turn. We have two terms that involve $(\mathbf{v}_e - \mathbf{v}_I)$, which can be replaced immediately with the current,

$$(\mathbf{v}_e-\mathbf{v}_I)=-\frac{\mathbf{J}}{en_e}. \tag{1.163}$$

The first term with $(\mathbf{v}_I - \mathbf{v})$ is exactly equal to $p_{In}/(\rho\rho_I\gamma)$. If, however, $|p_{In}| \gg |p_{en}|$, then Eq. (1.157) implies that, approximately,

$$(\mathbf{v}_I-\mathbf{v})\simeq\frac{\mathbf{J}\times\mathbf{B}}{c\rho\rho_I\gamma}. \tag{1.164}$$

Finally it can be shown (see Balbus [43] for details) that the final term with $(\mathbf{v}_I - \mathbf{v})$ can be consistently dropped. The version of Ohm's Law that we end up with is,

$$\mathbf{E}+\frac{\mathbf{v}\times\mathbf{B}}{c}-\frac{\mathbf{J}\times\mathbf{B}}{en_ec}+\frac{(\mathbf{J}\times\mathbf{B})\times\mathbf{B}}{c^2\rho\rho_I\gamma}-\frac{\nu_{ne}m_e}{e^2n_e}\mathbf{J}=0. \tag{1.165}$$

The non-ideal induction equation is then obtained by applying Faraday's law,

$$\nabla\times\mathbf{E}=-\frac{1}{c}\frac{\partial\mathbf{B}}{\partial t}, \tag{1.166}$$

to eliminate any explicit reference to the electric field. In its usual form,

$$\frac{\partial\mathbf{B}}{\partial t}=\nabla\times\left[\mathbf{v}\times\mathbf{B}-\eta\nabla\times\mathbf{B}-\frac{\mathbf{J}\times\mathbf{B}}{en_e}+\frac{(\mathbf{J}\times\mathbf{B})\times\mathbf{B}}{c\gamma\rho\rho_I}\right]. \tag{1.167}$$

We have defined the magnetic resistivity,

$$\eta=\frac{c^2}{4\pi\sigma} \tag{1.168}$$

where σ is (here) the electrical conductivity,

$$\sigma=\frac{e^2n_e}{m_e\nu_{en}}. \tag{1.169}$$

As before we can replace the current with the magnetic field via,

$$\mathbf{J}=\frac{c}{4\pi}\nabla\times\mathbf{B}, \tag{1.170}$$

so that the induction equation is solely a function of **B**. The terms on the right-hand-side are referred to as the inductive, Ohmic, Hall and ambipolar terms respectively.

The non-ideal terms in the induction equation depend upon the ionization state of the gas (through n_e and ρ_I) and upon the collision rates between the neutral and charged species (via η and γ). Standard values for these quantities are [68, 115],[13]

$$\eta = 234 \left(\frac{n}{n_e}\right) T^{1/2}\ \mathrm{cm^2\ s^{-1}},$$
$$\gamma = 3 \times 10^{13}\ \mathrm{cm^3\ s^{-1}\ g^{-1}}. \tag{1.171}$$

We are now ready to estimate the importance of the non-ideal terms in the environment of protoplanetary disks, and to ask what effect they have both on the MRI, and on the more general question of whether there is MHD turbulence or transport in disks.

1.5.4.2 Ohmic, Ambipolar and Hall Physics in Protoplanetary Disks

The non-ideal terms in Eq. (1.167) all depend inversely on the electron or ion density, so the strength of all non-ideal MHD effects relative to the inductive term increases with smaller ionization fraction. The three terms also have different dependencies on density, magnetic field strength and temperature, so the *relative* ordering of the non-ideal MHD effects varies with these parameters.

The Ohmic, Hall and ambipolar terms have different dependencies on the magnetic field geometry, and in a disk setting they influence the MRI in distinct ways (most importantly, the Hall effect differs from the others in being non-dissipative). There is therefore no model-independent way to precisely demarcate when each term will affect disk evolution. As a first guess, however, we can treat the magnetic field as a scalar and simply take the ratio of the Hall to the Ohmic term and the ambipolar to the Hall term,

$$\frac{\mathrm{H}}{\mathrm{O}} = \frac{cB}{4\pi e \eta n_e},$$
$$\frac{\mathrm{A}}{\mathrm{H}} = \frac{e n_e B}{c \gamma \rho \rho_I}. \tag{1.172}$$

Since $\eta \propto (n/n_e)$ and $n_e \propto \rho_I$ both of these ratios depend on (B/n). Substituting for η and γ, and taking the ion mass that enters into the ambipolar term as $30 m_H$, we

[13]Several assumptions are hardwired into these numbers. For the resistivity, it is assumed that currents are carried by the electrons, and that the conductivity is limited by electron-neutral collisions. For the drag coefficient we assume that the neutral gas is predominantly molecular hydrogen, and that the ions are moderately massive $m_i \simeq 30{-}40 m_{\mathrm{H}}$. It would be prudent to consult Blaes and Balbus [68], and references therein, should one encounter situations where these assumptions might fail.

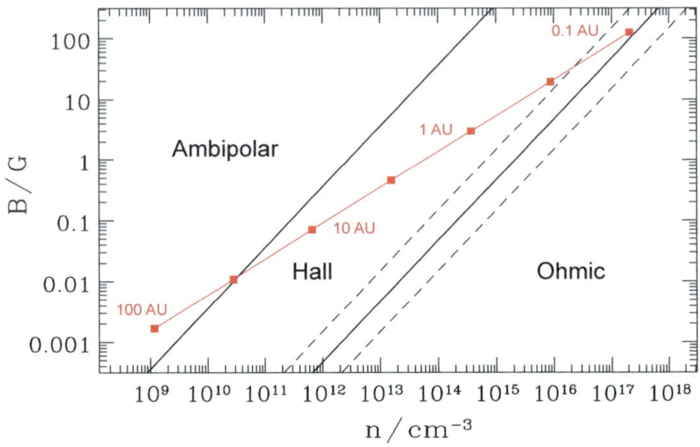

Fig. 1.20 Regions of the (n, B) parameter space in which different non-ideal terms are dominant. The boundary between the Ohmic and Hall regimes is plotted for $T = 100$ K (solid line) and also for temperatures of 10^3 K (upper dashed line) and 10 K (lower dashed line). The red line shows a very rough estimate of how the magnetic field in the disk might vary with density between the inner disk at 0.1 AU and the outer disk at 100 AU

can estimate the magnetic field strength for which the Ohmic and Hall terms have equal magnitude, and similarly for the Hall and ambipolar terms,

$$B_{\mathrm{O=H}} \approx 0.5 \left(\frac{n}{10^{15}\ \mathrm{cm}^{-3}} \right) \left(\frac{T}{100\ \mathrm{K}} \right)^{1/2} \mathrm{G},$$
$$B_{\mathrm{H=A}} \approx 4 \times 10^{-3} \left(\frac{n}{10^{10}\ \mathrm{cm}^{-3}} \right) \mathrm{G}. \tag{1.173}$$

Figure 1.20 shows these dividing lines in the (n, B) plane. Ohmic diffusion is dominant at high densities/low magnetic field strengths. Ambipolar diffusion dominates for low densities/high field strengths. The Hall effect is strongest for a fairly broad range of intermediate densities.

Estimating where protoplanetary disks fall in the (n, B) plane can be done in various ways. For a Solar System-motivated estimate we can start with the disk field inferred from laboratory measurements of chondrules in the Semarkona meteorite [144], which suggest that near the snow line ($r \simeq 3$ AU) the disk field was $B \simeq 0.5$ G. (There are caveats and a large systematic uncertainty associated with this measurement, all of which we ignore for now.) Let us assume that the surface density and temperature profiles are $\Sigma \propto r^{-3/2}$ and $T \propto r^{-1/2}$ respectively, and that the magnetic field pressure is the same fraction of the thermal pressure at all radii in the disk. Taking $\Sigma \simeq 300$ g cm^{-2} and $T = 150$ K at 3 AU, the inferred scalings of mid-plane density and magnetic field strength are then,

$$n \approx 2 \times 10^{13} \left(\frac{r}{3\ \mathrm{AU}}\right)^{-11/4} \mathrm{cm}^{-3},$$
$$B \approx 0.5 \left(\frac{r}{3\ \mathrm{AU}}\right)^{-13/8} \mathrm{G}. \tag{1.174}$$

The track defined by these relations is plotted in Fig. 1.20 for radii between 0.1 and 100 AU.

As we have emphasized, neither our approach to ranking the strength of the non-ideal terms, nor our estimate of the radial scaling of disk conditions, are anything more than crude guesses. Other approximations are equally valid (for example, one can order the terms in the (n, T) plane instead [19, 48, 231]). Nevertheless, because n and B vary by so many orders of magnitude across Fig. 1.20 the critical inferences we can draw are quite robust. We predict that the Hall effect is the dominant non-ideal MHD process at the disk mid-plane between (conservatively) 1 and 10 AU. Ohmic diffusion can become important as we approach the thermally ionized region interior to 1 AU. Ambipolar diffusion dominates at sufficiently large radii, of the order of 100 AU, and in the lower density gas away from the mid-plane.

1.5.4.3 The Dead Zone

The linear stability of Keplerian disk flow in non-ideal MHD has been extensively investigated (see, e.g. [48, 68, 110, 196, 231, 426]), and the reader interested in the non-ideal analogs to the MRI dispersion relation derived as Eq. (1.145) should start there. Proceeding less formally, we follow Gammie [148] to estimate the conditions under which Ohmic dissipation (ignoring for now the Hall term) would damp the MRI. The basic idea is to compare the time scale on which the ideal MRI would generate tangled magnetic fields to that on which Ohmic diffusion would smooth them out. We first note that diffusion erases small-scale structure in the field more efficiently than large scale features, so that the appropriate comparison is between growth and damping of the largest scale MRI models. Starting from the MRI dispersion relation (Eq. 1.145), for a Keplerian disk we consider the weak-field / long wavelength limit ($kv_A/\Omega \ll 1$). The growth rate of the MRI is,

$$|\omega| \simeq \sqrt{3} k v_A. \tag{1.175}$$

Writing this as a function of the spatial scale $\lambda = 2\pi/k$, we have,

$$|\omega| \simeq 2\pi\sqrt{3}\frac{v_A}{\lambda}. \tag{1.176}$$

Up to numerical factors the MRI on a given scale then grows on the Alfvén crossing time. Equating this growth rate to the Ohmic damping rate,

$$|\omega_\eta| \sim \frac{\eta}{\lambda^2} \tag{1.177}$$

we conclude that Ohmic dissipation will suppress the MRI on the largest available scale $\lambda \approx h$ provided that,

$$\eta > 2\pi\sqrt{3}v_A h. \tag{1.178}$$

We can express this result in a different form. Analogous to the fluid Reynolds number (Eq. 1.119) the magnetic Reynolds number Re_M is defined as,

$$\mathrm{Re}_M \equiv \frac{UL}{\eta} \tag{1.179}$$

where U is a characteristic velocity and L a characteristic scale. Taking $U = v_A$ and $L = h$ for a disk, the condition for Ohmic dissipation to suppress the MRI becomes

$$\mathrm{Re}_M < 1 \tag{1.180}$$

where order unity numerical factors have been omitted.

We now convert the condition for the suppression of the MRI into a limit on the ionization fraction $x \equiv n_e/n$. We make use of the formula for the magnetic resistivity (Eq. 1.171) and assume that Maxwell stresses transport angular momentum, so that $\alpha \sim v_A^2/c_s^2$ (this follows approximately from Eq. 1.120). The magnetic Reynolds number can then be estimated to be,

$$\mathrm{Re}_M = \frac{v_A h}{\eta} = \frac{\alpha^{1/2} c_s^2}{\eta \Omega}. \tag{1.181}$$

Substituting for η and c_s^2, the magnetic Reynolds number in a protoplanetary disk scales as,

$$\mathrm{Re}_M \approx 1.4 \times 10^{12} x \left(\frac{\alpha}{10^{-2}}\right)^{1/2} \left(\frac{r}{1\ \mathrm{AU}}\right)^{3/2} \left(\frac{T}{300\ \mathrm{K}}\right)^{1/2} \left(\frac{M_*}{M_\odot}\right)^{-1/2}. \tag{1.182}$$

For the given parameters, the critical ionization fraction below which Ohmic diffusion will quench the MRI is

$$x_{\mathrm{crit}} \sim 10^{-12}. \tag{1.183}$$

Clearly a very small ionization fraction suffices to couple the magnetic field to the gas and allows the MRI to operate, but there are large regions of the disk where even these ionization levels are not obtained and non-ideal effects are important.

Based on this analysis Gammie [148] noted, first, that the criterion for the MRI to operate under near-ideal MHD conditions in the inner disk coincides with the requirement that the alkali metals are thermally ionized (Fig. 1.7). The development of magnetized turbulence in the disk at radii where $T > 10^3$ K can therefore be modeled in ideal MHD. Second, he proposed that Ohmic diffusion would damp MHD turbulence in the low ionization environment near the disk mid-plane on scales of the order of 1 AU, creating a *dead zone* of sharply reduced turbulence and transport.

Gammie's original model is incomplete, as it did not include either ambipolar diffusion or the Hall effect, but the basic idea motivates much of the current work on MHD instabilities in disks.

1.5.4.4 Turbulence and Transport in Non-ideal MHD

Once we consider the full set of non-ideal terms, the first question is to assess the level of turbulence and transport that is expected as a function of their strengths. This is a well-defined but already difficult theoretical question, given that interesting values of the Ohmic, ambipolar and Hall terms span a broad range (depending physically on the temperature, density, ionization fraction and magnetic field strength). To define the problem in its most idealized form, we rewrite the non-ideal induction Eq. (1.167) as,

$$\frac{\partial \mathbf{B}}{\partial t} = \nabla \times \left[\mathbf{v} \times \mathbf{B} - \eta_O \nabla \times \mathbf{B} - \eta_H \frac{\mathbf{J} \times \mathbf{B}}{B} - \eta_A \frac{(\mathbf{J} \times \mathbf{B}) \times \mathbf{B}}{B^2} \right], \tag{1.184}$$

where *dimensionally* η_O, η_H and η_A are all diffusivities. There are different ways to construct dimensionless numbers from the diffusivities, but one useful set is,

$$\Lambda_O \equiv \frac{v_A^2}{\Omega \eta_O} \quad \text{(Ohmic Elsasser number)},$$
$$\text{Ha} \equiv \frac{v_A^2}{\Omega \eta_H} \quad \text{(Hall Elsasser number)},$$
$$\text{Am} \equiv \frac{v_A^2}{\Omega \eta_A} \quad \text{(ambipolar Elsasser number)}. \tag{1.185}$$

(Note that the ambipolar Elsasser number can also be written as $\text{Am} \equiv \gamma \rho_I / \Omega$, which is the number of ion-neutral collisions per dynamical time Ω^{-1}.) We further specify the net vertical magnetic field (if any) via the ratio of the mid-plane gas and magnetic field pressures (Eq. 1.149),

$$\beta_z = \frac{8\pi P}{B_z^2}. \tag{1.186}$$

Our question can then be rephrased; what is the level of angular momentum transport and turbulence in an MHD disk at radii where the non-ideal terms are characterized by the dimensionless parameters (Λ_O, Ha, Am, β_z).

Ohmic diffusion acts as a strictly dissipative process that stabilizes disks to magnetic field instabilities on scales below some critical value. Ambipolar diffusion is in principle more complex, because it does not dissipate currents that are parallel to the magnetic field. This distinction substantially impacts the linear stability of ambipolar-dominated disks [231], but appears to matter less for the non-linear evolution, whose properties are analogous to Ohmic diffusion. For both dissipative processes, simulations show that MRI-driven turbulence is strongly damped when the relevant dimensionless parameter (either Λ_0 or Am) drops below a critical value that depends upon the initial field geometry but is $\sim 1 - 10^2$ [38, 366, 377, 379, 414]. Consistent with the dead zone idea [148], we therefore expect substantial modification of the properties of MHD turbulence both in the mid-plane around 1 AU (where Ohmic diffusion is the dominant dissipative process) and in the disk atmosphere and at large radii $\sim 10^2$ AU (where ambipolar diffusion dominates).

The MRI dispersion relation is also modified by the Hall effect [48, 426], which differs from the other non-ideal terms in that it modifies the field structure without any attendant dissipation. In this respect the Hall term most closely resembles the inductive term $\nabla \times (\mathbf{v} \times \mathbf{B})$, and its strength can usefully be characterized by the ratio of the Hall to the inductive term. Non-linear simulations of the Hall effect in disks, which were pioneered by Sano and Stone [367, 368], have only recently been able to access the strongly Hall-dominated regime relevant to protoplanetary disks [32, 33, 232, 240, 381].[14] In vertically stratified disks with a net vertical field, Lesur et al. [240] find that the Hall effect has a controlling influence on disk dynamics on scales between 1 and 10 AU. For $\beta_z = 10^5$ a strong but *laminar* Maxwell stress (i.e. one dominated by large-scale radial and toroidal fields in Eq. 1.120) results when the net field is aligned with the rotation axis of the disk, whereas anti-alignment leads to extremely weak turbulence and transport.

The results from Hall-MHD simulations of protoplanetary disks are best interpreted not as a modification of the MRI, but rather as the signature of a distinct *Hall shear instability* [230]. In the presence of a net vertical magnetic field the Hall effect acts to *rotate* magnetic field vectors lying in the orbital plane (Fig. 1.21), with the sense of the rotation determined by whether the new field is aligned or anti-aligned to the rotation axis. In the aligned field case, the Hall-induced rotation allows the magnetic field to be amplified by the shear, while damping occurs in the anti-aligned limit. Unlike the MRI, the Hall shear instability does not depend on the Coriolis force, and is indifferent to the sign of the angular velocity gradient. By generating a radial field directly from an azimuthal one, the Hall effect (given a net field) supports a mean-field disk dynamo cycle [240, 407] that is qualitatively different from anything that is possible in ideal MHD.

Going beyond the idealized question of the effect of the non-ideal terms on turbulence and transport, our goal is to use the results described above to predict the structure and evolution of protoplanetary disks. This introduces new layers of uncertainty. To predict disk properties from first principles, we need at a minimum to know the strength of the different sources of ionization (Sect. 1.3.3.2), the rates of gas-phase and dust-induced recombination, and the global evolution of any net magnetic field (Sect. 1.4.5.1). We also need to model (or have good reasons to ignore)

[14] The numerical implementation of the Hall effect in simulation codes remains challenging, and the presence of large-scale fields in the saturated state suggests that local simulations may not be adequate to describe the outcome. Observationally important issues such as the level of fluid turbulence that accompanies the predominantly large-scale transport by Maxwell stresses remain to be fully understood.

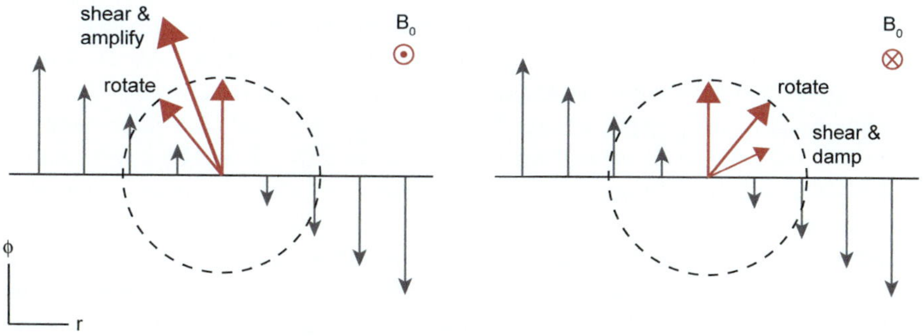

Fig. 1.21 An illustration (adapted from Kunz [230]) of the Hall shear instability. In the presence of a vertical field threading the disk, the Hall effect acts to rotate an initially toroidal field component either clockwise or anti-clockwise, depending upon the *sign* of the vertical field. The rotated field vector is then either amplified or damped by the shear. This instability differs from the MRI both technically (in that there is no reference to orbital motion, any shear flow suffices) and physically (via the dependence on the direction of the vertical magnetic field as well as its strength)

various non-MHD effects, including the hydrodynamic angular momentum transport processes already discussed and mass loss by photoevaporation [4] (Sect. 1.9). Given these uncertainties, the best that we can currently do is to highlight a number of qualitative predictions that receive support from numerical simulations:

- Net vertical magnetic fields are important for disk evolution. A vertical magnetic field enhances angular momentum transport by the MRI even in ideal MHD (roughly as $\alpha \propto \beta_z^{-1/2}$ [173]). In non-ideal MHD, local simulations by Simon et al. [376, 377] suggest that ambipolar damping of the MRI in the outer disk prevents resupply of the inner disk with gas *unless* a net field is present.[15] A net field with $\beta_z \simeq 10^4 - 10^5$ suffices, comparable to the fields expected in global models of flux evolution [165] but much weaker than the likely initial field left over from star formation. Accretion on scales of 30–100 AU occurs predominantly through a thin surface layer that is ionized by FUV photons [328, 376, 377], and is largely independent of the Hall effect [33].
- MHD winds and viscous transport can co-exist in disks. Local numerical simulations in ideal MHD by Suzuki and Inutsuka [392] showed that in the presence of net vertical field, the MRI was accompanied by mass and angular momentum loss in a disk wind. Winds are likewise seen in Bai and Stone's net field protoplanetary disk simulations at 1 AU that include Ohmic and ambipolar diffusion [39], in simulations at 30–100 AU where ambipolar diffusion dominates [376], and in simulations that include all non-ideal terms [240]. Caution is required before interpreting these

[15] It is not obvious that the inner disk *is* resupplied by gas, or, to put it more formally, that the disk attains a steady-state. Out to ~10 AU the viscous time scale is short enough that the disk will plausibly adjust to a steady state (provided only that a steady state is possible, see Sect. 1.6), but no such argument works out to 100 AU. Ultimately the question of whether gas at 100 AU ever reaches the star will need to be settled by observations as well as by theory.

local simulation results as quantitative predictions, because although the effective potential for wind launching is correctly represented (Eq. 1.108) there is a known and unphysical dependence of the mass loss rate on the vertical size of the simulation domain [142]. Outflows are also seen in recent *global* net field simulations [34, 64, 162], however, supporting the view that outflows driven by a combination of MHD and thermal processes could be a generic feature of protoplanetary disk accretion.

- Turbulence and angular momentum transport are not synonymous. In classical disk theory, the value of α determines not only the rate at which the disk evolves, but also the strength of turbulence and its effect on small solid particles. This link is doubly broken in more complete disk models. First, as already noted, angular momentum loss via winds (which need not be accompanied by turbulence) may be stronger than viscous transport at some radii. Second, even the internal component of transport may be primarily a large-scale "laminar" Maxwell stress, rather than small-scale turbulence [240, 376].
- The sign of the net field could lead to bimodality in disk properties. The Hall effect is the strongest non-ideal term interior to about 10 AU, and simulations [32, 33, 240] confirm the expectation from linear theory [48, 230, 426] that a disk with a weak field that is aligned to the rotation axis behaves quite differently from one with an anti-aligned field. Although there are possible confounding factors—for example the long-term evolution of the net field may itself differ with the sign of the field—it appears likely that the striking asymmetry seen in simulations introduces *some* observable bimodality in disk structure [381]. Hall MHD can also affect the properties of disks at an earlier time, during the collapse of molecular clouds and the formation of protostellar disks [224, 409], influencing for example their initial sizes.

Figure 1.22 illustrates a possible disk structure implied by the above results. The figure should be regarded as a work in progress; there is plenty of work remaining before we fully understand either the physics of potential angular momentum transport and loss processes, or how to tie that knowledge together into a consistent scenario for disk structure and evolution.

1.5.5 Transport in the Boundary Layer

The nature of transport in the boundary layer deserves a brief discussion. As discussed in Sect. 1.4.2.1 boundary layers are expected when the accretion rate is high enough to overwhelm the disruptive influence of the stellar magnetosphere (see Eq. 1.86 for a semi-quantitative statement of this condition). For most stars this requires high accretion rates, so the boundary layer and adjacent disk are hot enough to put us into the regime of thermal ionization and ideal MHD. In the disk we then expect angular momentum transport via the MRI. In the boundary layer, however, we face a problem. By definition, $d\Omega/dr > 0$ in the boundary layer (see Fig. 1.10), and this

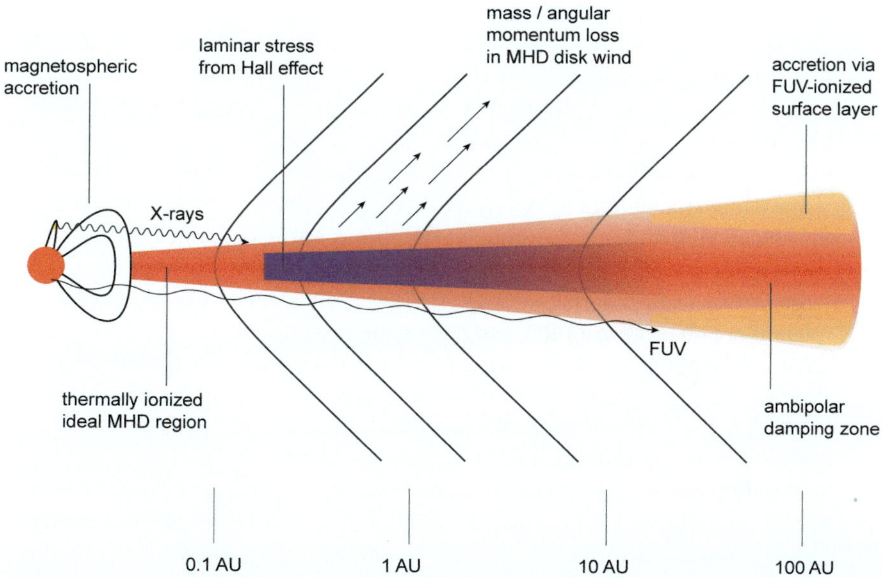

Fig. 1.22 A suggested structure for protoplanetary disks *if* MHD processes dominate over other sources of transport. The different regions are defined by the strength of non-ideal MHD terms (Ohmic diffusion, ambipolar diffusion and the Hall effect), and by mass and angular momentum loss in MHD disk winds. The Hall effect is predicted to behave differently if the net field threading the disk is anti-aligned to the rotation axis (here, alignment is assumed). Ionization by stellar X-rays and by FUV photons couples the stellar properties to those of the disk

angular velocity profile is *stable* against the MRI (Eq. 1.136). Something else much be responsible for transport in this region.

The angular velocity profile in the boundary layer is stable against the generation of turbulence by either the Rayleigh criterion or the MRI. It turns out to be *unstable*, however, to the generation of waves via a mechanism analogous to the hydrodynamic Papaloizou-Pringle instability of narrow tori [325]. Belyaev et al. [60, 61], using both analytic and numerical arguments, have shown that waves generated from the supersonic shear provide weak non-local transport of angular momentum (and energy) across the boundary layer. Magnetic fields are amplified by the shear [16] but do not play an essential role in boundary layer transport [61]. In protostellar systems boundary layers are present during eruptive accretion phases (see Sect. 1.6) when strong radiation fields are present [217]. Future work will need to combine the recent appreciation of the importance of wave angular momentum transport with radiation hydrodynamics for a full description of the boundary layer.

- Purely gas-phase turbulence in protoplanetary disks can be driven by MHD instabilities of Keplerian shear flow (the MRI and Hall-shear instability), by hydrodynamic instabilities that occur for specific vertical and/or radial entropy profiles, and by self-gravity.
- The magnetorotational instability probably dominates whenever the disk is thermally ionized, at temperatures exceeding about 10^3 K. Elsewhere, non-ideal MHD effects are important. Ohmic diffusion damps turbulence in high density regions, while ambipolar diffusion fulfills a similar role at large radii and in the disk atmosphere where the density is low. The Hall effect can enhance or inhibit angular momentum transport on $\sim 1-10$ AU scales, depending upon the sign of any net magnetic field relative to the disk angular momentum vector. The long-term evolution of the disk is thus coupled to the strength and evolution of net magnetic fields inherited from the star formation process.
- Self-gravity driven by ongoing envelope infall contributes to transport in young, massive disks. Fragmentation is expected to occur on sufficiently large scales (of the order of 10^2 AU), probably giving rise to objects with masses above the planetary mass regime.
- Entropy driven hydrodynamic instabilities could yield a significant level of transport in regions where MHD transport is suppressed. The operation of these instabilities is closely tied to the thermal structure and radiative properties of the disk.

1.6 Episodic Accretion

Young stellar objects (YSOs) are observed to be variable. The short time scale (lasting hours to weeks) component of that variability is complex [99], but can probably be attributed to a combination of turbulent inhomogeneities in the inner disk, stellar rotation [74], and the complex dynamics of magnetospheric accretion [3]. There is also longer time scale variability—lasting from years to (at least) many decades, that in some cases takes the form of well-defined outbursts in which the YSO brightens dramatically. The traditional classification of outbursting sources divides them into FU Orionis events [172, 178], characterized by a brightening of typically 5 magnitudes followed by a decay over decades, and EXors [179], which display repeated brightenings of several magnitudes over shorter time scales. The statistics on these uncommon long-duration outbursts (especially FU Orionis events [182]) are limited, and it is not even clear—either observationally or theoretically—whether FUOrs and EXors are variations on a theme or genuinely different phenomena [26]. Nonetheless, it is established that episodic accretion is common enough to matter for both stellar accretion and for planet formation processes occurring in the inner disk [26]. Our focus here is on the origin of these accretion outbursts.

Observations show that FU Orionis outbursts involve a large increase in the mass accretion rate through the inner disk on to the star [172]. During the outburst the inner disk will be relatively thick ($h/r \approx 0.1$), and hot enough to be thermally ionized. We therefore expect efficient angular momentum transport from the MRI, with $\alpha \approx 0.02$. Writing the viscous time scale (Eq. 1.69) in terms of these parameters,

$$t_\nu = \frac{1}{\alpha \Omega} \left(\frac{h}{r} \right)^{-2}, \tag{1.187}$$

we can estimate the disk radius associated with a (viscously driven) outburst of duration $t_{\rm burst}$,

$$r \simeq (GM_*)^{1/3} \alpha^{2/3} \left(\frac{h}{r} \right)^{4/3} t_{\rm burst}^{2/3}. \tag{1.188}$$

For a Solar mass star we find,

$$r \simeq 0.25 \left(\frac{\alpha}{0.02} \right)^{2/3} \left(\frac{h/r}{0.1} \right)^{4/3} \left(\frac{t_{\rm burst}}{100 \text{ yr}} \right)^{2/3} \text{AU}. \tag{1.189}$$

Disk-driven outbursts of broadly the right duration are thus likely to involve events on sub-AU scales, and could be associated physically with the magnetosphere, with the thermally ionized inner disk, or with the inner edge of the dead zone.

The physical origin of episodic accretion in YSOs has not been securely identified. Mooted ideas, illustrated in Fig. 1.23, fall into two categories. The first category invokes secular instabilities of protoplanetary disk structure that may occur on AU and sub-AU scales. The idea is that the inner disk may be intrinsically unable to

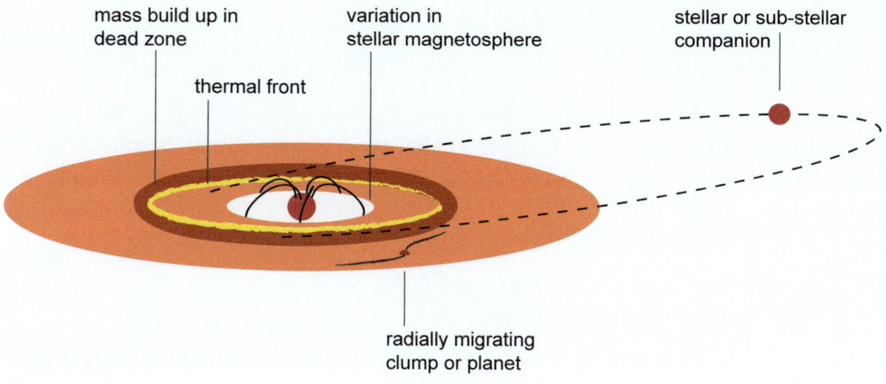

Fig. 1.23 An illustration of some of the processes suggested as the origin of episodic YSO accretion

accrete at a steady rate, and instead alternates between periods of high accretion rate when gas is draining on to the star and periods of low accretion rate when gas is accumulating in the inner disk. The instability could be a classical thermal instability [59], of the type accepted as causing dwarf nova outbursts [237], or a related instability of dead zone structure [23, 149]. The second category appeal to triggers independent of the inner disk to initiate the outburst. Ideas in this class are various and include perturbations from binary companions [72], the tidal disruption of radially migrating gas clumps/giant planets [301, 423], and disk variability linked to a stellar magnetic cycle [14, 21, 106]. Neither category of ideas is fully compelling (in the sense of being both fully worked out and consistent with currently accepted disk physics), so our discussion here will focus on a few key concepts that are useful for understanding current and future models of episodic accretion.

1.6.1 Secular Disk Instabilities

The classical instabilities that may afflict thin accretion disks are the thermal and viscous instabilities [338]. These are quite distinct from the basic fluid dynamical instabilities (the MRI, the VSI etc) that we discussed in Sect. 1.5, in that they address the stability of derived disk models rather than the fluid per se. Thus the thermal instability is an instability of the equilibrium vertical structure of the disk, while viscous instability in an instability of an (assumed) smooth radial structure under viscous evolution.

Before discussing how thermal or viscous instability might arise, we first define what these terms mean. Consider an annulus of the disk that is initially in hydrostatic and thermal equilibrium, such that the heating rate Q_+ matches the cooling rate Q_-. The heating rate per unit area depends upon the central temperature (Eq. 1.91), and can be written assuming the α-prescription as,

$$Q_+ = \frac{9}{4}\nu\Sigma\Omega^2 = \frac{9}{4}\alpha\frac{k_B T_c}{\mu m_H}\Sigma\Omega. \tag{1.190}$$

The cooling rate directly depends upon the effective temperature, $T_{\rm disk}$, but this can always be rewritten in terms of T_c using a calculation of the vertical thermal structure (Sect. 1.4.3). In the simple case when the disk is optically thick, for example, we have from Eq. (1.98) that $T_c^4/T_{\rm disk}^4 \simeq (3\tau/4)$, and hence,

$$Q_- = \frac{8\sigma}{3\tau}T_c^4. \tag{1.191}$$

Both α and τ may be functions of T_c. Now consider perturbing the central temperature on a time scale that is long compared to the dynamical time scale (so that hydrostatic equilibrium holds) but short compared to the viscous time scale (so that Σ remains fixed).[16] The disk will be unstable to runaway heating if an upward perturbation to the temperature increases the heating rate more than it increases the cooling rate, i.e. if,

$$\frac{\mathrm{d}\log Q_+}{\mathrm{d}\log T_c} > \frac{\mathrm{d}\log Q_-}{\mathrm{d}\log T_c}. \tag{1.192}$$

The same criterion predicts runaway cooling in the event of a downward perturbation. A disk that is unstable in this sense is described as *thermally unstable*. It would heat up (or cool down) until it finds a new structure in which heating and cooling again balance.

To determine the condition for viscous stability, we start by considering a steady-state solution $\Sigma(r)$ to the diffusive disk evolution Eq. (1.63). Following Pringle [338] we write $\mu \equiv \nu\Sigma$ and consider perturbations $\mu \to \mu + \delta\mu$ on a time scale long enough that *both* hydrostatic and thermal equilibrium hold (in this limit T_c is uniquely determined and $\nu = \nu(\Sigma)$). Substituting in Eq. (1.63) the perturbation $\delta\mu$ obeys,

$$\frac{\partial}{\partial t}(\delta\mu) = \frac{\partial\mu}{\partial\Sigma}\frac{3}{r}\frac{\partial}{\partial r}\left[r^{1/2}\frac{\partial}{\partial r}\left(r^{1/2}\delta\mu\right)\right]. \tag{1.193}$$

The perturbation $\delta\mu$ will grow if the diffusion coefficient, which is proportional to $\partial\mu/\partial\Sigma$, is negative. This is *viscous instability*, and it occurs if,

$$\frac{\partial}{\partial\Sigma}(\nu\Sigma) < 0. \tag{1.194}$$

A disk that is viscously unstable would tend to break up into rings, whose amplitude would presumably be limited by the onset of fluid instabilities that could be thought of as modifying the $\nu(\Sigma)$ relation.

1.6.1.1 The S-curve: A Toy Model

The instabilities of interest for YSO episodic accretion can broadly be considered to be thermal-type instabilities. Noting that $Q_+ \propto \alpha T_c$, and $Q_- \propto T_c^4/\tau \propto T_c^4/\kappa$ (where κ is the opacity at temperature T_c), we see that instability may occur according to Eq. (1.192) if,

- $\mathrm{d}\log Q_+/\mathrm{d}\log T_c$ is large, i.e. if α is strongly increasing with temperature.
- $\mathrm{d}\log Q_-/\mathrm{d}\log T_c$ is small, i.e. if κ is strongly increasing with temperature.

We expect α to increase rapidly with T_c at temperatures around 10^3 K, as we transition between damped non-ideal MHD turbulence at low temperature and the more vigorous ideal MHD MRI at higher temperature. We expect κ to increase most strongly at temperatures around 10^4 K, as hydrogen is becoming ionized and there

[16] This may seem to require fine tuning, but in fact the ordering of time scales in a geometrically thin disk always allows for such a choice [338].

is a strong contribution to the opacity from H^- scattering (in this regime κ can vary as something like T^{10} [59]). Either of these changes can result in instability.

Before detailing the specifics of possible thermal and dead zone instabilities in protoplanetary disks, it is useful to analyze a toy model that displays their essential features. We consider an optically thick disk described by the usual classical equations [139] whose angular momentum transport efficiency α and opacity κ are both piece-wise constant functions of central temperature T_c. Specifically,

$$\begin{aligned} T_c < T_{\rm crit} &: \alpha = \alpha_{\rm low}, \kappa = \kappa_{\rm low}, \\ T_c > T_{\rm crit} &: \alpha = \alpha_{\rm high}, \kappa = \kappa_{\rm high}, \end{aligned} \tag{1.195}$$

with $\alpha_{\rm low} \leq \alpha_{\rm high}$ and $\kappa_{\rm low} \leq \kappa_{\rm high}$. Our goal is to calculate the explicit form of the $\dot{M}(\Sigma)$ relation in the "low" and "high" states below and above the critical temperature $T_{\rm crit}$. For a steady-state disk, at $r \gg R_*$, heated entirely by viscous dissipation, the equations we need (mostly from Sect. 1.4) read,

$$\begin{aligned} T_{\rm disk}^4 &= \frac{3\Omega^2}{8\pi\sigma}\dot{M}, \\ \frac{T_c^4}{T_{\rm disk}^4} &= \frac{3}{4}\tau, \\ \tau &= \frac{1}{2}\kappa\Sigma, \\ \nu\Sigma &= \frac{\dot{M}}{3\pi}, \\ \nu &= \alpha\frac{c_s^2}{\Omega} = \frac{\alpha}{\Omega}\frac{k_B T_c}{\mu m_H}. \end{aligned} \tag{1.196}$$

Note that the stellar mass M_* and radius in the disk r enter these formulae only via their combination in the Keplerian angular velocity Ω. Eliminating T_c, $T_{\rm disk}$, τ and ν between these equations, we obtain a solution for $\dot{M}(\Sigma)$,

$$\dot{M} = \frac{9\pi}{4}\left(\frac{k_B^4}{\mu^4 m_H^4 \sigma}\right)^{1/3} \kappa^{1/3}\alpha^{4/3}\Omega^{-2/3}\Sigma^{5/3}, \tag{1.197}$$

valid on either the low or the high branch when the appropriate values for α and κ are inserted. A solution on the low branch is possible provided that $\Sigma \leq \Sigma_{\rm max}$, where $\Sigma_{\rm max}$ is defined by the condition that $T_c = T_{\rm crit}$. Similarly, a high branch solution requires $\Sigma \geq \Sigma_{\rm min}$ with $T_c = T_{\rm crit}$ at $\Sigma_{\rm min}$. The limiting surface densities are given by,

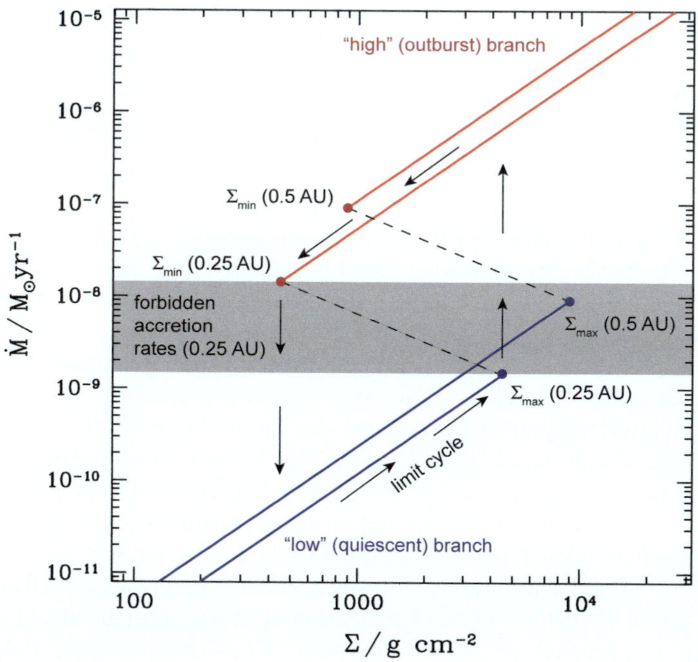

Fig. 1.24 Example S-curves in the accretion rate-surface density plane from the toy model described in the text. For these curves we take $\kappa_{\rm low} = \kappa_{\rm high} = 1$ cm^2 g^{-1}, $\alpha_{\rm low} = 10^{-4}$, $\alpha_{\rm high} = 10^{-2}$, and $T_{\rm crit} = 10^3$ K. The lower of the two curves is for $\Omega = 1.6 \times 10^{-6}$ s^{-1} (0.25 AU for a Solar-mass star), the upper for $\Omega = 5.6 \times 10^{-7}$ s^{-1} (0.5 AU)

$$\Sigma_{\rm max} = \frac{8}{3^{3/2}} \left(\frac{\mu m_H \sigma}{k_B} \right)^{1/2} \Omega^{-1/2} T_{\rm crit}^{3/2} \kappa_{\rm low}^{-1/2} \alpha_{\rm low}^{-1/2},$$
$$\Sigma_{\rm min} = \frac{8}{3^{3/2}} \left(\frac{\mu m_H \sigma}{k_B} \right)^{1/2} \Omega^{-1/2} T_{\rm crit}^{3/2} \kappa_{\rm high}^{-1/2} \alpha_{\rm high}^{-1/2}. \quad (1.198)$$

If $\kappa_{\rm high} > \kappa_{\rm low}$ and/or $\alpha_{\rm high} > \alpha_{\rm low}$, then $\Sigma_{\rm max} > \Sigma_{\rm min}$ and there will be a range of surface densities where accretion rates corresponding to either the low or the high branch are possible.

Figure 1.24 shows, for a fairly arbitrary choice of the model parameters, the thermal equilibrium solutions that correspond to the low and high states of the disk annulus. One should not take the results of such a toy model very seriously, but it captures several features of more realistic models,

- The solution has stable thermal equilibrium solutions on two branches, a low state branch where $\dot{M}$ for a given surface density is small, and a high state branch where it is substantially larger. In the toy model these branches are entirely separate, but in more complete models they are linked by an *unstable* middle branch (giving the plot the appearance of an "S"-curve).

- There is a range of surface densities for which either solution is possible.
- There is a band of accretion rates for which no stable equilibrium solutions exist.
- The position of the S-curve in the $\Sigma - \dot{M}$ plane is a function of radius, with the band of forbidden accretion rates moving to higher $\dot{M}$ further from the star.

The S-curve is derived from a local analysis, and the existence of annuli whose thermal equilibrium solutions have this morphology is a necessary but not sufficient condition for a global disk outburst. That time dependent behavior of some sort is inevitable can be seen by supposing that the annulus at 0.25 AU in Fig. 1.24 is fed with gas from outside at a rate that falls into the forbidden band. No stable thermal equilibrium solution with this accretion rate is possible. If the disk is initially on the lower branch, the rate of gas supply exceeds the transport rate through the annulus, and the surface density increases. This continues until Σ reaches and exceeds $\Sigma_{\max}$, at which point the only available solution lies on the high branch at much higher accretion rate. The annulus transitions to the high branch, where the transport rate is now larger than the supply rate, and the surface density starts to drop. The cycle is completed when Σ falls below $\Sigma_{\min}$, triggering a transition back to the low state.

The evolution of a disk that is potentially unstable (i.e. one that has some annuli with S-curve thermal equilibria) is critically dependent upon the *radial* flow of mass and heat, which are the key extra ingredients needed if an unstable disk is to "organize" itself and produce a long-lived outburst. To see this, imagine a disk where annuli outside r_f are already on the high branch of the S-curve, while those inside remain on the low branch. The strong radial gradient of T_c implies a similarly rapid change in ν, which leads to a large mass flux from the annuli that are already in outburst toward those that remain quiescent (Eq. 1.64). The resulting increase in surface density, along with the heat that goes with it, can push the neighboring annulus on to the high branch, initiating a propagating "thermal front" that triggers a large scale transition of the disk into an outburst state.

The quantitative modeling of global disk evolution including these thermal processes is well-developed within the classical α disk formalism [237]. For a minimal model, all that is needed is to supplement the disk evolution Eq. (1.63) with a model for the vertical structure (conceptually as described for the toy model above) and an equation for the evolution of the central temperature. This takes the form [79, 237],

$$\frac{\partial T_c}{\partial t} = \frac{Q_+ - Q_-}{c_p \Sigma} - \frac{\mathscr{R} T_c}{\mu c_p} \frac{1}{r} \frac{\partial}{\partial r} (r v_r) - v_r \frac{\partial T_c}{\partial r} + \cdots \qquad (1.199)$$

Here c_p is the specific heat capacity at constant pressure, and $\mathscr{R}$ is the gas constant. The first term on the right hand side describes the direct heating and cooling due to viscosity and radiative losses, while the second and third terms describe $P\mathrm{d}V$ work and the advective transport of heat associated with radial mass flows. In general there should be additional terms to represent the radial flow of heat due to radiative and/or turbulent diffusion (these effects are small in most thin disk situations, but become large when there is an abrupt change in T_c at a thermal front). The treatment of these additional terms is somewhat inconsistent in published models, though they can significantly impact the character of derived disk outbursts [315].

1.6.1.2 Classical Thermal Instability

In physical rather than toy models for episodic accretion α and κ are smooth rather than discontinuous functions of the temperature. A local instability, with a resulting S-curve, occurs if one or both of these functions changes sufficiently rapidly with T_c (so that Eq. 1.192 is satisfied). No simple condition specifies when a disk that has some locally unstable annuli will generate well-defined global outbursts, but loosely speaking outbursts occur provided that the branches of the S-curve (and the values of $\Sigma_{\rm max}$ and $\Sigma_{\rm min}$) are well separated.

The classical cause of disk thermal instability is the rapid increase in opacity associated with the ionization of hydrogen, at $T \simeq 10^4$ K. Around this temperature κ can rise as steeply as T^{10}, and the disk will invariably satisfy at least the condition for local thermal instability. The evolution of disks subject to a hydrogen ionization thermal instability was first investigated as a model for dwarf novae (eruptive disk systems in which a white dwarf accretes from a low-mass companion star) [168, 290], and subsequently applied to low-mass X-ray binaries. Thermal instability models provide a generally good match to observations of outbursts in these systems (which are of shorter duration that YSO outbursts, and correspondingly better characterized empirically), and are accepted as the probable physical cause. Good fits to data require models to include not only the large change in κ that is the cause of classical thermal instability, but also changes in α between the low and high branches of the S-curve. Typically $\alpha_{\rm low} \leq 10^{-2}$, whereas $\alpha_{\rm high} \sim 0.1$ [215]. MHD simulations that include radiation transport have shown that the S-curve derived from α disk models can be approximately recovered as a consequence of the MRI, and that the difference in stress between the quiescent and outburst states may be attributable in part to the development of vertical convection within the hot disk [100, 183].

By eye, the light curves of FU Orionis events look quite similar to scaled versions of dwarf novae outbursts, so the success of thermal instability models in the latter sphere makes them a strong candidate for YSO accretion outbursts. The central temperature of some outbursting FUOr disks, moreover, almost certainly does exceed the 10^4 K needed to ionize hydrogen, making it inevitable that thermal instability physics will play some role in the phenomenon. Detailed thermal instability models of FU Orionis events were constructed by Bell and Lin [59], who combined a one-dimensional (in r) treatment of the global evolution with detailed α model vertical structure calculations. They were able to find periodic solutions that describe "self-regulated" disk outbursts (i.e., requiring no external perturbation or trigger), with the disk alternating between quiescent periods (with $\dot{M} = 10^{-8}$–10^{-7} $M_\odot$ yr^{-1}) and outburst states (with $\dot{M} \geq 10^{-5}$ $M_\odot$ yr^{-1}). These properties, and the inferred outburst duration of $\sim 10^2$ yr, are in as good an agreement with observations as could reasonably be expected given the simplicity of the model.

The weakness of classical thermal instability models as an explanation for YSO accretion outbursts is that they require unnatural choices of the viscosity parameter α [15]. Thermal instability is immutably tied to the hydrogen ionization temperature, which exceeds even the mid-plane temperatures customarily attained in protoplanetary disks. The required temperatures can be reached (if at all) only extremely close to the star, and the radial region affected by instability extends to no more than 0.1–0.2 AU. The viscous time scale on these scales is short, so matching the century-long outbursts seen in FUOrs requires a very weak viscosity—Bell and Lin [59] adopt $\alpha_{\rm high} = 10^{-3}$. This is at least an order of magnitude lower than the expected efficiency of MRI transport under ideal MHD conditions [107, 378]. Moreover, in the specific case of FU Orionis itself, radiative transfer models suggest that the region of high accretion rate during the outburst extends out to 0.5–1 AU [441], substantially larger than would be expected in the thermal instability scenario.

1.6.1.3 Instabilities of Dead Zones

A dead zone in which the MRI is suppressed by Ohmic resistivity (Sect. 1.5.4.3) supports a related type of instability whose high and low states are distinguished primarily by different values of α, rather than by the thermal instability's different values of κ. The origin of instability is clear within Gammie's original conception [148] of the dead zone, which has a simple two-layer structure. The surface of the disk, ionized by X-rays,[17] is MRI-active and supports accretion with a local $\alpha \sim 10^{-2}$. Below a critical column density Ohmic resistivity completely damps MRI-induced turbulence (according to Eq. 1.181), and the disk is dead with $\alpha = 0$. This structure can be bistable if the surface density exceeds that of the ionized surface layer. The low accretion rate state corresponds to a cool, externally heated disk with a dead zone; the high accretion rate state to a hot thermally ionized disk at the same Σ.

Martin and Lubow [282, 284] have shown that the local physics of Ohmic dead zone instability can be analyzed in an manner closely analogous to thermal instability. The sole difference lies in the reason why the lower state ceases to exist above a critical surface density ($\Sigma_{\rm max}$ in Fig. 1.24). For thermal instability $\Sigma_{\rm max}$ is set by the onset of ionization at the disk mid-plane, and the attendant rise in opacity. A simple dead zone, however, *does not get hotter with increasing* Σ, because the heating (either from irradiation, or viscous dissipation in the ionized surface layer) occurs at low optical depth at a rate that is independent of the surface density. It is then not obvious how even a very thick dead zone can be heated above 10^3 K, "ignited", and induced to transition to the high state of the S-curve.

One way to trigger a jump to the high state is to postulate that some source of turbulence other than the MRI is present to heat the disk. Gammie [149] suggested that inefficient transport through the ionized surface layer would lead to the build of mass in the dead zone [148] until $Q \approx 1$ and self-gravity sets in (see Sect. 1.5.2). We can readily estimate the properties we might expect for an instability triggered

[17] Cosmic rays in the original model, though this is an unimportant distinction.

in this manner by the onset of self-gravity at small radii. Suppose that at 1 AU the temperature in the quiescent (externally irradiated) disk state is 150 K. Then to reach $Q = 1$ requires gas to build up until $\Sigma \simeq 7 \times 10^4$ g cm^{-2}, at which point the mass interior is $M \sim \pi r^2 \Sigma \sim 0.025\, M_\odot$. This is comparable to the amount of mass accreted per major FU Orionis-like outburst. In the outburst state the disk will be moderately thick (say $h/r = 0.3$ [58]), and the viscous time scale $(1/\alpha\Omega)(h/r)^{-2}$ works out to be about 200 yr for $\alpha = 0.01$, again similar to inferred FUOr time scales. At this crude level of estimate, it therefore seems possible that a self-gravity triggered dead zone instability could be consistent both with the inferred size of the outbursting region [441] and with theoretical best guesses as to the strength of angular momentum transport in ideal MHD conditions.

Time-dependent models for outbursts arising from a dead zone instability were computed by Armitage et al. [23], and subsequently by several groups in both one-dimensional [282, 315, 442, 444] and two-dimensional models [29, 443]. The more recent studies show that a self-gravity triggered instability of an Ohmic dead zone can give rise to outbursts whose properties are broadly consistent with those of observed FUOrs. Instability persists even if there is a small residual viscosity within the dead zone [28, 284], which could arise hydrodynamically in response to the "stirring" from the overlying turbulent surface layer [131]. How well such models work when either or both of the Hall effect and MHD winds contribute to the dynamics close to the star remains unclear. In principle these effects could also lead to entirely different types of eruptive behavior.

1.6.2 Triggered Accretion Outbursts

Accretion variability, including (perhaps) the large scale outbursts of FUOrs, can also be triggered by processes largely independent of the inner disk itself. Stellar activity cycles, binaries with small periastron distances, and tidal disruption of gaseous clumps or planets may all contribute.

1.6.2.1 Stellar Activity Cycles

As discussed in Sect. 1.4.2.2 the inner disk is expected [219] and observed [73] to be disrupted by the stellar magnetosphere. The complex dynamics of the interaction between the field—which may be misaligned to the stellar spin axis and have non-dipolar components [206]—and the disk [233] is likely the dominant cause of T Tauri variability on time scales comparable to the stellar rotation period (i.e. days to weeks). If the strength of the field also varies systematically due to the presence of activity cycles analogous to the Solar cycle, these could trigger longer time scale (years to decades) accretion variability. The simplest mechanism is modulation of the magnetospheric radius across corotation [94]. When the field is strong and $r_m > r_{\rm co}$ the linkage between the stellar field lines and the disks adds angular momentum to the disk, impeding accretion in the same way as gravitational torques from a binary (Sect. 1.4.2.3).[18] Gas then accumulates just outside the magnetospheric radius, and can subsequently be accreted in a burst when the field weakens.

The viability of such magnetically "gated" accretion as an origin for large scale variability is limited by the short viscous time scale of the disk at $r \approx r_m$, which makes it hard to accumulate large masses of gas if the stellar fields are, as expected, of no more than kG strength. Models [14] suggest that significant decade-long variability could be associated with protostellar activity cycles, but there is no clear path to generating FU Orionis outbursts. Activity cycles are more promising as an explanation for lower amplitude, periodic EXor outbursts [105, 106]. The interaction between the time-variable stellar magnetic field and the non-ideal physics of the inner disk, just outside the thermally ionized region, could be significant [21].

1.6.2.2 Binaries

An eccentric binary with an AU-scale periastron distance could funnel gas into the inner disk, increasing the accretion rate and leading to an outburst if the increased surface density is high enough to trigger thermal or dead zone instability. This mechanism was proposed by Bonnell and Bastien [72], and has subsequently been studied with higher resolution simulations [135, 332]. Although there are some differences between the predicted outbursts and those observed (this is almost inevitable, as the limited sample of FUOrs is already quite diverse), it is clear that close encounters from binary or cluster companions induce episodes of substantially enhanced accretion. The obvious prediction—that FUOrs ought to be found with observable binary companions or preferentially associated with higher-density star forming regions—is neither confirmed nor ruled out given the small sample of known objects.

1.6.2.3 Clump Tidal Disruption

A final possibility is that accretion outbursts could be triggered by the tidal disruption of a bound object (a planet or gas cloud) that migrates too close to the star. The necessary condition for this to occur is given by the usual argument for the Roche limit. If we consider a planet with radius R_p and mass M_p, orbiting a star of mass M_* at radius r, the differential (tidal) gravitational force between the center of the planet and its surface is,

$$F_{\rm tidal} = \frac{GM_*}{r^2} - \frac{GM_*}{(r+R_p)^2} \simeq \frac{2GM_*}{r^3} R_p. \tag{1.200}$$

Equating the tidal force to the planet's own self-gravity, $F_{\rm self} = GM_p/R_p^2$, we find that tidal forces will disrupt the planet at a radius $r_{\rm tidal}$ given approximately by,

[18] In compact object accretion, this is described as the "propeller" regime of accretion [189].

$$r_{\rm tidal} = \left(\frac{M_*}{M_p} \right)^{1/3} R_p. \tag{1.201}$$

An equivalent condition is that tidal disruption occurs when the mean density of the planet $\bar{\rho} < M_*/r^3$.

It is difficult to tidally disrupt a mature giant planet. A Jupiter mass planet has a radius of $R_p \simeq 1.5\ R_J$ at an age of 1 Myr [278], and will not be disrupted outside the photospheric radius of a typical young star. (Though such planets, if present in the inner disk, could alter the course of thermal or dead zone instability [97, 253].) If tidal disruption is to be relevant to episodic accretion we require, first, that the outer disk is commonly gravitationally unstable to fragmentation, and, second, that the clumps that form migrate rapidly inward (in the Type 1 regime discussed by Kley in this volume) without contracting too rapidly. Numerical evidence supports the idea that clump migration can be rapid [54, 84, 291, 423], though it is at best unclear whether contraction can be deferred sufficiently to deliver clumps that would be tidally disrupted on sub-AU scales [147]. Assuming that these pre-conditions are satisfied, however, Nayakshin and Lodato [301] studied the tidal mass loss from the close-in planets and its impact on the disk. They found that the tidal disruption of $\sim$20 R_J clumps, interior to 0.1 AU, led to accretion outbursts consistent with the basic properties of FUOrs.

The primary theoretical doubts about tidal disruption as a source of outbursts concern the relative rates of inward migration and clump contraction, which are both hard to calculate at substantially better than order of magnitude level. Observationally, this process would produce outbursts in systems whose disks were young, massive, and probably still being fed by envelope infall.

- Episodic accretion on to Young Stellar Objects occurs in several different flavors, including high-amplitude outbursts such as FU Orionis events that would certainly have an important influence on disk chemistry and planet formation close to the star.
- A possible explanation for episodic accretion is an instability in the equilibrium disk structure on AU or sub-AU scales, related to the classical thermal instability that explains outbursts in dwarf nova systems. A potential instability of dead zone structure has been demonstrated in one- and two-dimensional models, but may need to be modified to take account of additional non-ideal MHD physics and disk winds.
- Outbursts could also be *internally* triggered, for example by changes in the stellar magnetic field, or *externally* triggered, if dense clumps of gas migrate quickly into the inner disk and are tidally disrupted.

1.7 Single and Collective Particle Evolution

The evolution of solid particles within disks differs from that of gas because solid bodies are unaffected by pressure gradients but do experience aerodynamic forces. We discuss here how these differences affect the motion of single particles orbiting within the gas disk, and how we can describe the evolution of a "fluid" made up of small solid particles interacting aerodynamically with the gas. Issues such as the rate and outcome of particle collisions, that are central to early stage planet formation, are treated in the accompanying part by Kley, and elsewhere [18].

The key parameter describing the aerodynamic coupling between solid particles and gas is the *stopping time*. For a particle of mass m that is moving with velocity Δv relative to the local gas, the stopping time is defined as,

$$t_s \equiv \frac{m\Delta v}{|F_{\rm drag}|}, \tag{1.202}$$

where $F_{\rm drag}$ is the magnitude of the aerodynamic drag force that acts in the opposite direction to Δv. Very frequently, what matters most is how the stopping time compares to the orbital time at the location of the particle. We therefore define a dimensionless stopping time by multiplying t_s by the orbital frequency $\Omega_{\rm K}$,

$$\tau_s \equiv t_s \Omega_{\rm K}. \tag{1.203}$$

The dimensionless stopping time is also called the *Stokes number*.

For our immediate purposes it largely suffices to describe aerodynamic effects in terms of the stopping time, but eventually you will want to translate the results into concrete predictions for how particles of various sizes and material properties behave. This requires specifying $F_{\rm drag}$, whose form depends on the size of the particle relative to the mean free path of gas molecules, and (in the fluid regime) on the Reynolds number of the flow around the particle [431]. If the particle radius s is small compared to the mean free path λ ($s < 9\lambda/4$) the particle experiences *Epstein drag*, with a drag force,

$$\mathbf{F}_{\rm drag} = -\frac{4\pi}{3}\rho s^2 v_{\rm th} \Delta \mathbf{v}. \tag{1.204}$$

Here ρ is the density of the surrounding gas, and the thermal speed of the molecules,

$$v_{\rm th} = \sqrt{\frac{8k_B T}{\pi \mu m_H}}, \tag{1.205}$$

is roughly the same as the sound speed. Because Epstein drag is proportional to the velocity difference Δv (rather than the more familiar square of the velocity difference), the stopping time is a function of the particle properties that is independent of the velocity difference. For a spherical particle of material density ρ_m,

$$t_s = \frac{\rho_m}{\rho} \frac{s}{v_{\rm th}}. \quad (1.206)$$

The mean free path in protoplanetary disks is of the order of cm (larger in the outer disk), so the Epstein regime is relevant for particles that range from dust to those of small macroscopic dimensions. Drag laws appropriate for larger bodies, which fall into the Stokes regime of drag, are given by Whipple [431].

1.7.1 *Radial Drift*

The most important consequence of aerodynamic forces is the phenomenon of *radial drift*. In Sect. 1.3.1.2 we showed that radial pressure gradients result in a gas orbital velocity that differs from the Keplerian value by 50–100 m s^{-1} (using typical disk parameters at 1 AU). Most commonly, the gas is partially supported against gravity by the pressure gradient, and so rotates more slowly than the Keplerian value. We will consider how this velocity differential affects the evolution of solids in various limits. We will start by ignoring both turbulence and the feedback of aerodynamic forces *on the gas*, before revisiting the problem with these processes included.

1.7.1.1 Particle Drift Without Feedback

Large bodies ($\tau_s \gg 1$) orbit at close to the Keplerian speed, and the effect of gas on their evolution can be considered as a simple "headwind" if the disk is sub-Keplerian. Suppose that the gas orbits at a speed $v_K - \Delta v$, with $\Delta v \ll v_K$. The drag force $|F_{\rm drag}| = m \Delta v \Omega_{\rm K} / \tau_s$ does work at a rate,

$$\dot{E} \simeq -|F_{\rm drag}| v_K, \quad (1.207)$$

that leads to a change in the orbital energy $E = -GM_* m / 2a$, where a is the radius of the orbit. Noting that,

$$\dot{E} = \frac{GM_* m}{2a^2} \frac{{\rm d}a}{{\rm d}t}, \quad (1.208)$$

and equating the two expressions for $\dot{E}$, we find that the orbit decays at a speed $v_r = {\rm d}a/{\rm d}t$ that is given by,

$$v_r = -\frac{2}{\tau_s} \Delta v. \quad (1.209)$$

The radial drift of large bodies is inversely proportional to their Stokes number.

The simple headwind argument fails for small particles with $\tau_s \ll 1$, which instead are forced to orbit *at the gas speed* by the strong aerodynamic coupling. The particles do not feel the pressure gradient, so their non-Keplerian orbital motion results in a net radial force,

$$\frac{F_r}{m} = \frac{(v_K - \Delta v)^2}{a} - \frac{GM_*}{a^2} \simeq -\frac{2v_K \Delta v}{a}. \tag{1.210}$$

Equating this to the drag force for *radial* motion at speed v_r, $|F_{\rm drag}|/m = v_r \Omega_{\rm K}/\tau_s$, we find that radial drift for small particles occurs at the terminal drift speed,

$$v_r = -2\tau_s \Delta v. \tag{1.211}$$

This is the speed relative to the gas, so for a disk that is accreting there is an additional component given by the gas' radial velocity.

Intermediate-sized particles orbit at some speed between that of the gas and that given by the Keplerian velocity. To derive the general rate of radial drift [395, 428], we consider a gas disk whose orbital velocity is,

$$v_{\phi,{\rm gas}} = v_K (1 - \eta)^{1/2}. \tag{1.212}$$

The parameter $\eta \propto (h/r)^2$. For example, if the disk has $\Sigma \propto r^{-1}$ and central temperature $T_c \propto r^{-1/2}$, we showed in Sect. 1.3.1.2 that $\eta = (11/4)(h/r)^2$. Defining the particle radial and azimuthal velocities to be v_r and v_ϕ respectively, the equations of motion are,

$$\frac{{\rm d}v_r}{{\rm d}t} = \frac{v_\phi^2}{r} - \Omega_{\rm K}^2 r - \frac{1}{t_{\rm s}}\left(v_r - v_{r,{\rm gas}}\right) \tag{1.213}$$

$$\frac{\rm d}{{\rm d}t}\left(r v_\phi\right) = -\frac{r}{t_{\rm s}}\left(v_\phi - v_{\phi,{\rm gas}}\right). \tag{1.214}$$

The azimuthal equation can be simplified by noting that the specific angular momentum remains close to Keplerian,

$$\frac{\rm d}{{\rm d}t}\left(r v_\phi\right) \simeq v_r \frac{\rm d}{{\rm d}r}(r v_K) = \frac{1}{2} v_r v_K. \tag{1.215}$$

This yields,

$$v_\phi - v_{\phi,{\rm gas}} \simeq -\frac{1}{2}\frac{t_{\rm s} v_r v_K}{r}. \tag{1.216}$$

We now substitute for Ω_K in the radial equation using Eq. (1.212). Discarding higher order terms we obtain,

$$\frac{{\rm d}v_r}{{\rm d}t} = -\eta \frac{v_K^2}{r} + \frac{2v_K}{r}\left(v_\phi - v_{\phi,{\rm gas}}\right) - \frac{1}{t_{\rm s}}\left(v_r - v_{r,{\rm gas}}\right). \tag{1.217}$$

The ${\rm d}v_r/{\rm d}t$ term is negligible. Dropping that, we eliminate $(v_\phi - v_{\phi,{\rm gas}})$ between Eqs. (1.216) and (1.217) to obtain,

$$v_r = \frac{(r/v_K)t_s^{-1}v_{r,\text{gas}} - \eta v_K}{(v_K/r)t_s + (r/v_K)t_s^{-1}}. \tag{1.218}$$

In terms of the Stokes number the final result for the particle radial velocity is,

$$v_r = \frac{\tau_s^{-1}v_{r,\text{gas}} - \eta v_K}{\tau_s + \tau_s^{-1}}. \tag{1.219}$$

The previously derived results for very small and very large particles are recovered by taking the appropriate limits.

The speed of the radial drift implied by Eq. (1.219) is shown in Fig. 1.25. Very rapid drift is predicted for particles with $\tau_s \sim 1$. For our fiducial disk model with $\Sigma \propto r^{-1}$, $T_c \propto r^{-1/2}$ and $h/r = 0.03$, the radial drift time scale $t_{\text{drift}} = r/|v_r|$ is just 10^3 orbital periods—one thousand years at 1 AU! Particles with this stopping time are very roughly meter-sized, and their rapid drift is the origin of the "meter-sized barrier" that severely constrains models of planetesimal formation [89, 197].

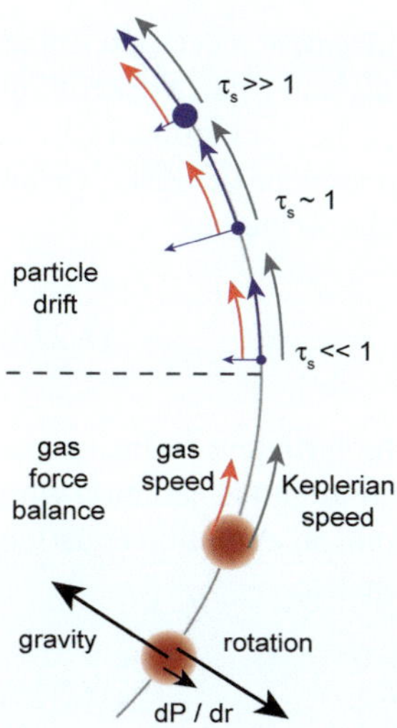

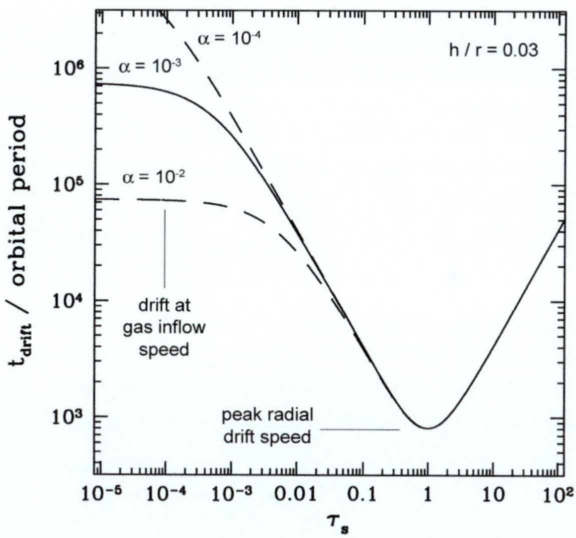

Fig. 1.25 Particles drift inwards in a disk wherever $\mathrm{d}P/\mathrm{d}r$, due to the combination of aerodynamic forces and sub-Keplerian gas rotation. The radial drift time scale $t_{\text{drift}} = r/|v_r|$, in units of the local orbital period, is plotted as a function of the dimensionless stopping time τ_s. The fastest radial drift occurs for $\tau_s = 1$. The specific numbers shown in the plot are appropriate for a disk with $h/r = 0.03$ and α values of 10^{-2} (lowest curve on the left-hand side of the figure), 10^{-3} and 10^{-4}

1.7.1.2 Drift with Diffusion

If the disk is turbulent, small dust particles that are aerodynamically well-coupled to the gas will diffuse radially and vertically. (The same physics will apply to trace gas species, such as water or CO molecules.) Diffusion will tend to equalize the *concentration* of the dust or trace gas species relative to the dominant gaseous component in the disk. To derive an equation for the evolution of the trace species [96], we initially ignore any radial drift due to aerodynamic effects and write the concentration of the trace gas or dust (generically the "contaminant") as,

$$f = \frac{\Sigma_d}{\Sigma}, \tag{1.220}$$

where Σ_d is the surface density of the contaminant. If the contaminant is neither created nor destroyed within the region of the disk under consideration, continuity demands that,

$$\frac{\partial \Sigma_d}{\partial t} + \nabla \cdot \mathbf{F}_d = 0, \tag{1.221}$$

where $\mathbf{F}_d$, the flux, can be decomposed into two parts: an advective piece describing transport of the dust or gas with the mean disk flow, and a diffusive piece describing the tendency of turbulence to equalize the concentration of the contaminant across the disk. For $f \ll 1$ we can assume that the diffusive properties of the disk depend only on the *gas* surface density, in which case the flux can be written as,

$$\mathbf{F}_d = \Sigma_d \mathbf{v} - D\Sigma\nabla\left(\frac{\Sigma_d}{\Sigma}\right). \tag{1.222}$$

Here $\mathbf{v}$ is the mean velocity of gas in the disk and D is the turbulent diffusion co-efficient. The diffusive term vanishes if f is constant. Combining this equation with the continuity equation for the gaseous component, we obtain an evolution equation for f in an axisymmetric disk. In cylindrical polar co-ordinates,

$$\frac{\partial f}{\partial t} = \frac{1}{r\Sigma}\frac{\partial}{\partial r}\left(Dr\Sigma\frac{\partial f}{\partial r}\right) - v_r\frac{\partial f}{\partial r}. \tag{1.223}$$

This result has the form of an advection-diffusion equation, with the advective component being due to the radial flow of the disk gas. It is easy to generalize this equation to account for the radial drift of larger particles that are imperfectly coupled to the gas, by adding an additional flux representing the radial drift speed [5].

Determining what is the appropriate value for the turbulent diffusion co-efficient D involves many of the same uncertainties that afflict the determination of the turbulent viscosity. For trace gas species and very small dust particles the zeroth-order expectation is that $D \approx \nu$ [297], though simulations of non-ideal MHD disk turbulence show both significant deviations from ν and anisotropy between the vertical and radial directions [447]. For larger bodies there is a well-determined analytic scaling with the Stokes number of the particles [438].

1.7.1.3 Particle Pile-Up

Solids that experience significant radial drift tend to become concentrated ("pile-up") in the inner disk [436, 439]. The basic effect is present in the simplest case where diffusion and inward drag by the mean flow are small, and feedback of the particles on the gas can be neglected. The radial drift speed is then, approximately,

$$v_r \simeq -\tau_s \eta v_K, \tag{1.224}$$

with $\eta \propto (h/r)^2$. For particles that are in the Epstein regime of drag, the stopping time in the mid-plane is (from Eqs. 1.206 and 1.16),

$$t_s = \frac{\rho_m}{\rho_0} \frac{s}{v_{th}} = \sqrt{2\pi} \frac{\rho_m h}{\Sigma} \frac{s}{v_{th}}. \tag{1.225}$$

Since $h = c_s/\Omega$, and c_s and v_{th} differ only by a numerical factor, we obtain,

$$\tau_s = \frac{\pi}{2} \frac{\rho_m}{\Sigma} s. \tag{1.226}$$

Suppose now (perhaps not very realistically) that the surface density profile of solids has attained a steady-state, such that the mass flux is constant with radius. Then,

$$\dot{M}_d = -2\pi r \Sigma_d v_r = \text{constant}, \tag{1.227}$$

and substituting for v_r we find,

$$\frac{\Sigma_d}{\Sigma} \propto \left(\frac{h}{r}\right)^{-2} r^{-1/2}. \tag{1.228}$$

For a disk with constant (h/r) the steady-state concentration of solids increases closer to the star as $r^{-1/2}$. In the more realistic case of a flaring disk with mid-plane temperature profile, say, $T_c \propto r^{-1/2}$, the effect is stronger. A constant α model of such a disk has a steady-state gas surface density profile $\Sigma \propto r^{-1}$, with the solids following $\Sigma_d \propto r^{-2}$.

1.7.1.4 The Nakagawa-Sekiya-Hayashi Equilibrium

Up till now we have implicitly assumed that the evolution of the gas is unaffected by the evolution of the solids within it. Obviously this can never be strictly correct.

If a population of solid particles are losing angular momentum to the gas through aerodynamic forces and spiraling inward, the gas must gain a corresponding amount of angular momentum. If the solid to gas ratio is only of the order of 1%, however, one might suppose that the effect of the angular momentum exchange on the gas would be small and, perhaps, ignorable. This is only partially true. First, a number of processes, including vertical particle settling [117], gas loss via photo-evaporation [5, 402] or MHD winds, and radial drift itself [436], can boost the solid to gas ratio, at least locally. Second, the equilibrium solution for radial drift in the presence of back reaction on to the gas can be unstable to the *streaming instability* [437], which can result in strong localized clumping of the solids.

The generalization of the radial drift formula (Eq. 1.219) to account for the back reaction of the drift on the gas is known as the Nakagawa-Sekiya-Hayashi (NSH) equilibrium [300]. The NSH solution is derived by considering the interaction between solids and gas in a simple disk model that ignores the effects of turbulence and vertical gravity. Both the gas, with density ρ_g, pressure P and velocity $\mathbf{v}_g$, and the solid particles, with density density ρ_p and velocity $\mathbf{v}_p$, are treated as fluids that interact with each other via aerodynamic drag. They obey continuity and momentum equations of the form [437],

$$\frac{\partial \rho_p}{\partial t} + \nabla \cdot (\rho_p \mathbf{v}_p) = 0, \tag{1.229}$$

$$\nabla \cdot \mathbf{v}_g = 0, \tag{1.230}$$

$$\frac{\partial \mathbf{v}_p}{\partial t} + \mathbf{v}_p \cdot \nabla \mathbf{v}_p = -\Omega_{\mathrm{K}}^2 \mathbf{r} - \frac{1}{t_s} \left(\mathbf{v}_p - \mathbf{v}_g \right), \tag{1.231}$$

$$\frac{\partial \mathbf{v}_g}{\partial t} + \mathbf{v}_g \cdot \nabla \mathbf{v}_g = -\Omega_{\mathrm{K}}^2 \mathbf{r} + \frac{1}{t_s} \frac{\rho_p}{\rho_g} \left(\mathbf{v}_p - \mathbf{v}_g \right) - \frac{\nabla P}{\rho_g}. \tag{1.232}$$

We have replaced the continuity equation for the gas with the condition for incompressibility, which is valid provided that velocities remain highly subsonic. The only other differences between the equations for the two species are the presence of a pressure gradient term in the gas momentum equation, and the pre-factor in the aerodynamic drag term expressing the differing inertia of the two fluids.

It is straightforward to derive the steady-state axisymmetric solution for the drift of solids and gas from the above equations [300, 437]. To illustrate the difference between the NSH and no-feedback solutions, we quote here just the result for the relative radial velocity of the solids and the gas. With our definition of η (Eq. 1.212) this takes the form,

$$v_{r,\mathrm{rel}} = -\frac{\rho_g}{\rho} \frac{\eta \tau_s v_K}{1 + (\tau_s \rho_g / \rho)^2}, \tag{1.233}$$

where $\rho = \rho_g + \rho_p$. Equation (1.219) is recovered in the case where $\rho_p \ll \rho_g$ and feedback can be neglected (note that we have ignored any radial gas motions due to processes *other* than particle-gas coupling in this version of the NSH solution).

The predicted relative radial velocity between the solid and gas components is plotted in Fig. 1.26 for various values of the solid to gas ratio. For $\rho_p / \rho_g = 10^{-2}$–$10^{-1}$

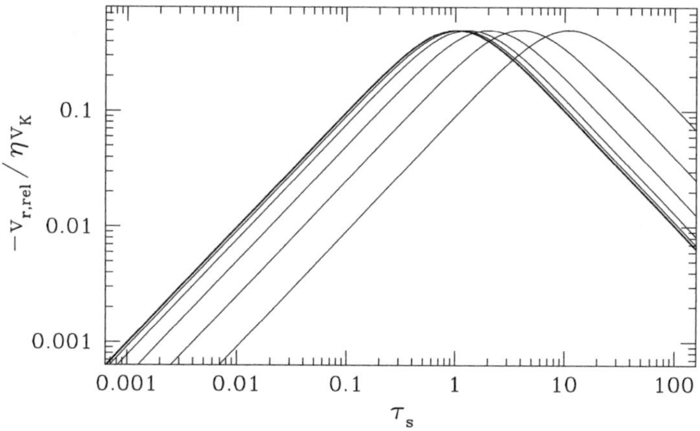

Fig. 1.26 The relative velocity of solids compared to gas (normalized by ηv_K, the parameter specifying departure of the gas disk from Keplerian rotation) is plotted as a function of the Stokes number τ_s. From left to right the curves show the NSH equilibrium solutions for $\rho_p/\rho_g = 10^{-2}$ (effectively identical to the no-feedback solution), 3×10^{-2}, 0.1, 0.3, 1, 3 and 10

the rate of drift at a given value of the stopping time is very similar to the no-feedback solution given by Eq. (1.219). For large values of ρ_p/ρ_g, however, the peak of the relative velocity curve (which in this limit is predominantly outward motion of the *gas*) shifts to somewhat higher values of τ_s.

1.7.2 Vertical Settling

Aerodynamic drag also modifies the vertical distribution of solids relative to gas. We first consider the forces acting on a small particle of mass m at height z above the mid-plane of a laminar disk. The vertical component of stellar gravity exerts a force,

$$|F_{\rm grav}| = m\Omega_{\rm K}^2 z. \tag{1.234}$$

The gas in the disk is supported against this force by the vertical pressure gradient, but solid particles are not. If started at rest a particle will accelerate until the gravitational force is balanced by drag. In the Epstein regime (given by Eq. 1.204) the resulting terminal velocity is,

$$v_{\rm settle} = \frac{\rho_m}{\rho}\frac{s}{v_{\rm th}}\Omega_{\rm K}^2 z. \tag{1.235}$$

Using numerical values appropriate for a 1 μm particle at $z \sim h$ at 1 AU ($\rho = 6 \times 10^{-10}$ g cm^{-3}, $z = 3 \times 10^{11}$ cm, $v_{\rm th} = 10^5$ cm s^{-1}) gives a settling speed $v_{\rm settle} \approx 0.06$ cm s^{-1}. The settling time, defined as,

$$t_{\rm settle} = \frac{z}{|v_{\rm settle}|}, \tag{1.236}$$

is about 1.5×10^5 yr. In the absence of turbulence micron-sized particles ought to settle out of the upper layers of the disk on a time scale that is shorter than the disk lifetime.

Turbulent diffusion acts to counteract the effects of settling. If the particle fluid with density ρ_p behaves as a trace species (i.e. $\rho_p/\rho \ll 1$) then it obeys an advection-diffusion equation [117, 143],

$$\frac{\partial \rho_p}{\partial t} = D\frac{\partial}{\partial z}\left[\rho\frac{\partial}{\partial z}\left(\frac{\rho_p}{\rho}\right)\right] + \frac{\partial}{\partial z}\left(\Omega_{\rm K}^2 t_s \rho_p z\right). \tag{1.237}$$

Steady-state solutions to this equation can be found in the limit where the particle layer is thin enough that the *gas* density is approximately constant across the particle scale height. The dimensionless friction time τ_s is then independent of z and we find,

$$\frac{\rho_p}{\rho} = \left(\frac{\rho_p}{\rho}\right)_{z=0} \exp\left[-\frac{z^2}{2h_p^2}\right], \tag{1.238}$$

where h_p, the scale height of the particle concentration ρ_p/ρ, is,

$$h_p = \sqrt{\frac{D}{\Omega_{\rm K}^2 t_s}}. \tag{1.239}$$

If the turbulent diffusivity is comparable to the turbulent viscosity, i.e. $D \sim \nu$, the ratio of the concentration scale height to the gas scale height is just,

$$\frac{h_p}{h} \simeq \sqrt{\frac{\alpha}{\tau_s}}. \tag{1.240}$$

Solid particles become strongly concentrated toward the disk mid-plane whenever their dimensionless friction time substantially exceeds α. For reasonable values of α this requires substantial particle growth.

1.7.3 Streaming Instability

Youdin and Goodman [437] demonstrated that the aerodynamically coupled system of gas and solids, described by the NSH equilibrium, is linearly unstable to the growth of perturbations. The instability, known as the *streaming instability* by (very rough) analogy with the two-stream instability of plasmas, taps the free energy present in the relative motion between the solid and gaseous fluids, which is ultimately sustained by the background gradient in the pressure. It provides a physically plausible route to forming planetesimals (km or larger bodies that are largely decoupled from the gas) rapidly from smaller solids with $\tau_s \sim 1$ or less.

The pre-requisites for the existence of the streaming instability are two-way aerodynamic coupling between gas and dust within a rotating system (with shear and Coriolis force). A minimal mathematical analysis can be performed in the "terminal velocity approximation", in which the relative velocity between gas and dust is,

$$\Delta \mathbf{v} = -\frac{\nabla P}{\rho} t_s. \quad (1.241)$$

This approximation yields a third-order dispersion relation [437], while a more complete analysis (still neglecting vertical stratification) results in a sixth-order system. The linear growth rates, plotted in Youdin and Goodman [437], are functions of τ_s and the ratio of solid to gas density, ρ_d/ρ_g. Growth is typically substantially slower than dynamical, with the most unstable modes having scales $\ll h$. For $\tau_s \sim 10^{-2}$ and $\rho_p/\rho_g \sim 0.1$, for example, the linear growth time scale is a few hundred orbits.

A simple physical (as opposed to mathematical) explanation of the streaming instability is frustratingly elusive. (Analogies to "traffic jams", or to the drag reducing properties of pelotons in bicycle races, are more relevant to the strong clustering that *results* from the streaming instability than to its existence as a linear instability.) The reader distressed by this state of affairs may seek solace in papers by Chiang and Youdin [89], and by Jacquet et al. [195], who discuss the origin of instabilities in simplified or related physical systems.

The relationship between the saturated state of the streaming instability (which is of greatest interest when the fluctuations in particle density are very strongly non-linear) and the linear growth phase is non-obvious, and requires numerical simulations. Starting with the work of Johansen and Youdin [201], several authors have quantified the outcome of the instability in protoplanetary disks (for recent examples, see e.g. [36, 198, 434]). Simplifying greatly, the streaming instability depends on the particle stopping time τ_s, the local solid to gas ratio (or metallicity, with super-Solar metallicities $Z > 10^{-2}$ being favored [36, 203]) and on the magnitude of the deviation from Keplerian velocity $\eta v_K/c_s$ (with *small* values of this parameter promoting clumping [37]). Using two-dimensional simulations Carrera et al. [81] and Yang et al. [435] find that strong clumping occurs for $Z > Z_{\rm crit}(\tau_s)$, where the critical metallicity is fit by a piece-wise function,

$$\begin{aligned} \log Z_{\rm crit} &= 0.10(\log \tau_s)^2 + 0.20 \log \tau_s - 1.76 \;\; (\tau_s < 0.1), \\ \log Z_{\rm crit} &= 0.30(\log \tau_s)^2 + 0.59 \log \tau_s - 1.57 \;\; (\tau_s > 0.1). \end{aligned} \quad (1.242)$$

These results suggest that the sweet spot where the lowest metallicity is required for strong clustering occurs for $\tau_s \approx 0.1$ at $Z \approx 0.015$.

The existence of a strongly inhomogenous distribution of solid particles has important implications for particle growth and planetesimal formation, irrespective of whether the over-densities are strong or very strong. There is particular interest, however, in determining whether the streaming instability can yield over-densities that exceed the Roche density (Sect. 1.6.2.3), given approximately by,

$$\rho_p \sim \frac{M_*}{r^3}. \tag{1.243}$$

Particle clumps whose density exceeds the Roche density can collapse gravitationally into planetesimals, whose properties (such as size and binarity [303]) will depend upon the statistics of the particle density field generated by the streaming instability. Collapse is a likely outcome in regions of the disk where the streaming instability is strong [198, 200]. Current simulation results suggest that collapse forms a population of planetesimals with an initial mass function that can be fit with a power-law [199, 369, 374],

$$\frac{\mathrm{d}N}{\mathrm{d}M_p} \propto M_p^{-1.6}, \tag{1.244}$$

whose exponent displays at most a weak dependence on the stopping time of the particles that participate in the instability [375]. This predicted mass function is top-heavy, in the sense that most of the mass resides in a handful of the largest objects (in contrast to the stellar IMF, which has the opposite property [55]). The role of intrinsic gas-phase disk turbulence (which is not included in most of the recent calculations) in modifying the predicted mass function has not, however, been fully established.

The circumstances that lead to the formation of streaming-unstable regions within protoplanetary disks, along with the outcome of the instability when it occurs, are by no means definitively established. Figure 1.27 illustrates the flavor of theoretical models now under investigation [197], which invoke the single particle processes discussed in this section as essential elements. Vertical settling and radial drift, operating on particles that have grown through collisions [70] to be imperfectly coupled to the gas, act to enhance the local metallicity toward the values where the streaming instability would operate. Settling and pile-up, however, may not always be sufficient, and the next section is devoted to processes that can generate structure and additional enhancement in the local metallicity within the disk.

- Particles with sizes from microns up to centimeters are coupled to the disk gas by aerodynamic forces, operating in the Epstein regime (where the particle size is smaller than the molecular mean-free-path, leading to a linear dependence of the drag on relative velocity).
- Aerodynamic forces acting on particles lead to radial drift, because the gas in the disk feels radial pressure forces that are not experienced by the particles. Drift is directed toward pressure maxima. In a disk with a monotonically declining pressure profile drift is inward, extremely rapid, and can lead to a transient pile-up of solids in the inner disk regions.

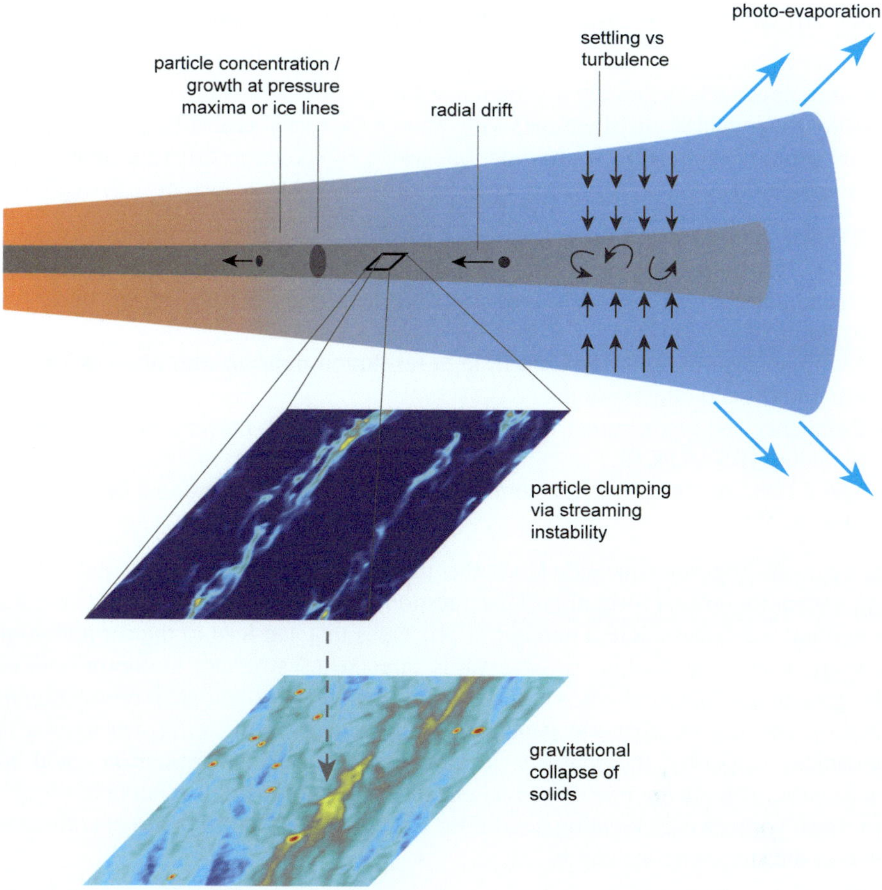

Fig. 1.27 Illustration of some of the processes that can lead to streaming instability in the aerodynamically coupled particle-gas system. Simulations of the streaming instability and gravitational collapse by Jake Simon [374]

- Vertical settling also occurs on a time scale that is short compared to the disk lifetime. It can be inhibited by even modest levels of fluid turbulence, and is usually accompanied by coagulation and fragmentation processes.
- When the feedback of a radially drifting particle fluid on the gas is accounted for, the resulting coupled system of gas and particles is often linearly unstable to the streaming instability. Given appropriate conditions this can lead to very strong clustering of the particles, followed by gravitational collapse.

1.8 Structure Formation in Protoplanetary Disks

Up until now we have largely assumed that the gas and dust in protoplanetary disks follow axisymmetric distributions, with smooth (but very probably different [12]) radial profiles. This is an approximation, which is known to fail spectacularly in some observed systems. As discussed in Sect. 1.2 disks show a variety of structures:

- Classification of a significant fraction of protoplanetary disks as *transitional disks* [124], based on evidence of inner cavities in (at least) the dust distribution.
- Radial structure in molecular emission linked to the presence of ice lines, for example of CO [342].
- Multiple rings of emission seen in high resolution mm/sub-mm observations of HL Tau [8] and other systems.
- Pronounced non-axisymmetric ("horseshoe"-shaped) sub-mm emission in systems including IRS 48 [420].
- Spiral arms and other non-axisymmetric structures seen in scattered light images of disks [161].

An open and important question is whether these structures are a consequence of—or a precursor to—planet formation. That question cannot yet be answered, but keeping it in mind we discuss here a number of processes that can lead to the formation of directly observable (and hence necessarily large-scale) structure in one or both of the gas and dust distributions within disks. Independent of the topical observational interest, any process that can generate inhomogeneity in the solid distribution is potentially important theoretically. In particular planetesimal formation could be made easier if there are processes that enhance the ratio of solids relative to the gas (in rings, vortices etc), creating conditions more favorable for both direct collisions and for the streaming instability.

1.8.1 Ice Lines

The water snow line, together with the silicate sublimation front and various ice lines in the outer disk, are potentially critical locations for planet formation. Most often, this importance is quantified by noting that the equilibrium chemical composition of a Solar abundance gas has a substantially larger mass of condensible solids outside the snow line than inside (by about a factor of 4 in the classical Minimum Mass Solar Nebula [175], rather less than that using more modern calculations of the chemical equilibrium [257]). The likelihood that this leads to a jump in solid surface density at the snow line is then invoked as the reason why the Solar System has only terrestrial planets at smaller radii, and giants beyond.

These arguments are valid but incomplete. First, the equilibrium chemical composition is only linked directly to the solid surface density in the limit where the solid particles remain small and well-coupled to the gas. Particles that grow to be large enough that radial drift becomes significant will instead develop a surface density profile that is both different from that of the gas (Sect. 1.7.1.3, [439]), and dependent on the size distribution. If icy particles are typically substantially larger than silicates (as is frequently suggested) their more rapid radial drift could lead to an instantaneously *lower* surface density of solids outside the snow line than inside. Second, although it is *possible* to construct models in which an assumed jump in planetesimal surface density at the snow line contributes to efficient core formation, the compositional effect is not the whole story. The greater area of planetary feeding zones at larger radii, along with more complex effects such as Type I migration and pebble accretion, affect the outcome of planet formation at different radii to a similar extent. The most important role of ice lines may instead be as a preferential site for planetesimal formation, or perhaps as a location where Type I migration stalls.

The pressure in the protoplanetary disk is substantially below that of the triple point of water, and hence the snow line marks a radial transition between ice and water vapor. Under mid-plane conditions, the corresponding temperature is typically $T = 150-180$ K. Where this isotherm lies in the disk is a function of the stellar luminosity and accretion rate—neither of which are constant over time—and of the disk opacity which may change due to coagulation. Theoretical models [151, 239, 292] suggest that when $\dot{M} \approx 10^{-7}\ M_\odot\ \mathrm{yr}^{-1}$ r_{snow} is at $\approx$3 AU, before moving inward to within 1 AU as the accretion drops to $\dot{M} \approx 10^{-9}\ M_\odot\ \mathrm{yr}^{-1}$. At still lower $\dot{M}$ the inner disk becomes optically thin, and the resultant rise in temperature pushes the snow line back out to 2–3 AU.

The above estimates, calculated within the framework of relatively simple disk models that include viscous heating and irradiation, are significantly modified [281] if the true disk structure instead resembles Gammie's [148] layered model with a mid-plane dead zone. The calculation of snow line evolution may require further revision if winds, which potentially change both the surface density profile and the fraction of potential energy that goes into disk heating, are important on AU-scales.

Figure 1.28 illustrates some of the key physical processes occurring near ice lines. *Where* ice lines occur can be calculated by use of the Clausius-Clapeyron relation, which gives the saturated vapor pressure P_{eq} at temperature T in terms of the latent heat L of the phase transition,

$$P_{\mathrm{eq}} = C_L e^{-L/\mathscr{R}T}. \tag{1.245}$$

Here $\mathscr{R}$ is the gas constant, and C_L is a constant that depends upon the species involved. For water, $L/\mathscr{R} = 6062$ K and $C_L = 1.14 \times 10^{13}$ g cm^{-1} s^{-2} [91]. We can compare this pressure to the actual partial pressure of vapor in the disk. Using water (molecular weight $\mu_{\mathrm{H_2O}} = 18$) as an example, if the surface density of vapor is Σ_v, the mid-plane pressure is,

$$P_v = \frac{1}{\sqrt{2\pi}} \frac{\Sigma_v}{h} \frac{k_B T}{\mu_{\mathrm{H_2O}} m_H}. \tag{1.246}$$

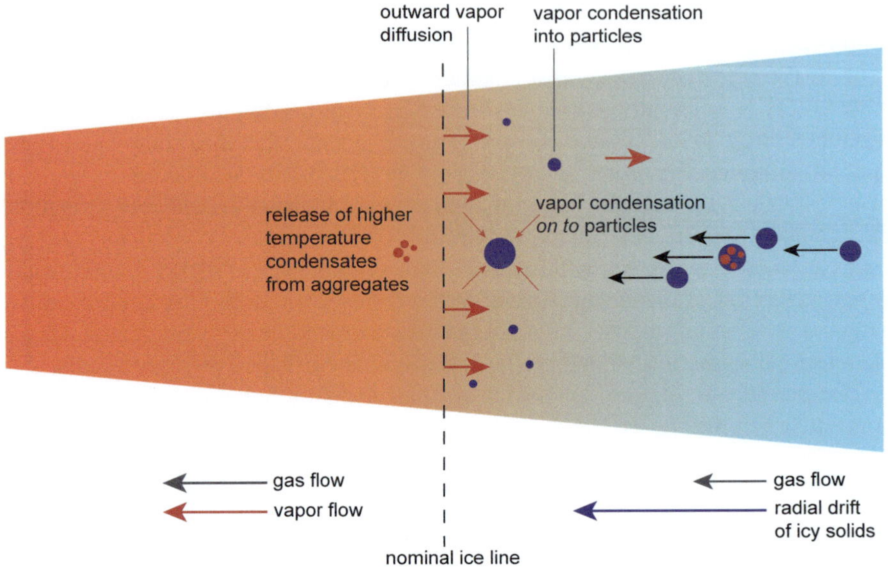

Fig. 1.28 Illustration of some of the physical processes occurring near ice lines in the protoplanetary disk. Icy materials drifting radially inward sublimate when they reach the ice line, releasing any higher temperature materials that were embedded into aggregates [91]. The resulting vapor flows toward the star at the same speed as the rest of the gas, but also diffuses outward down the steep gradient in concentration [388]. It may then recondense, either into new particles or on to pre-existing particles [361]. Some combination of these effects may feedback upon the gas physics via changes to either the opacity or, in models where MHD processes dominate angular momentum transport, the ionization state [228]

If $P_v < P_{\rm eq}$ water ice will sublimate, whereas if $P_v > P_{\rm eq}$ vapor will condense into solid form. For small particles whose sizes are measured in mm or cm sublimation is rapid [91], and hence to a reasonable approximation sublimation and condensation processes act to maintain the vapor pressure close to the equilibrium value.

The large value of $L/\mathscr{R}$ implies that the snow line is a sharply defined transition within the disk. At 150 K, the characteristic temperature interval over which the equilibrium vapor pressure varies, $P_{\rm eq}/({\rm d}P_{\rm eq}/{\rm d}T)$, is just a few Kelvin. If sublimation of icy particles that drift through the snow line is fast, this has the consequence of imposing a sharp radial gradient in water vapor concentration at the snow line which, in a turbulent disk, will in turn result in a diffusive outward flux of vapor (Eq. 1.222). Re-condensation of the vapor can then lead to an enhancement of the solid surface density immediately outside the snow line.

Stevenson and Lunine [388] proposed that a vapor diffusion/condensation cycle of this type could enhance the surface density of ice enough to promote the rapid formation of Jupiter's core. The strength of the effect depends upon the size of the icy solids drifting in toward the snow line, and upon what is assumed about processes including disk turbulence, condensation and coagulation/fragmentation.

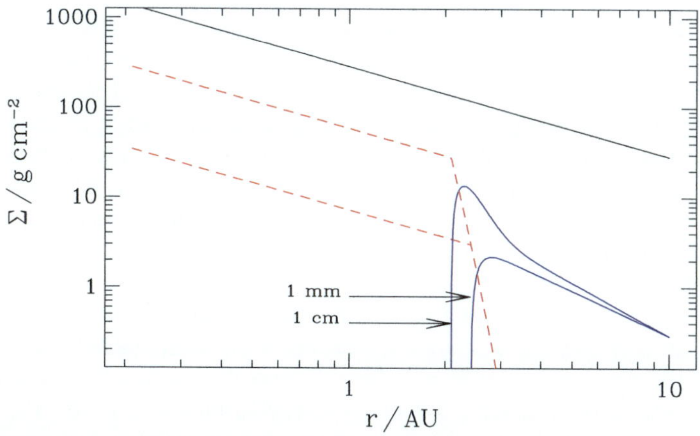

Fig. 1.29 Example steady-state profiles of gas (upper solid line), radially drifting icy particles (solid blue lines) and water vapor (red dashed lines) in a turbulent protoplanetary disk. The assumed disk model has an accretion rate of $10^{-8}\ M_\odot\ \mathrm{yr}^{-1}$, a temperature $T = 150(r/3\ \mathrm{AU})^{-1/2}$, and an α parameter of 5×10^{-3}. The ratio of the turbulent diffusivity to the turbulent viscosity is taken to be unity, and the concentration of icy solids is set (arbitrarily) to be 10^{-2} at 10 AU. The rapid radial drift of cm-sized particles leads to a high concentration of water vapor in the inner disk. Outward diffusion and re-condensation of this vapor—assumed here to form particles of a single fixed size—leads to an enhancement of solids just outside the snow line [388]. Note also the more elementary conclusion that the vapor concentration in the inner disk is a direct probe of the mass flux of radially drifting solids encountering the snow line

Figure 1.29 shows the results of a particularly simple calculation ([22] after [91, 96]), which assumes that vapor condenses (or, condenses and rapidly coagulates) into solid particles of a single size that *matches* the size of icy solids drifting in from larger radii. For the adopted disk model, mm-sized icy solids have relatively low drift velocities, and these particles sublimate into vapor without generating any local enhancement in the surface density of solids. Larger cm-sized particles, conversely, are enhanced by a factor of several outside the snow line as a consequence of the diffusive transport of vapor followed by condensation. Surface density enhancements of this magnitude could be important, particularly in models where planetesimal formation depends upon the disk locally exceeding a threshold value of metallicity (as is the case for the streaming instability, see Sect. 1.7.3, [203]).

Ros and Johansen [361] investigated a related possibility. Instead of assuming that vapor condenses into new particles (or, on to very small grains released when aggregates break up at the snow line), they modeled the growth of pre-existing solids as vapor condenses on to their surfaces. A simple collisional argument gives the growth rate due to vapor condensation/sublimation as,

$$\frac{\mathrm{d}m}{\mathrm{d}t} = 4\pi s^2 v_{\mathrm{th}} \rho_v \left(1 - \frac{P_{\mathrm{eq}}}{P_v}\right), \tag{1.247}$$

for a particle of mass m and radius s, surrounded by vapor of density ρ_v and thermal speed v_{th}. (As noted by Supulver and Lin [391], this is not an exact expression [176].) Using a Monte Carlo approach, Ros and Johansen [361] found that condensation on to particle surfaces could provide an efficient growth mechanism up to sizes of the order of 10 cm. This could aid planetesimal formation by boosting the stopping time of particles into the range preferred by the streaming instability [81]. Moreover, by removing mass from the directly observable mm-size regime, condensation-driven growth could suppress the mm and sub-mm flux from disks in the vicinity of ice lines [440].

The aforementioned physics affects only trace components of the disk—the icy solids and the resulting vapor. It is easy, however, to contemplate feedback processes that couple the evolution of solids at ice lines to the bulk of the gas disk. At a minimum the opacity will vary depending upon the radial distribution of solids. Beyond that, we have already noted that the efficiency of angular momentum transport (in MHD models) is expected to be a function of the local ionization state, and that the ionization balance is affected by the abundance of small grains. Kretke and Lin [228] suggested that the enhanced abundance of solids near the snow line would act to suppress the rate of angular momentum transport, and that this could lead to the formation of a local pressure maximum that would act to trap particles (producing, in principle, a positive feedback loop). This argument (which has been invoked in some models of collisional growth [76]) is highly plausible, though the breadth and complexity of the physics involved makes quantitative investigation challenging.

Although we have focused here on the snow line, analogous considerations carry over to other ice lines. The CO ice line corresponds to a temperature of $T = 17-19$ K [66], and is somewhat more complicated to model because the CO is typically mixed with N_2 and water ices. The silicate "ice line" (or sublimation front) could also be important, since it lies at radii where a high fraction of stars are observed to host short-period planetary systems.

1.8.2 Particle Traps

In a disk with a monotonically declining pressure profile radial drift is always inward. More generally, however, aerodynamic drift is directed toward pressure maxima, and can be outward if there is a local pressure maximum within the disk. This possibility was recognized in a prescient paper by Whipple [431], who appealed to it as part of a model for the formation of comets.[19] The tendency for solids to be aerodynamically enhanced in the vicinity of pressure maxima is often described as particle "trapping", though this term is somewhat misleading; in a turbulent disk small particles are

[19]Quoting from his paper, "should it be possible for a toroid of higher density to occur in the Solar nebula, the growing planetesimals would be drawn toward it from the inside as well as from the outside …".

at most temporarily detained by pressure maxima rather than being permanently trapped.

The effect of local pressure maxima on the radial distribution of solids can be derived, in the limit where the solids remain a trace contaminant, following the methods described in Sects. 1.7.1.1 and 1.7.1.2. For an axisymmetric disk with an arbitrary mid-plane pressure profile, the radial velocity v_r of particles under the action of aerodynamic forces remains as given by Eq. (1.219), with the parameter η describing the deviation from Keplerian velocity becoming [395],

$$\eta = -\left(\frac{h}{r}\right)^2 \left[\frac{\mathrm{d}\ln\Sigma}{\mathrm{d}\ln r} + (q-3)\right]. \tag{1.248}$$

In this formula q is defined as the local power-law index describing the flaring of the disk,

$$\frac{h}{r} \propto r^{q-1}, \tag{1.249}$$

such that a non-flaring disk has $q = 1$. In axisymmetry and steady-state, Eqs. (1.221) and (1.222) can be immediately integrated to give,

$$r(F_{\mathrm{diff}} + F_{\mathrm{adv}}) = k, \tag{1.250}$$

where the diffusive and advective fluxes are,

$$\begin{aligned} F_{\mathrm{diff}} &= -D\Sigma\frac{\mathrm{d}}{\mathrm{d}r}\left(\frac{\Sigma_d}{\Sigma}\right), \\ F_{\mathrm{adv}} &= \Sigma_d v_r, \end{aligned} \tag{1.251}$$

and the constant k is just the radial flux of solid material. Written out explicitly, the concentration of particles $f \equiv \Sigma_d/\Sigma$ obeys a first-order differential equation,

$$\frac{\mathrm{d}f}{\mathrm{d}r} - \frac{v_r}{D}f = \frac{k}{D\Sigma}\frac{1}{r}. \tag{1.252}$$

Analytic solutions to this equation are possible for simple choices of v_r, D and Σ. A straightforward quadrature gives the solution for the concentration profile given more realistic choices of these functions.

Figure 1.30 shows an illustrative numerical solution to Eq. (1.252). For this example, we have modeled a trap by assuming (arbitrarily) that the viscosity in the gas disk is reduced across a moderately narrow annulus. The lower viscosity leads to a higher surface density, producing a pressure maximum which in turn concentrates particles. The concentration effect is strongly size-dependent. Small particles (in this example those with radii of 0.1 and 1 mm) have a radial velocity that is similar to that of the gas, and are largely unaffected by the pressure maximum. Larger cm-sized particles, on the other hand, have a larger magnitude of radial drift, and can be

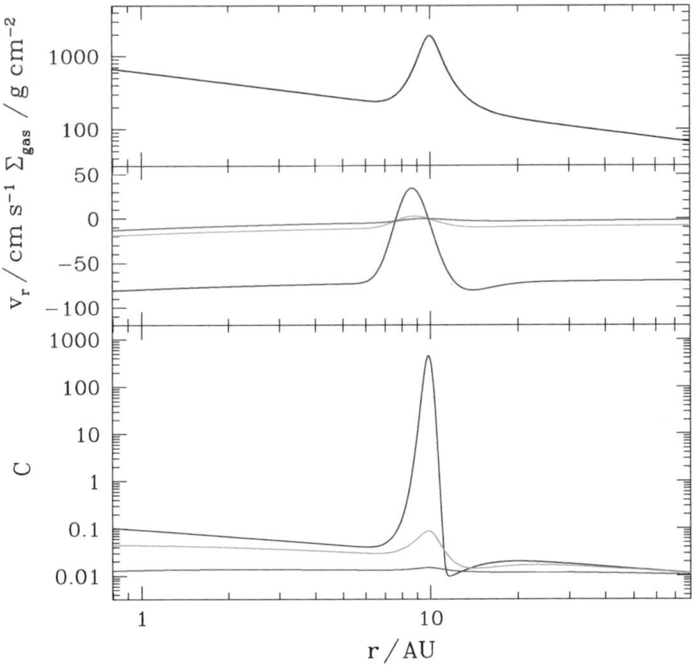

Fig. 1.30 The steady-state radial distribution of solids in a turbulent disk with an axisymmetric local pressure maximum (a particle "trap"). Upper panel: the surface density of the gas. Middle panel: the radial velocity of 0.1 mm (red), 1 mm (green) and 1 cm (blue) particles. The smallest particles have a radial velocity that is almost indistinguishable from that of the gas, while the larger particles experience rapid radial drift that can be outward near the location of the pressure maximum. Lower panel: the concentration $C = \Sigma_d / \Sigma_{\rm gas}$, normalized to an arbitrary value of 10^{-2} at 100 AU. The assumed disk model for this calculation has $\dot{M} = 10^{-8} \, M_\odot \, {\rm yr}^{-1}$, $M_* = M_\odot$, $h/r = 0.5$, $\alpha = 10^{-3}$ and $D = \nu$ (c.f. Eq. 1.223). The trap is modeled as a gaussian-shaped reduction in α to a minimum of 10^{-4}, with a width of $4h$. The particles are assumed to be spherical, with a material density of $1 \, {\rm g \, cm^{-3}}$, and to follow the Epstein drag law

strongly concentrated at the location of the pressure maximum. The enhancement in the local particle density can reach several orders of magnitude, depending both on the particle size and on the "strength" (radial width and amplitude) of the pressure maximum within the gas disk.

It should be noted that in a turbulent disk (and ignoring particle feedback on the gas) it is *always* possible to find a steady-state solution for the radial particle concentration in which the particle mass flux is a constant at all radii. Physically, this is because particles accumulate near pressure maxima until the radial gradient in concentration becomes large enough for turbulent diffusion to allow them to leak out [425, 445]. Pressure maxima only act as a "filter" [353], *permanently* removing large particles from the inward radial drift flow, when additional physical effects are included [445]. If the local concentration of large particles becomes large enough, for example, planetesimal formation [169] could cause some fraction of the solid material to drop out of the radial flow.

A number of physical effects have been identified that could lead to the formation of local pressure maxima within protoplanetary disks. These include the exterior edges of planet-carved gaps [320, 353, 445], the outer edges of cavities created by photoevaporation (Sect. 1.9.1, [5]), and a photoelectric heating instability that may operate in gas-poor systems (primarily debris disks) [63, 270]. Spiral arms in self-gravitating disks [153, 355] and the inner edges of dead zones [269] can also concentrate solids via closely related physical processes, though these environments typically involve significant non-axisymmetry. With the exception of self-gravity, these possible locations for pressure maxima either form at specific places within the disk (e.g. at the inner dead zone edge, defined by a characteristic mid-plane temperature), or at a time after when we expect planets to have formed (during photoevaporative disk clearing, or in *response* to pre-existing massive planets). It is possible, however, for the turbulence within the gas disk to be generically unstable to the formation of pressure maxima within zonal flows. This would be interesting because it would imply the (possibly transient) existence of multiple particle traps within the disk, that could play a role in early-stage planet formation [333].

1.8.3 Zonal Flows

It is evident from Eq. (1.23) that an equilibrium can be set up in which radial forces from a complex pressure profile (that may include local pressure maxima) are balanced by radial variations in azimuthal velocity. In a local description (Eq. 1.138, but here ignoring magnetic fields) the balance is between the pressure gradient term $\nabla P/\rho$ and that describing the Coriolis force $2\Omega_0 \times \mathbf{v}$. When these terms balance, such that

$$\frac{\mathrm{d}P}{\mathrm{d}x} = 2\Omega_0 \rho v_y, \tag{1.253}$$

the system is said to be in *geostrophic balance* (here x is the radial direction, and y the azimuthal). The pressure gradient is compensated by variations in the orbital velocity, creating *zonal flows* analogous to the banded structure of winds in giant planet atmospheres.

A disk zonal flow is an equilibrium solution to the fluid equations, but that equilibrium may not be stable. Too pronounced a deviation from Keplerian rotation results in a shear profile that is unstable to Rossby wave instability [259]. We will discuss this instability, which is similar to Kelvin-Helmholtz instability, in Sect. 1.8.5. Even if the rotation profile is stable, the diffusive nature of a classical viscosity (Eq. 1.62) would tend to erase any small-scale perturbations in the pressure that are sourced from surface density fluctuations. Persistent zonal flows are thus not expected in classical disk theory. They have been observed, however, in local numerical simulations of MHD turbulent disks [202]. The key to their formation appears to be the ability of MHD disk turbulence to generate large-scale structure in the magnetic field [378], which could be viewed as an inverse cascade of turbulent power. The details of how and when zonal flows form are not entirely clear, though Johansen et al. [202] describe a simplified dynamical model in which large-scale variations in the Maxwell stress lead first to azimuthal velocity perturbations and thence to axisymmetric structure in the pressure and density.

The lifetime and radial scale of zonal flows in protoplanetary disks depend upon the same factors that determine the properties of MHD disk turbulence more generally (Sect. 1.5.4.4), namely the strength of non-ideal terms in the induction equation and the presence of net vertical magnetic field. In ideal MHD, lifetimes of tens of orbital periods and radial scales of the order of 10 h appear to be typical [202, 378], though these results require a double dose of caveats—first because the inferred scales are not much smaller than the size of the local simulation domains used, and second because they are large enough that curvature terms neglected in local models may be important. Nonetheless, the amplitude, scale and lifetime of zonal flows under ideal MHD conditions plausibly lead to strong local enhancements in the dust to gas ratio for particles with stopping times $\tau_s \sim 0.1-1$ [111]. The presence of net vertical fields substantially enhances the amplitude of zonal flows [40].

Zonal flows are also found in MHD disk simulations that include ambipolar diffusion, and can be comparable in amplitude to the ideal MHD case if the net field is sufficient to stimulate a significant $\alpha \sim 10^{-2}$ [373]. However, both inferences from local simulations [373], and explicit tracking of particles in global simulations [447], suggest that zonal flows in the outer regions of protoplanetary disks, where ambipolar diffusion dominates, have properties close to the boundary beyond which strong particle enhancement would be expected.

In summary, zonal flows are likely to be present in the inner disk, where ideal MHD is a good approximation, though these flows would only strongly influence the dynamics of relatively large solid particles. In the outer disk, where ambipolar diffusion is important and even mm-sized particles have significant stopping times, zonal flows could introduce observable large-scale axisymmetric structure and may contribute to particle concentration. In the Hall-dominated regime that prevails around 1 AU theoretical expectations are less clear. Extremely strong zonal structures were observed in local vertically unstratified Hall-MHD simulations [232], whereas comparable stratified runs instead led to large-scale Maxwell stress [240]. It is therefore unclear whether there are circumstances in which zonal flows on AU-scales could contribute to particle concentration and planetesimal formation.

1.8.4 Vortices

Few issues in planet formation are as long-debated as the possible role of vortices. Very general arguments suggest that large-scale vortices could be present in protoplanetary disks and play an important role in planetesimal formation. We note first that disks are (approximately) two dimensional fluid systems, and in contrast to three dimensional systems they therefore support an inverse cascade of turbulent energy toward large scales [223]. Second, for a barotropic disk (i.e. $P = P(\rho)$ only) the vortensity ω/ρ is conserved (Eq. 1.124). Taken together, these properties imply that vortices within disks have the potential to form persistent long-lived structures. Indeed, simulations of strictly two dimensional flows show that disks seeded with small-scale vorticity perturbations evolve to form a small number of large and persistent anticyclonic vortices [154, 205]. Anticyclonic vortices are high pressure regions that attract marginally coupled solids [51, 398], potentially catalyzing the subsequent formation of planetesimals. Even absent planetesimal formation, the natural tendency of vortices to form large-scale non-axisymmetric dust features makes it tempting to identify them with observed disk asymmetries [420].

The basic properties of disk vortices are well-established. What is much trickier is to determine (1) whether vortices form spontaneously in disks or only *after* planet formation (for the reasons already mentioned, spontaneous vortex formation generally requires non-barotropic processes), and (2) whether three dimensional instabilities and/or particle feedback are fatal impediments to their survival. Observations as well as theory are probably needed to resolve these issues.

1.8.4.1 The Kida Solution

The magnitude of the vorticity in a strictly Keplerian disk is $\omega_K = -(3/2)\Omega_K$. A vortex can be modeled as a spatially localized elliptical perturbation, within which the vorticity $\omega = \omega_K + \omega_v$, with ω_v a constant. Other types of vortex are possible, but rather remarkably this type can be described by an exact non-linear solution [213] that is useful for both analytic and numerical studies.

The Kida solution [213] describes a vortex within a shearing-sheet approximation to disk flow. Following Lesur and Papaloizou [243] we define a cartesian co-ordinate system (x, y) that is centered at radius r_0 and which co-rotates with the background disk flow at angular velocity $\Omega_K = \Omega_K(r_0)$,

$$\begin{aligned} x &= r_0\phi, \\ y &= -(r - r_0). \end{aligned} \tag{1.254}$$

Kida considered time-dependent vortex solutions, but here we will worry only about vortices that are steady.[20] Time-independent solutions are possible if the semi-major axis of the vortex is aligned with the azimuthal direction (x in the shearing sheet model), and the vorticity perturbation satisfies,

$$\frac{\omega_v}{\omega_K} = \frac{1}{\chi}\left(\frac{\chi + 1}{\chi - 1}\right). \tag{1.255}$$

[20] For a derivation of the steady Kida solution, see e.g. the appendix of Chavanis [88].

Here $\chi = a/b$ is the aspect ratio of the vortex, which forms an elliptical patch with semi-major axis a and semi-minor axis b. The right-hand-side is evidently positive, which implies that the only steady Kida vortices in Keplerian disks are anticyclonic (with ω_v having the opposite sign to Ω_K).

The complete Kida solution is written in terms of a streamfunction ψ in an elliptic co-ordinate systems (μ, ν), where,

$$x = f \cosh(\mu) \cos(\nu),$$
$$y = f \sinh(\mu) \sin(\nu), \tag{1.256}$$

and $f = a\sqrt{(\chi^2 - 1)/\chi^2}$. The solution can be split into a core and an exterior part,

$$\begin{aligned}\psi_{\rm core} &= -\frac{3\Omega_K f^2}{4(\chi - 1)}\left[\chi^{-1}\cosh^2(\mu)\cos^2(\nu) + \chi \sinh^2(\mu)\sin^2(\nu)\right],\\ \psi_{\rm ext} &= -\frac{3\Omega_K f^2}{8(\chi - 1)^2}\left[1 + 2(\mu - \mu_0) + 2(\chi - 1)^2 \sinh^2(\mu)\sin^2(\nu)\right.\\ &\qquad \left.+ \frac{\chi - 1}{\chi + 1}\exp[-2(\mu - \mu_0)]\cos(2\nu)\right],\end{aligned} \tag{1.257}$$

which match at $\mu = \mu_0 = \tanh^{-1}(\chi^{-1})$. The cartesian velocity field is then given by $v_x = -\partial\psi/\partial y$, $v_y = \partial\psi/\partial x$. In general the cartesian representation of the velocity has no simple form, but within the core it is,

$$\begin{aligned}v_{x,\rm core} &= \frac{3\Omega_K \chi}{2(\chi - 1)} y,\\ v_{y,\rm core} &= -\frac{3\Omega_K}{2\chi(\chi - 1)} x,\end{aligned} \tag{1.258}$$

describing simple elliptical motion. Figure 1.31 shows the contours (logarithmically spaced) of the full streamfunction for Kida vortices of varying aspect ratio.

Barge and Sommeria [51] studied the trajectories of aerodynamically coupled solids that encounter vortices (using a different and approximate vortex model). The high pressure within anticyclonic vortex cores acts as an attractor for solids. The cross-section for capture is maximized for particles with dimensionless stopping time $\tau_s = 1$ [88]. Since vortices can potentially grow to have radial extents $\Delta r \approx h$ (the supersonic velocity perturbations of larger vortices would radiate sound waves sapping their energy) a single large vortex can potentially trap a substantial mass of solids flowing radially toward it due to ordinary radial drift.

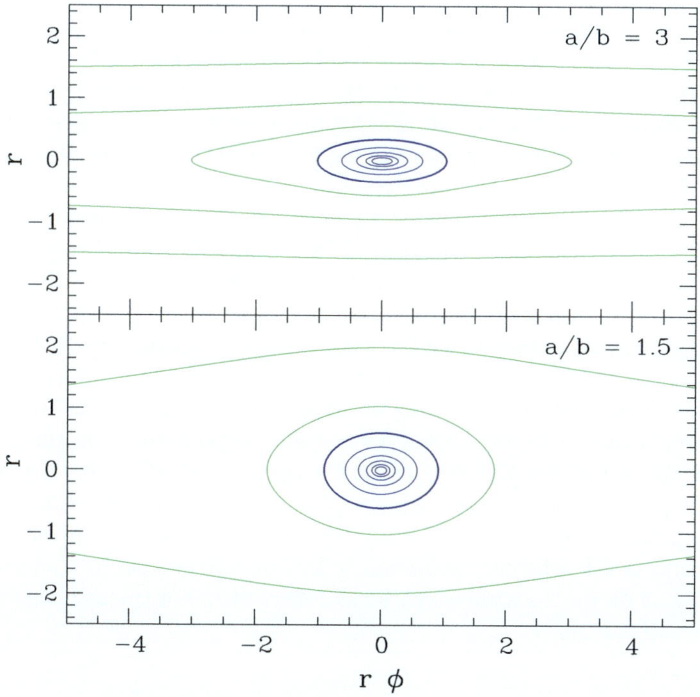

Fig. 1.31 Contours of the Kida vortex streamfunction $\psi(x, y)$ are shown for different values of the vortex aspect ratio $\chi \equiv a/b$. Within the vortex core, delineated by the bold contour, the streamlines defined by the solution are elliptical, with fixed aspect ratio. Outside the core, the vortex merges smoothly into the background shear flow of the disk

1.8.4.2 Stability of Vortices

Although the geometry of protoplanetary disks is approximately two dimensional, the fact that disk vortices are limited in radial size to $\Delta r \approx h$ means that they are three dimensional objects that notice the vertical stratification. Barranco and Marcus [53] and Shen et al. [371], using three dimensional simulations, found that mid-plane vortices that would be highly stable in 2D are rapidly destroyed by three dimensional instabilities. The origin of these observed instabilities, at least in part, appears to be the *elliptical instability* [212], which occurs whenever there is a resonance between the vortex rotation period and inertial waves within the disk. In a disk environment, Lesur and Papaloizou [243] find that purely gaseous vortices are unstable for almost all choices of the vortex aspect ratio and degree of vertical stratification, though these parameters strongly affect the linear growth time of the instability. The instability that afflicts 3D vortices, however, is typically slow-growing and of small radial scale. Numerically this makes studies of vortex survival particularly challenging. Physically it means that the questions of vortex formation and vortex survival are closely linked, except perhaps at very large radii (where "primordial vortices" might persist for interesting periods of time) the vortex population in a disk is expected to reflect an equilibrium between formation and destruction processes.

A significant loading of solids will also impact vortex longevity, generally for the worse. Railton and Papaloizou [347] studied the stability of generalized Kida vortices, containing both gas and dust in the limit of strongly aerodynamically coupled particles. They found that these configurations were vulnerable to parametric instabilities in the same way as gas-only vortices. This is consistent with a wide range of other analytic and numerical work [86, 145, 190], which suggests that dust to gas ratios in the range between 0.1 and 1 are sufficient to imperil the survival of disk vortices. This limitation may not, however, preclude vortices playing a role in planet formation. If we assume that there is a continual source of vorticity within the disk, then vortex particle concentration up to $\rho_p \sim \rho_g$ could be sufficient to initiate planetesimal formation via a variation of the streaming instability [343].

Observations have identified a number of systems (primarily transition disks) that show a non-axisymmetric distribution of sub-mm emission [83, 194, 420], consistent with that expected if aerodynamically coupled solids are accumulating in a vortex [446]. The putative vortices in these examples may all be *caused* by planets. It would be interesting to ask whether useful constraints on vortices could be derived from observations of the *most-axisymmetric* disks, where there is no independent suspicion that planets already exist.

1.8.5 Rossby Wave Instability

The Rossby wave instability (RWI) [247, 248, 259] is a well-characterized mechanism for producing vortices within protoplanetary disks. The RWI is a linear instability that grows whenever there is "sufficiently sharp" radial structure in the disk. Specifically, for a two dimensional disk model with angular velocity $\Omega(r)$, vertically integrated pressure $P(r)$, and adiabatic index γ, we define the entropy S and epicyclic frequency κ via,

$$\begin{aligned} S &= \frac{P}{\Sigma^\gamma}, \\ \kappa^2 &= \frac{1}{r^3}\frac{\partial}{\partial r}(r^4\Omega^2). \end{aligned} \tag{1.259}$$

In terms of these quantities, the stability of the disk to the RWI is determined by the radial profile of a generalized potential vorticity,

$$\mathscr{L}(r) = \frac{\kappa^2}{2\Omega\Sigma} \times S^{-2/\gamma}. \tag{1.260}$$

A necessary condition for RWI is that $\mathscr{L}$ have an extremum. A precise sufficient condition is not known, but variations in the potential vorticity of the order of 10%

over radial scales $\approx h$ appear to be enough to trigger instability, which leads to the formation of anticyclonic vortices on time scales that can be rapid—of the order of $10\Omega^{-1}$ [248]. The instability, which can be understood in terms of the local trapping of waves in the vicinity of the vortensity perturbation [155, 416], has similarities to the Papaloizou-Pringle [325] instability of accretion tori. The RWI is essentially a two dimensional instability [251, 288], though the vortices that it forms are as vulnerable to unrelated three dimensional instabilities as any others.

We have already remarked that the edges of dead zones (and ice lines [228]) are places where local pressure maxima may form. The RWI criterion (Eq. 1.260) is not necessarily satisfied at every local pressure maximum, but it is nonetheless true that dead zone edges are plausible locations where the RWI may occur. Hydrodynamic models [190, 268, 421], and MHD simulations that include Ohmic diffusion as a simple model of dead zones [126, 132, 271, 272], support this expectation, and show that the radial disk structure introduced by dead zones is likely to be unstable to the RWI and subsequent formation of vortices. The vortices in turn act as sites of particle concentration [268]. Depending upon the radii involved, a single vortex generated by the RWI may have a lifetime that is very short as compared to the disk lifetime (this will be particularly true at the inner edge of the dead zone). However, in this scenario fresh generations of vortices may be expected to form as long as the non-ideal disk physics that sustains the RWI-unstable dead zone structure persists.

A second location where RWI may occur is at the edge of a gap created dynamically by a massive planet [108, 220]. Figure 1.32 shows the evolution of a disk in a two-dimensional simulation of this scenario. The formation of a massive planet

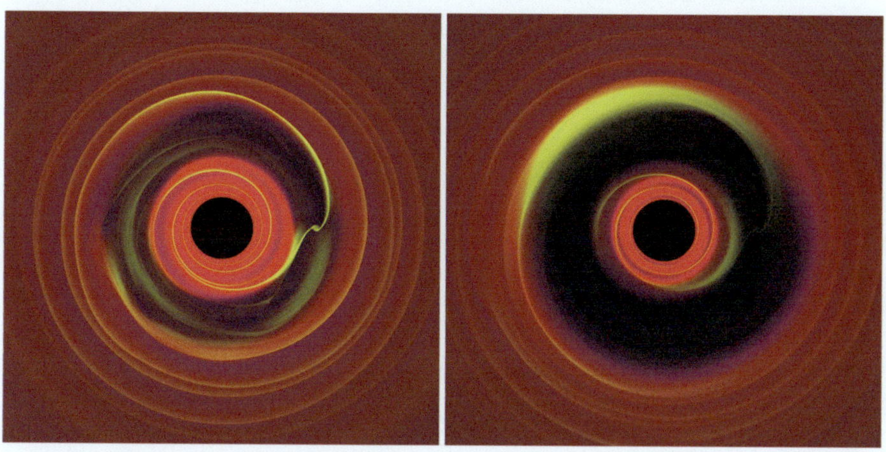

Fig. 1.32 Snapshots showing the hydrodynamic evolution of an almost inviscid disk containing a massive planet (from [20]). The planet rapidly clears an annular gap within the disk, whose edge is unstable to the generation of vortices. The system then evolves through a phase when the outer gap edge hosts a single large vortex, which can be an efficient trap for solid particles. The disk has $\alpha = 10^{-4}$ and $h/r = 0.05$ at the location of the planet, which has a mass ratio to the star of 5×10^{-3}

creates an approximately axisymmetric gap within the disk, whose edges can be unstable to the RWI. The vortices that are generated fairly rapidly merge, creating a single large vortex at either edge of the gap then can subsequently trap particles. This process works best if the disk in the vicinity of the planet's orbit has a low viscosity (roughly $\alpha \sim 10^{-3}$ or lower), and is at least partially a transient effect—the vortices form during the phase when the planet accretes its gaseous envelope. For these reasons, observable structure from planet-initiated vortices is likely to be easiest to see in the outer disk, where ambipolar diffusion damps turbulence [377] and the absolute lifetime of a single generation of vortices against disruptive instabilities is long. The particles trapped efficiently in the outer disk include those with $s \sim$ mm that can be seen in sub-mm observations of protoplanetary disks [446, 448].

- Condensation and sublimation processes are expected to modify the surface density and size distribution of particles in annuli that are adjacent to ice lines (of water, silicates, and possibly species such as CO with cooler condensation temperatures).
- Pressure maxima, whether caused by planets or by intrinsic disk processes such as zonal flows, lead to local axisymmetric enhancements in the particle surface density.
- Vortices can also trap particles. A population of vortices in a disk tend to merge to form a small number of large vortices, which unless replenished have a finite lifetime.
- Sharp features in the radial disk structure can trigger the Rossy Wave Instability, which generates vortices.

1.9 Disk Dispersal

There is no leading order mystery as to where the gas in protoplanetary disks goes to. If we divide the mean disk mass estimated from sub-mm studies in Taurus ($M_d \sim 5 \times 10^{-3}\ M_\odot$ [9]) by the median accretion rate ($\dot{M} \sim 10^{-8}\ M_\odot\ \mathrm{yr}^{-1}$ [167]) estimated in the same region, we obtain a characteristic evolution timescale of 0.5 Myr. The lifetime of detectable disks might be expected to be a small multiple of this timescale—say a few Myr—which is indeed what is observed [181].

The above exercise establishes that, from a purely observational perspective, most or all of the gas in protoplanetary disks *could* be lost via accretion on to the star. Beyond that, it proves nothing. Disk mass estimates and measurements of stellar accretion rates are subject to uncertainties, which when combined probably allow us to shift the inferred characteristic lifetime by an order of magnitude in either direction. In particular, if disk masses are systematically *under-estimated* from sub-mm continuum observations (because of particle growth to sizes too large to contribute to the sub-mm opacity) the characteristic lifetime could approach or exceed the observed one.

The next order observational diagnostic probes the time-dependence of the decay of disk signatures. Here a puzzle does emerge. The classical disk evolution Eq. (1.63) admits a self-similar solution [267] that ought to approximate the evolution at sufficiently late times. For a disk with a viscosity that scales with radius as $\nu \propto r^{\gamma}$, the late-time evolution of the surface density close to the star is predicted to follow (e.g. [18]),

$$\Sigma \propto t^{-(5/2-\gamma)/(2-\gamma)}. \tag{1.261}$$

If $\gamma = 1$, for example, as would be the case for a disk with a steady-state surface density profile $\Sigma \propto r^{-1}$, then the predicted late-time decay goes as $t^{-3/2}$. This is relatively slow, and would probably lead to a population of stars with weak disk signatures (in gas and dust tracers) that are not observed (for a review of these observational arguments, see Alexander et al. [4]). One might argue, of course, that this is an illusory problem that we have created for ourselves by placing unwarranted faith in the validity of classical viscous disk theory. A resolution along these lines is possible. More commonly, however, the discrepancy between the simple theory and observations is taken to imply that some distinct process acts to rapidly disperse the disk and terminate accretion. Photoevaporation is almost certainly part of the story, on account both of robust theoretical estimates that suggest it should be important, and observations that are consistent with photoevaporation occurring in some relatively extreme situations.

1.9.1 Photoevaporation

Disk photoevaporation is a purely hydrodynamic process that occurs when molecular gas in the disk is dissociated or ionized by high energy (UV or X-ray) photons. If the gas is heated sufficiently to become unbound, it accelerates away from the disk under the influence of pressure gradient forces to form a thermally driven wind. Early models for photoevaporative flows were developed in the 1980s by Bally and Scoville [50] in the context of massive stars surrounded by neutral disks, and in more quantitative detail by Begelman et al. [57] who studied X-ray heated disks around compact objects. The essential physics is thus very well-established. We will begin by considering how photoevaporation works if the disk is exposed to extreme ultraviolet (EUV) photons that have sufficient energy ($h\nu > 13.6$ eV) to ionize hydrogen. This is probably *not* the dominant driver of photoevaporation from protoplanetary disks around low mass stars, but it is amenable to an analytic treatment that exposes the main principles [186].

1.9.1.1 Thermal Winds from Disks

We consider a disk whose surfaces are illuminated by a source of high energy photons, that may come either from the central star or from other luminous stars within a cluster. The radiation heats a surface layer of the disk to a temperature T, with a sound speed c_s. A characteristic scale r_g (the "gravitational radius") can be defined by asking where the sound speed equal the local Keplerian velocity,

$$r_g = \frac{GM_*}{c_s^2}. \tag{1.262}$$

Noting that the thermal energy per particle is $\sim k_B T$, the radius r_g is approximately equivalent to the smallest radius where the *total* energy of the gas in the heated surface layer (i.e. thermal plus gravitational) is zero. For radii $r > r_g$ the total energy is positive, and it is energetically possible for the surface gas to flow away in a thermal wind.

This is all quite rough, and we should really consider both the hydrodynamic structure of the wind (which is similar to the textbook example of a Parker wind [427]) and the role of rotation [250]. Doing so results in an improved estimate of the critical radius beyond which a thermal wind is launched, which scales with but is significantly smaller than r_g,

$$r_c \approx 0.2\frac{GM_*}{c_s^2} \approx 1.8\left(\frac{M_*}{M_\odot}\right)\left(\frac{c_s}{10 \text{ km s}^{-1}}\right)^{-2} \text{ AU}. \tag{1.263}$$

We have picked 10 km s^{-1} as a fiducial sound speed because this is approximately the sound speed in EUV-ionized gas, which has a temperature near 10^4 K. X-rays or far ultraviolet (FUV) photons (those with 6 eV $< h\nu <$ 13.6 eV that dissociate but do not ionize hydrogen) heat the surface to lower temperatures, $\sim 10^3$ K, resulting in correspondingly larger critical radii.

Interior to r_g the hot gas is bound, and unless some other process intervenes (such as a stellar wind or an MHD disk wind) it will form a static isothermal atmosphere with a scale height that varies with radius as $h \propto r^{3/2}$ (see Sect. 1.3.1.1). Outside r_g it will flow away, at a speed of the order of the sound speed. In the case of EUV illumination the hot and ionized surface layer is separated from the underlying cool gas by a sharp ionization front, which gives a clearly defined "base" to the wind. If the number density at the base is n_0, we expect a mass loss rate per unit area of the disk that is given by,

$$\dot{\Sigma}_w \simeq 2\mu m_H n_0 c_s, \tag{1.264}$$

up to factors of order unity that depend again on the detailed hydrodynamic structure of the flow. In the EUV case the mass loss profile due to photoevaporation is then determined by the radial scaling of the base density $n_0(r)$. Noting that the integrated mass loss rate is just,

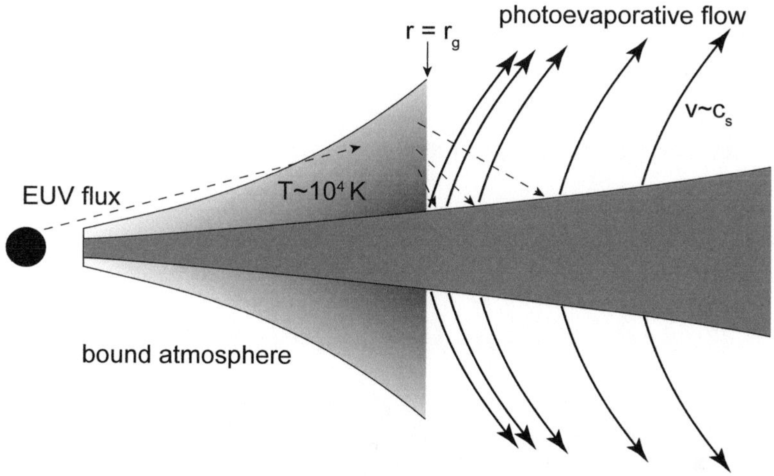

Fig. 1.33 Illustration of the simplest model of internal photoevaporation driven by extreme ultraviolet (EUV) radiation (based on the "weak stellar wind" case from Hollenbach et al. [186]). Stellar EUV radiation ionizes and heats the surface layers of the disk. Where the thermal energy of the surface layer remains small compared to the binding energy, the hot gas forms a bound atmosphere. At larger radii, where the gas is more weakly bound, the hot gas flows away in a thermally driven wind. Details of the radiative transfer and heating processes differ depending upon the nature of the high energy radiation, but a qualitatively similar scenario applies also to X-ray and FUV-driven photoevaporation. Taken from Fig. 3.8 in [18], © Cambridge University Press

$$\dot{M}_w = \int_{r_c}^{r_{\rm out}} 2\pi r \dot{\Sigma}_w {\rm d}r, \tag{1.265}$$

we conclude that if $n_0(r)$ declines more steeply that r^{-2} the mass loss is predominantly from the inner disk (near r_c), whereas a shallower profile leads to most of the mass being lost from the outer disk.

Actually finding $n_0(r)$ for photoevaporation driven by EUV radiation from the central star requires the solution of a radiative transfer problem, whose geometry is illustrated in Fig. 1.33. The base density is determined by equating the rate of ionization to that of recombination, which occurs at a rate per unit volume $\alpha n_e n_p$ with α, the recombination co-efficient, given by $\alpha \approx 3 \times 10^{-13}$. The main difficulty is determining the radial scaling of the ionization rate. This in principle has two components, a "direct"flux from the star and a"diffuse" field that originates from the fraction (about one third) of recombinations within the bound atmosphere that go to the ground state and hence regenerate an ionizing photon. Hollenbach et al. [186] presented analytic and approximate numerical solutions to the radiative transfer problem that imply $n_0(r) \propto r^{-5/2}$, and this scaling has been widely adopted in subsequent work. Important features of the Hollenbach et al. solution have been verified in more detailed radiation hydrodynamic simulations (including the $\Phi^{1/2}$ of the mass loss with the ionizing photon flux [356]), though the slope of $n_0(r)$ at $r > r_g$

remains to be confirmed (indeed, a recent radiative transfer calculation is inconsistent with the canonical slope [397]).

1.9.1.2 Drivers of Photoevaporation

Photoevaporation driven by a photon flux of EUV radiation from the central star Φ is estimated to result in a mass loss rate [4, 134],

$$\dot{M}_w \simeq 1.6 \times 10^{-10} \left(\frac{M_*}{M_\odot} \right)^{1/2} \left(\frac{\Phi}{10^{41}\ \mathrm{s}^{-1}} \right)^{1/2} M_\odot \mathrm{yr}^{-1}, \tag{1.266}$$

provided that the disk is present and optically thick at all radii. The dominant uncertainty in applying this estimate to specific systems comes from lack of knowledge of Φ, which can be constrained but which remains hard to pin down precisely [6, 326]. For low mass stars ($M \leq M_\odot$) reasonable estimates imply that EUV photoevaporation rates are negligible when compared to the median accretion rate of T Tauri stars [167], but large enough to matter for disk dispersal if no stronger mass loss processes are operative. (EUV photon fluxes are of course vastly larger for massive stars, for which the theory was originally developed.)

Low mass pre-main-sequence stars are strong emitters of FUV and X-ray radiation [137, 163, 336]. The FUV luminosity has a base level that is set by chromospheric activity, on top of which there is a potentially much larger component from accretion [191]. The X-ray luminosity scales linearly with the bolometric luminosity, $\log(L_X/L_{\mathrm{bol}}) = -3.6$ [336], but with a large scatter. Qualitatively, these photons affect the disk in the same way as the EUV. The surfaces of the disk are heated, albeit to a somewhat lower temperature than the 10^4 K that is characteristic of HII regions, and where this heating results in unbound gas a wind ensues. Quantitatively, the main difference is that X-ray and FUV heated layers are not separated from the cool underlying disk by any analog of a sharp ionization front, and this makes modeling of FUV and X-ray photoevaporation more difficult. State of the art calculations [159, 316, 317], however, suggest that X-rays and FUV radiation drive mass loss rates that are substantially higher than the EUV prediction—with values of the order of $10^{-8}\ M_\odot\ \mathrm{yr}^{-1}$ being possible—with X-rays likely dominating in the inner region. The exact mass loss rates have a significant dependence on the adopted thermochemistry within the disk [424].

1.9.1.3 Disk Evolution Including Photoevaporation

Including photoevaporation in classical viscous models for disk evolution is particularly simple, because thermal winds exert no torque on the disk (Eq. 1.74). The rate and radial dependence of the mass loss, moreover, is primarily a function of the spectral energy distribution of the irradiation. This may depend upon the stellar

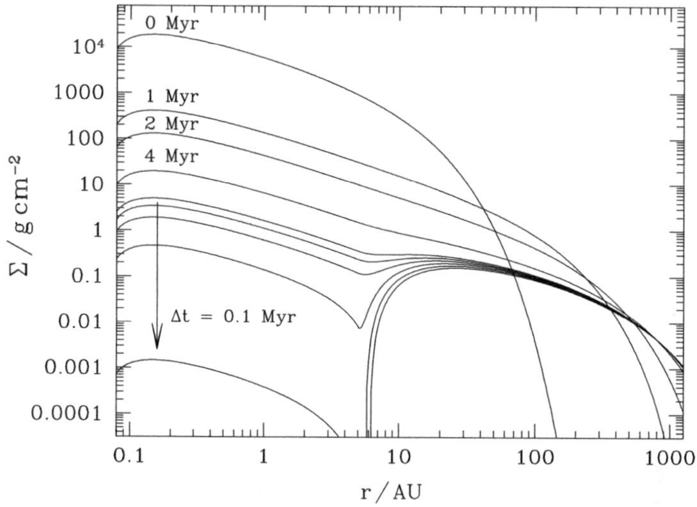

Fig. 1.34 An illustrative calculation of disk evolution including photoevaporative mass loss. The model plotted is based on a disk with $\nu \propto r$, and wind mass loss that scales as $\dot{\Sigma}_w \propto r^{-5/2}$ outside 5 AU. The disk displays the two time scale evolutionary behavior that is reasonably generic to internal photoevaporation models, with a long period of slow "viscous" evolution being followed by rapid inside-out dispersal. Taken from Fig. 3.10 in [18], © Cambridge University Press

accretion rate (if the FUV luminosity is an important driver of photoevaporation) but it is not coupled at leading order to details of the disk structure.[21] These properties mean that disks evolving under the joint action of viscosity and photoevaporation exhibit two distinct phases of evolution, an early phase in which viscosity dominates and a short subsequent phase in which the wind results in rapid disk dispersal [95].

Figure 1.34, based on the original calculations by Clarke et al. [95], shows how a disk evolves under the action of viscosity and EUV photoevaporation from the central star. In addition to the two time scale evolutionary behavior, the concentration of EUV mass loss toward the inner disk leads to a characteristic radial structure of disk dispersal. As the disk accretion rate drops, photoevaporation first dominates the evolution near to the innermost radius where mass loss is possible (in the illustrated example, this is taken to be 5 AU). A gap opens at this location, separating the inner disk (with a short viscous time scale) from the outer disk (where the viscous time scale is much longer). The inner disk then drains on to the star, stellar accretion ceases, and the disk is dispersed from the inside-out. The final dispersal is rapid, because the formation of a hole in the disk allows EUV [7] or X-ray [318] radiation to directly illuminate the inner edge of the disk, which typically accelerates mass loss. The dust to gas ratio in whatever is left of the disk increases during the dispersal phase [5, 160, 402]. Although the details depend on the radial profile of the photoevaporative

[21] In more detail, however, the grain population within the disk will affect the absorption of high energy photons and hence the local mass loss rate [160].

wind, broadly similar evolution is predicted in most models [158]. The observed lifetimes of disks are broadly consistent with theoretically estimates that are based on "best-guess" values of photoevaporative mass loss rates [27].

There is no universal form describing how the action of photoevaporation cuts off accretion in the inner disk. For simple models of EUV photoevaporation, however, it can be shown that the inner accretion rate $\dot{M}(t)$ is related to the accretion rate $\dot{M}_{\rm SS}$ that would be predicted by a self-similar model [267] without mass loss via [363],

$$\dot{M} = \left[1 - \left(\frac{t}{t_0}\right)^{3/2}\right] \dot{M}_{\rm SS}, \tag{1.267}$$

where t_0 is the time at which accretion ceases. This formula is derived for a specific viscosity law ($\nu \propto r$) and photoevaporation model, but provides a qualitative idea of how the inner disk drains under more general circumstances.

The inside-out character of photoevaporative dispersal applies only in the limit where radiation from the central star is dominant. In sufficiently rich stellar clusters, photoevaporation due to intense FUV radiation fields from other (massive) stars is more important. Unlike in the case of central star photoevaporation, for which the observational evidence is indirect [4], photoevaporative flows driven by external UV fields can sometimes be seen directly, most spectacularly in the core of the Orion Nebula [49]. Adams et al. [1] and Clarke [92] have modeled the evolution of viscous disks under external photoevaporation, and shown that it results in destruction of the disk from the outside-in. For the fraction of stars that form in such clusters, this process evidently limits the time over which gas-rich disks would be present on scales comparable to the Solar System's Kuiper Belt.

1.9.2 MHD Winds

The theoretical and observational arguments for photoevaporation being an important component of disk evolution are strong, but other processes may also contribute to disk dispersal. The obvious alternate candidate is MHD winds, which are likely to be present if mature disks retain a dynamically significant net magnetic field (see the discussion in Sect. 1.5.4.4). MHD disk winds have obvious qualitative differences from their photoevaporative cousins,

- Their strength depends upon the disk's net flux, rather than on the stellar radiation field.
- They can be accelerated from arbitrarily small radii, where even EUV-ionized gas would be bound, with a velocity proportional to the Keplerian velocity at the magnetic field footpoint.
- The local mass loss rate is (roughly) expected to scale with the disk surface density, rather than being (approximately) a constant independent of the underlying column.

- Predominantly neutral or molecular gas can, at least in principle, be accelerated (though it might subsequently be dissociated or ionized by stellar radiation).

If MHD disk winds are sufficiently strong, they can affect disk dispersal via some combination of mass and angular momentum loss. The resultant evolution can be quite similar to the photoevaporative case. In particular, if mass rather than angular momentum loss is dominant, Suzuki et al. [393, 394] showed that MHD winds lead to the formation of a shallow surface density profile at small radii and, eventually, an expanding inner hole. In the opposite limit where angular momentum loss is strong (and mass loss negligible) Armitage et al. [24] suggested that dispersal could occur through the late onset of magnetic braking. Whether this is possible depends entirely on how the mass to flux ratio of the disk changes over time, and hence on the uncertain question of how net flux is transported and lost (Sect. 1.4.5.1). The most realistic scenario is one in which winds are driven by a combination of thermal (i.e. photoevaporative) and MHD processes, with the rate and radial profile of the mass and angular momentum loss depending jointly on the magnetic field structure and on the strength of impinging high energy photons [41].

- High-energy radiation in the form of FUV, EUV and X-ray photons heats the upper layers of protoplanetary disks, and drives a thermal wind from radii where the hot gas is unbound.
- Photoevaporative winds lead to distinct dispersal pathways for relatively isolated disks (where the high energy flux comes from the central star) and disks in rich clusters where FUV from massive stars dominates.
- Both thermal (photoevaporative) and MHD driving may contribute to disk evolution and dispersal, though the latter depends on the persistence of a significant net magnetic flux to late times.

Acknowledgements My work on protoplanetary disk physics and planet formation has been supported by the National Science Foundation, by NASA under the Origins of Solar Systems, Exoplanet Research and Astrophysics Theory programs, and by the Space Telescope Science Institute. I acknowledge the hospitality of the IIB at the University of Liverpool, where much of this chapter was written, and thank Kaitlin Kratter for an informal review of the manuscript.

References

1. Adams, F.C., Hollenbach, D., Laughlin, G., Gorti, U.: Photoevaporation of circumstellar disks due to external far-ultraviolet radiation in stellar aggregates. ApJ **611**, 360–379 (2004). https://doi.org/10.1086/421989
2. Adams, F.C., Lada, C.J., Shu, F.H.: Spectral evolution of young stellar objects. ApJ **312**, 788–806 (1987). https://doi.org/10.1086/164924

3. Alencar, S.H.P., Teixeira, P.S., Guimarães, M.M., McGinnis, P.T., Gameiro, J.F., Bouvier, J., Aigrain, S., Flaccomio, E., Favata, F.: Accretion dynamics and disk evolution in NGC 2264: a study based on CoRoT photometric observations. A&A **519**, A88 (2010). https://doi.org/10.1051/0004-6361/201014184
4. Alexander, R., Pascucci, I., Andrews, S., Armitage, P., Cieza, L.: The Dispersal of Protoplanetary Disks. Protostars and Planets VI, pp. 475–496 (2014). https://doi.org/10.2458/azu_uapress_9780816531240-ch021
5. Alexander, R.D., Armitage, P.J.: Dust dynamics during protoplanetary disc clearing. MNRAS **375**, 500–512 (2007). https://doi.org/10.1111/j.1365-2966.2006.11341.x
6. Alexander, R.D., Clarke, C.J., Pringle, J.E.: Constraints on the ionizing flux emitted by T Tauri stars. MNRAS **358**, 283–290 (2005). https://doi.org/10.1111/j.1365-2966.2005.08786.x
7. Alexander, R.D., Clarke, C.J., Pringle, J.E.: Photoevaporation of protoplanetary discs—I. Hydrodynamic models. MNRAS **369**, 216–228 (2006). https://doi.org/10.1111/j.1365-2966.2006.10293.x
8. ALMA Partnership, Brogan, C.L., Pérez, L.M., Hunter, T.R., Dent, W.R.F., Hales, A.S., Hills, R.E., Corder, S., Fomalont, E.B., Vlahakis, C., Asaki, Y., Barkats, D., Hirota, A., Hodge, J.A., Impellizzeri, C.M.V., Kneissl, R., Liuzzo, E., Lucas, R., Marcelino, N., Matsushita, S., Nakanishi, K., Phillips, N., Richards, A.M.S., Toledo, I., Aladro, R., Broguiere, D., Cortes, J.R., Cortes, P.C., Espada, D., Galarza, F., Garcia-Appadoo, D., Guzman-Ramirez, L., Humphreys, E.M., Jung, T., Kameno, S., Laing, R.A., Leon, S., Marconi, G., Mignano, A., Nikolic, B., Nyman, L.A., Radiszcz, M., Remijan, A., Rodón, J.A., Sawada, T., Takahashi, S., Tilanus, R.P.J., Vila Vilaro, B., Watson, L.C., Wiklind, T., Akiyama, E., Chapillon, E., de Gregorio-Monsalvo, I., Di Francesco, J., Gueth, F., Kawamura, A., Lee, C.F., Nguyen Luong, Q., Mangum, J., Pietu, V., Sanhueza, P., Saigo, K., Takakuwa, S., Ubach, C., van Kempen, T., Wootten, A., Castro-Carrizo, A., Francke, H., Gallardo, J., Garcia, J., Gonzalez, S., Hill, T., Kaminski, T., Kurono, Y., Liu, H.Y., Lopez, C., Morales, F., Plarre, K., Schieven, G., Testi, L., Videla, L., Villard, E., Andreani, P., Hibbard, J.E., Tatematsu, K.: The 2014 ALMA long baseline campaign: first results from high angular resolution observations toward the HL Tau region. ApJ, **808**, L3 (2015). https://doi.org/10.1088/2041-8205/808/1/L3
9. Andrews, S.M., Williams, J.P.: Circumstellar dust disks in Taurus-Auriga: the submillimeter perspective. ApJ **631**, 1134–1160 (2005). https://doi.org/10.1086/432712
10. Andrews, S.M., Wilner, D.J., Espaillat, C., Hughes, A.M., Dullemond, C.P., McClure, M.K., Qi, C., Brown, J.M.: Resolved images of large cavities in protoplanetary transition disks. ApJ **732**, 42 (2011). https://doi.org/10.1088/0004-637X/732/1/42
11. Andrews, S.M., Wilner, D.J., Hughes, A.M., Qi, C., Dullemond, C.P.: Protoplanetary disk structures in Ophiuchus. ApJ **700**, 1502–1523 (2009). https://doi.org/10.1088/0004-637X/700/2/1502
12. Andrews, S.M., Wilner, D.J., Hughes, A.M., Qi, C., Rosenfeld, K.A., Öberg, K.I., Birnstiel, T., Espaillat, C., Cieza, L.A., Williams, J.P., Lin, S.Y., Ho, P.T.P.: The TW Hya disk at 870 μm: comparison of CO and dust radial structures. ApJ **744**, 162 (2012). https://doi.org/10.1088/0004-637X/744/2/162
13. Andrews, S.M., Wilner, D.J., Zhu, Z., Birnstiel, T., Carpenter, J.M., Pérez, L.M., Bai, X.N., Öberg, K.I., Hughes, A.M., Isella, A., Ricci, L.: Ringed substructure and a gap at 1 au in the nearest protoplanetary disk. ApJ **820**, L40 (2016). https://doi.org/10.3847/2041-8205/820/2/L40
14. Armitage, P.J.: Magnetic cycles and photometric variability of T Tauri stars. MNRAS **274**, 1242–1248 (1995)
15. Armitage, P.J.: Turbulence and angular momentum transport in a global accretion disk simulation. ApJ **501**, L189–L192 (1998). https://doi.org/10.1086/311463
16. Armitage, P.J.: Magnetic activity in accretion disc boundary layers. MNRAS **330**, 895–900 (2002). https://doi.org/10.1046/j.1365-8711.2002.05152.x
17. Armitage, P.J.: Lecture notes on the formation and early evolution of planetary systems. ArXiv Astrophysics e-prints (2007)

18. Armitage, P.J.: Astrophysics of Planet Formation, 294 pp. Cambridge University Press, Cambridge, UK (2010). https://doi.org/10.1017/CBO9780511802225. ISBN 978-0-521-88745-8 (hardback)
19. Armitage, P.J.: Dynamics of protoplanetary disks. ARA&A **49**, 195–236 (2011). https://doi.org/10.1146/annurev-astro-081710-102521
20. Armitage, P.J.: A trap for planet formation. Science **340**, 1179–1180 (2013). https://doi.org/10.1126/science.1239404
21. Armitage, P.J.: EXor outbursts from disk amplification of stellar magnetic cycles. ApJ **833**, L15 (2016). https://doi.org/10.3847/2041-8213/833/2/L15
22. Armitage, P.J., Eisner, J.A., Simon, J.B.: Prompt planetesimal formation beyond the snow line. ApJ **828**, L2 (2016). https://doi.org/10.3847/2041-8205/828/1/L2
23. Armitage, P.J., Livio, M., Pringle, J.E.: Episodic accretion in magnetically layered protoplanetary discs. MNRAS **324**, 705–711 (2001). https://doi.org/10.1046/j.1365-8711.2001.04356.x
24. Armitage, P.J., Simon, J.B., Martin, R.G.: Two timescale dispersal of magnetized protoplanetary disks. ApJ **778**, L14 (2013). https://doi.org/10.1088/2041-8205/778/1/L14
25. Artymowicz, P., Lubow, S.H.: Mass flow through gaps in circumbinary disks. ApJ **467**, L77 (1996). https://doi.org/10.1086/310200
26. Audard, M., Ábrahám, P., Dunham, M.M., Green, J.D., Grosso, N., Hamaguchi, K., Kastner, J.H., Kóspál, Á., Lodato, G., Romanova, M.M., Skinner, S.L., Vorobyov, E.I., Zhu, Z.: Episodic Accretion in Young Stars. Protostars and Planets VI, pp. 387–410 (2014)
27. Bae, J., Hartmann, L., Zhu, Z., Gammie, C.: The long-term evolution of photoevaporating protoplanetary disks. ApJ **774**, 57 (2013). https://doi.org/10.1088/0004-637X/774/1/57
28. Bae, J., Hartmann, L., Zhu, Z., Gammie, C.: Variable accretion outbursts in protostellar evolution. ApJ **764**, 141 (2013). https://doi.org/10.1088/0004-637X/764/2/141
29. Bae, J., Hartmann, L., Zhu, Z., Nelson, R.P.: Accretion outbursts in self-gravitating protoplanetary disks. ApJ **795**, 61 (2014). https://doi.org/10.1088/0004-637X/795/1/61
30. Baehr, H., Klahr, H., Kratter, K.M.: The fragmentation criteria in local vertically stratified self-gravitating disk simulations. ApJ **848**, 40 (2017). https://doi.org/10.3847/1538-4357/aa8a66
31. Bai, X.N.: Magnetorotational-instability-driven accretion in protoplanetary disks. ApJ **739**, 50 (2011). https://doi.org/10.1088/0004-637X/739/1/50
32. Bai, X.N.: Hall-effect-controlled gas dynamics in protoplanetary disks. I. Wind solutions at the inner disk. ApJ **791**, 137 (2014). https://doi.org/10.1088/0004-637X/791/2/137
33. Bai, X.N.: Hall effect controlled gas dynamics in protoplanetary disks. II. Full 3D simulations toward the outer disk. ApJ **798**, 84 (2015). https://doi.org/10.1088/0004-637X/798/2/84
34. Bai, X.N.: Global simulations of the inner regions of protoplanetary disks with comprehensive disk microphysics. ApJ **845**, 75 (2017). https://doi.org/10.3847/1538-4357/aa7dda
35. Bai, X.N., Goodman, J.: Heat and dust in active layers of protostellar disks. ApJ **701**, 737–755 (2009). https://doi.org/10.1088/0004-637X/701/1/737
36. Bai, X.N., Stone, J.M.: Dynamics of solids in the midplane of protoplanetary disks: implications for planetesimal formation. ApJ **722**, 1437–1459 (2010). https://doi.org/10.1088/0004-637X/722/2/1437
37. Bai, X.N., Stone, J.M.: The effect of the radial pressure gradient in protoplanetary disks on planetesimal formation. ApJ **722**, L220–L223 (2010). https://doi.org/10.1088/2041-8205/722/2/L220
38. Bai, X.N., Stone, J.M.: Effect of ambipolar diffusion on the nonlinear evolution of magnetorotational instability in weakly ionized disks. ApJ **736**, 144 (2011). https://doi.org/10.1088/0004-637X/736/2/144
39. Bai, X.N., Stone, J.M.: Wind-driven accretion in protoplanetary disks. I. Suppression of the magnetorotational instability and launching of the magnetocentrifugal wind. ApJ **769**, 76 (2013). https://doi.org/10.1088/0004-637X/769/1/76
40. Bai, X.N., Stone, J.M.: Magnetic flux concentration and zonal flows in magnetorotational instability turbulence. ApJ **796**, 31 (2014). https://doi.org/10.1088/0004-637X/796/1/31

41. Bai, X.N., Ye, J., Goodman, J., Yuan, F.: Magneto-thermal disk winds from protoplanetary disks. ApJ **818**, 152 (2016). https://doi.org/10.3847/0004-637X/818/2/152
42. Bakes, E.L.O., Tielens, A.G.G.M.: The photoelectric heating mechanism for very small graphitic grains and polycyclic aromatic hydrocarbons. ApJ **427**, 822–838 (1994). https://doi.org/10.1086/174188
43. Balbus, S.A.: Magnetohydrodynamics of Protostellar Disks, pp. 237–282 (2011)
44. Balbus, S.A., Hawley, J.F.: A powerful local shear instability in weakly magnetized disks. I—Linear analysis. II—Nonlinear evolution. ApJ **376**, 214–233 (1991). https://doi.org/10.1086/170270
45. Balbus, S.A., Hawley, J.F.: Instability, turbulence, and enhanced transport in accretion disks. Rev. Mod. Phys. **70**, 1–53 (1998). https://doi.org/10.1103/RevModPhys.70.1
46. Balbus, S.A., Hawley, J.F., Stone, J.M.: Nonlinear stability, hydrodynamical turbulence, and transport in disks. ApJ **467**, 76 (1996). https://doi.org/10.1086/177585
47. Balbus, S.A., Papaloizou, J.C.B.: On the dynamical foundations of α disks. ApJ **521**, 650–658 (1999). https://doi.org/10.1086/307594
48. Balbus, S.A., Terquem, C.: Linear analysis of the Hall effect in protostellar disks. ApJ **552**, 235–247 (2001). https://doi.org/10.1086/320452
49. Bally, J., O'Dell, C.R., McCaughrean, M.J.: Disks, microjets, windblown bubbles, and outflows in the Orion Nebula. AJ **119**, 2919–2959 (2000). https://doi.org/10.1086/301385
50. Bally, J., Scoville, N.Z.: Structure and evolution of molecular clouds near H II regions. II—The disk constrained H II region, S106. ApJ **255**, 497–509 (1982). https://doi.org/10.1086/159850
51. Barge, P., Sommeria, J.: Did planet formation begin inside persistent gaseous vortices? A&A **295**, L1–L4 (1995)
52. Barker, A.J., Ogilvie, G.I.: Hydrodynamic instability in eccentric astrophysical discs. MNRAS **445**, 2637–2654 (2014). https://doi.org/10.1093/mnras/stu1939
53. Barranco, J.A., Marcus, P.S.: Three-dimensional vortices in stratified protoplanetary disks. ApJ **623**, 1157–1170 (2005). https://doi.org/10.1086/428639
54. Baruteau, C., Meru, F., Paardekooper, S.J.: Rapid inward migration of planets formed by gravitational instability. MNRAS **416**, 1971–1982 (2011). https://doi.org/10.1111/j.1365-2966.2011.19172.x
55. Bastian, N., Covey, K.R., Meyer, M.R.: A universal stellar initial mass function? A critical look at variations. ARA&A **48**, 339–389 (2010). https://doi.org/10.1146/annurev-astro-082708-101642
56. Beckwith, S.V.W., Sargent, A.I.: Particle emissivity in circumstellar disks. ApJ **381**, 250–258 (1991). https://doi.org/10.1086/170646
57. Begelman, M.C., McKee, C.F., Shields, G.A.: Compton heated winds and coronae above accretion disks. I Dynamics. ApJ **271**, 70–88 (1983). https://doi.org/10.1086/161178
58. Bell, K.R., Cassen, P.M., Klahr, H.H., Henning, T.: The structure and appearance of protostellar accretion disks: limits on disk flaring. ApJ **486**, 372–387 (1997)
59. Bell, K.R., Lin, D.N.C.: Using FU Orionis outbursts to constrain self-regulated protostellar disk models. ApJ **427**, 987–1004 (1994). https://doi.org/10.1086/174206
60. Belyaev, M.A., Rafikov, R.R., Stone, J.M.: Angular momentum transport by acoustic modes generated in the boundary layer. I. Hydrodynamical theory and simulations. ApJ **770**, 67 (2013). https://doi.org/10.1088/0004-637X/770/1/67
61. Belyaev, M.A., Rafikov, R.R., Stone, J.M.: Angular momentum transport by acoustic modes generated in the boundary layer. II. Magnetohydrodynamic simulations. ApJ **770**, 68 (2013). https://doi.org/10.1088/0004-637X/770/1/68
62. Bergin, E.A., Cleeves, L.I., Gorti, U., Zhang, K., Blake, G.A., Green, J.D., Andrews, S.M., Evans II, N.J., Henning, T., Öberg, K., Pontoppidan, K., Qi, C., Salyk, C., van Dishoeck, E.F.: An old disk still capable of forming a planetary system. Nature **493**, 644–646 (2013). https://doi.org/10.1038/nature11805
63. Besla, G., Wu, Y.: Formation of Narrow dust rings in circumstellar debris disks. ApJ **655**, 528–540 (2007). https://doi.org/10.1086/509495

64. Béthune, W., Lesur, G., Ferreira, J.: Global simulations of protoplanetary disks with net magnetic flux. I. Non-ideal MHD case. A&A **600**, A75 (2017). https://doi.org/10.1051/0004-6361/201630056
65. Birnstiel, T., Klahr, H., Ercolano, B.: A simple model for the evolution of the dust population in protoplanetary disks. A&A **539**, A148 (2012). https://doi.org/10.1051/0004-6361/201118136
66. Bisschop, S.E., Fraser, H.J., Öberg, K.I., van Dishoeck, E.F., Schlemmer, S.: Desorption rates and sticking coefficients for CO and N_2 interstellar ices. A&A **449**, 1297–1309 (2006). https://doi.org/10.1051/0004-6361:20054051
67. Bjorkman, J.E., Wood, K.: Radiative equilibrium and temperature correction in Monte Carlo radiation transfer. ApJ **554**, 615–623 (2001). https://doi.org/10.1086/321336
68. Blaes, O.M., Balbus, S.A.: Local shear instabilities in weakly ionized, weakly magnetized disks. ApJ **421**, 163–177 (1994). https://doi.org/10.1086/173634
69. Blandford, R.D., Payne, D.G.: Hydromagnetic flows from accretion discs and the production of radio jets. MNRAS **199**, 883–903 (1982)
70. Blum, J., Wurm, G.: The growth mechanisms of macroscopic bodies in protoplanetary disks. ARA&A **46**, 21–56 (2008). https://doi.org/10.1146/annurev.astro.46.060407.145152
71. Bollard, J., Connelly, J.N., Whitehouse, M.J., Pringle, E.A., Bonal, L., Jørgensen, J.K., Nordlund, Å., Moynier, F., Bizzarro, M.: Early formation of planetary building blocks inferred from Pb isotopic ages of chondrules. Sci. Adv. **3**, e1700,407 (2017). https://doi.org/10.1126/sciadv.1700407
72. Bonnell, I., Bastien, P.: A binary origin for FU Orionis stars. ApJ **401**, L31–L34 (1992). https://doi.org/10.1086/186663
73. Bouvier, J., Alencar, S.H.P., Harries, T.J., Johns-Krull, C.M., Romanova, M.M.: Magnetospheric Accretion in Classical T Tauri Stars. Protostars and Planets V, pp. 479–494 (2007)
74. Bouvier, J., Cabrit, S., Fernandez, M., Martin, E.L., Matthews, J.M.: Coyotes-I—the photometric variability and rotational evolution of T-Tauri stars. A&A **272**, 176 (1993)
75. Bouvier, J., Matt, S.P., Mohanty, S., Scholz, A., Stassun, K.G., Zanni, C.: Angular Momentum Evolution of Young Low-Mass Stars and Brown Dwarfs: Observations and Theory. Protostars and Planets VI, pp. 433–450 (2014)
76. Brauer, F., Henning, T., Dullemond, C.P.: Planetesimal formation near the snow line in MRI-driven turbulent protoplanetary disks. A&A **487**, L1–L4 (2008). https://doi.org/10.1051/0004-6361:200809780
77. Burke, J.R., Hollenbach, D.J.: The gas-grain interaction in the interstellar medium—thermal accommodation and trapping. ApJ **265**, 223–234 (1983). https://doi.org/10.1086/160667
78. Calvet, N., D'Alessio, P., Watson, D.M., Franco-Hernández, R., Furlan, E., Green, J., Sutter, P.M., Forrest, W.J., Hartmann, L., Uchida, K.I., Keller, L.D., Sargent, B., Najita, J., Herter, T.L., Barry, D.J., Hall, P.: Disks in transition in the Taurus population: Spitzer IRS spectra of GM Aurigae and DM Tauri. ApJ **630**, L185–L188 (2005). https://doi.org/10.1086/491652
79. Cannizzo, J.K.: The accretion disk limit cycle model: toward an understanding of the long-term behavior of SS Cygni. ApJ **419**, 318 (1993). https://doi.org/10.1086/173486
80. Cao, X., Spruit, H.C.: Instability of an accretion disk with a magnetically driven wind. A&A **385**, 289–300 (2002). https://doi.org/10.1051/0004-6361:20011818
81. Carrera, D., Johansen, A., Davies, M.B.: How to form planetesimals from mm-sized chondrules and chondrule aggregates. A&A **579**, A43 (2015). https://doi.org/10.1051/0004-6361/201425120
82. Casassus, S., Marino, S., Pérez, S., Roman, P., Dunhill, A., Armitage, P.J., Cuadra, J., Wootten, A., van der Plas, G., Cieza, L., Moral, V., Christiaens, V., Montesinos, M.: Accretion kinematics through the warped transition disk in HD142527 from resolved CO(6–5) observations. ApJ **811**, 92 (2015). https://doi.org/10.1088/0004-637X/811/2/92
83. Casassus, S., van der Plas, G., M, S.P., Dent, W.R.F., Fomalont, E., Hagelberg, J., Hales, A., Jordán, A., Mawet, D., Ménard, F., Wootten, A., Wilner, D., Hughes, A.M., Schreiber, M.R., Girard, J.H., Ercolano, B., Canovas, H., Román, P.E., Salinas, V.: Flows of gas through a protoplanetary gap. Nature **493**, 191–194 (2013). https://doi.org/10.1038/nature11769

84. Cha, S.H., Nayakshin, S.: A numerical simulation of a 'Super-Earth' core delivery from 100 to 8 AU. MNRAS **415**, 3319–3334 (2011). https://doi.org/10.1111/j.1365-2966.2011.18953.x
85. Chandrasekhar, S.: Hydrodynamic and hydromagnetic stability (1961)
86. Chang, P., Oishi, J.S.: On the stability of dust-laden protoplanetary vortices. ApJ **721**, 1593–1602 (2010). https://doi.org/10.1088/0004-637X/721/2/1593
87. Chauvin, G., Vigan, A., Bonnefoy, M., Desidera, S., Bonavita, M., Mesa, D., Boccaletti, A., Buenzli, E., Carson, J., Delorme, P., Hagelberg, J., Montagnier, G., Mordasini, C., Quanz, S.P., Segransan, D., Thalmann, C., Beuzit, J.L., Biller, B., Covino, E., Feldt, M., Girard, J., Gratton, R., Henning, T., Kasper, M., Lagrange, A.M., Messina, S., Meyer, M., Mouillet, D., Moutou, C., Reggiani, M., Schlieder, J.E., Zurlo, A.: The VLT/NaCo large program to probe the occurrence of exoplanets and brown dwarfs at wide orbits. II. Survey description, results, and performances. A&A **573**, A127 (2015). https://doi.org/10.1051/0004-6361/201423564
88. Chavanis, P.H.: Trapping of dust by coherent vortices in the solar nebula. A&A **356**, 1089–1111 (2000)
89. Chiang, E., Youdin, A.N.: Forming planetesimals in solar and extrasolar nebulae. Annu. Rev. Earth Planet. Sci. **38**, 493–522 (2010). https://doi.org/10.1146/annurev-earth-040809-152513
90. Chiang, E.I., Goldreich, P.: Spectral energy distributions of T Tauri stars with passive circumstellar disks. ApJ **490**, 368–376 (1997)
91. Ciesla, F.J., Cuzzi, J.N.: The evolution of the water distribution in a viscous protoplanetary disk. Icarus **181**, 178–204 (2006). https://doi.org/10.1016/j.icarus.2005.11.009
92. Clarke, C.J.: The photoevaporation of discs around young stars in massive clusters. MNRAS **376**, 1350–1356 (2007). https://doi.org/10.1111/j.1365-2966.2007.11547.x
93. Clarke, C.J.: Pseudo-viscous modelling of self-gravitating discs and the formation of low mass ratio binaries. MNRAS **396**, 1066–1074 (2009). https://doi.org/10.1111/j.1365-2966.2009.14774.x
94. Clarke, C.J., Armitage, P.J., Smith, K.W., Pringle, J.E.: Magnetically modulated accretion in T Tauri stars. MNRAS **273**, 639–642 (1995)
95. Clarke, C.J., Gendrin, A., Sotomayor, M.: The dispersal of circumstellar discs: the role of the ultraviolet switch. MNRAS **328**, 485–491 (2001). https://doi.org/10.1046/j.1365-8711.2001.04891.x
96. Clarke, C.J., Pringle, J.E.: The diffusion of contaminant through an accretion disc. MNRAS **235**, 365–373 (1988)
97. Clarke, C.J., Syer, D.: Low-mass companions to T Tauri stars: a mechanism for rapid-rise FU Orionis outbursts. MNRAS **278**, L23–L27 (1996)
98. Cleeves, L.I., Bergin, E.A., Qi, C., Adams, F.C., Öberg, K.I.: Constraining the X-ray and cosmic-ray ionization chemistry of the TW Hya protoplanetary disk: evidence for a sub-interstellar cosmic-ray rate. ApJ **799**, 204 (2015). https://doi.org/10.1088/0004-637X/799/2/204
99. Cody, A.M., Stauffer, J., Baglin, A., Micela, G., Rebull, L.M., Flaccomio, E., Morales-Calderón, M., Aigrain, S., Bouvier, J., Hillenbrand, L.A., Gutermuth, R., Song, I., Turner, N., Alencar, S.H.P., Zwintz, K., Plavchan, P., Carpenter, J., Findeisen, K., Carey, S., Terebey, S., Hartmann, L., Calvet, N., Teixeira, P., Vrba, F.J., Wolk, S., Covey, K., Poppenhaeger, K., Günther, H.M., Forbrich, J., Whitney, B., Affer, L., Herbst, W., Hora, J., Barrado, D., Holtzman, J., Marchis, F., Wood, K., Medeiros Guimarães, M., Lillo Box, J., Gillen, E., McQuillan, A., Espaillat, C., Allen, L., D'Alessio, P., Favata, F.: CSI 2264: simultaneous optical and infrared light curves of young disk-bearing stars in NGC 2264 with CoRoT and Spitzer—evidence for multiple origins of variability. AJ **147**, 82 (2014). https://doi.org/10.1088/0004-6256/147/4/82
100. Coleman, M.S.B., Kotko, I., Blaes, O., Lasota, J.P., Hirose, S.: Dwarf nova outbursts with magnetorotational turbulence. MNRAS **462**, 3710–3726 (2016). https://doi.org/10.1093/mnras/stw1908
101. Connolly, H.C., Jones, R.H.: Chondrules: the canonical and noncanonical views. J. Geophys. Res. (Planets) **121**, 1885–1899 (2016). https://doi.org/10.1002/2016JE005113
102. Cossins, P., Lodato, G., Clarke, C.: The effects of opacity on gravitational stability in protoplanetary discs. MNRAS **401**, 2587–2598 (2010). https://doi.org/10.1111/j.1365-2966.2009.15835.x

103. Cossins, P., Lodato, G., Clarke, C.J.: Characterizing the gravitational instability in cooling accretion discs. MNRAS **393**, 1157–1173 (2009). https://doi.org/10.1111/j.1365-2966.2008.14275.x
104. Curry, C., Pudritz, R.E.: On the global stability of magnetized accretion disks. II. Vertical and Azimuthal magnetic fields. ApJ **453**, 697 (1995). https://doi.org/10.1086/176431
105. D'Angelo, C.R., Spruit, H.C.: Episodic accretion on to strongly magnetic stars. MNRAS **406**, 1208–1219 (2010). https://doi.org/10.1111/j.1365-2966.2010.16749.x
106. D'Angelo, C.R., Spruit, H.C.: Accretion discs trapped near corotation. MNRAS **420**, 416–429 (2012). https://doi.org/10.1111/j.1365-2966.2011.20046.x
107. Davis, S.W., Stone, J.M., Pessah, M.E.: Sustained magnetorotational turbulence in local simulations of stratified disks with zero net magnetic flux. ApJ **713**, 52–65 (2010). https://doi.org/10.1088/0004-637X/713/1/52
108. de Val-Borro, M., Artymowicz, P., D'Angelo, G., Peplinski, A.: Vortex generation in protoplanetary disks with an embedded giant planet. A&A **471**, 1043–1055 (2007). https://doi.org/10.1051/0004-6361:20077169
109. Deng, H., Mayer, L., Meru, F.: Convergence of the critical cooling rate for protoplanetary disk fragmentation achieved: the key role of numerical dissipation of angular momentum. ApJ **847**, 43 (2017). https://doi.org/10.3847/1538-4357/aa872b
110. Desch, S.J.: Linear analysis of the magnetorotational instability, including ambipolar diffusion, with application to protoplanetary disks. ApJ **608**, 509–525 (2004). https://doi.org/10.1086/392527
111. Dittrich, K., Klahr, H., Johansen, A.: Gravoturbulent planetesimal formation: the positive effect of long-lived zonal flows. ApJ **763**, 117 (2013). https://doi.org/10.1088/0004-637X/763/2/117
112. D'Orazio, D.J., Haiman, Z., MacFadyen, A.: Accretion into the central cavity of a circumbinary disc. MNRAS **436**, 2997–3020 (2013). https://doi.org/10.1093/mnras/stt1787
113. Draine, B.T.: Photoelectric heating of interstellar gas. ApJs **36**, 595–619 (1978). https://doi.org/10.1086/190513
114. Draine, B.T.: On the submillimeter opacity of protoplanetary disks. ApJ **636**, 1114–1120 (2006). https://doi.org/10.1086/498130
115. Draine, B.T., Roberge, W.G., Dalgarno, A.: Magnetohydrodynamic shock waves in molecular clouds. ApJ **264**, 485–507 (1983). https://doi.org/10.1086/160617
116. Draine, B.T., Sutin, B.: Collisional charging of interstellar grains. ApJ **320**, 803–817 (1987). https://doi.org/10.1086/165596
117. Dubrulle, B., Morfill, G., Sterzik, M.: The dust subdisk in the protoplanetary nebula. Icarus **114**, 237–246 (1995). https://doi.org/10.1006/icar.1995.1058
118. Dutrey, A., Semenov, D., Chapillon, E., Gorti, U., Guilloteau, S., Hersant, F., Hogerheijde, M., Hughes, M., Meeus, G., Nomura, H., Piétu, V., Qi, C., Wakelam, V.: Physical and Chemical Structure of Planet-Forming Disks Probed by Millimeter Observations and Modeling. Protostars and Planets VI, pp. 317–338 (2014)
119. Eckhardt, B., Schneider, T.M., Hof, B., Westerweel, J.: Turbulence transition in pipe flow. Annu. Rev. Fluid Mech. **39**, 447–468 (2007). https://doi.org/10.1146/annurev.fluid.39.050905.110308
120. Edlund, E.M., Ji, H.: Nonlinear stability of laboratory quasi-Keplerian flows. Phys. Rev. E **89**(2), 021004 (2014). https://doi.org/10.1103/PhysRevE.89.021004
121. Eisner, J.A., Hillenbrand, L.A., Carpenter, J.M., Wolf, S.: Constraining the evolutionary stage of Class I protostars: multiwavelength observations and modeling. ApJ **635**, 396–421 (2005). https://doi.org/10.1086/497161
122. Ercolano, B., Barlow, M.J., Storey, P.J.: The dusty MOCASSIN: fully self-consistent 3D photoionization and dust radiative transfer models. MNRAS **362**, 1038–1046 (2005). https://doi.org/10.1111/j.1365-2966.2005.09381.x
123. Ercolano, B., Glassgold, A.E.: X-ray ionization rates in protoplanetary discs. MNRAS **436**, 3446–3450 (2013). https://doi.org/10.1093/mnras/stt1826

124. Espaillat, C., Muzerolle, J., Najita, J., Andrews, S., Zhu, Z., Calvet, N., Kraus, S., Hashimoto, J., Kraus, A., D'Alessio, P.: An Observational Perspective of Transitional Disks. Protostars and Planets VI, pp. 497–520 (2014)
125. Evans II, N.J., Dunham, M.M., Jørgensen, J.K., Enoch, M.L., Merín, B., van Dishoeck, E.F., Alcalá, J.M., Myers, P.C., Stapelfeldt, K.R., Huard, T.L., Allen, L.E., Harvey, P.M., van Kempen, T., Blake, G.A., Koerner, D.W., Mundy, L.G., Padgett, D.L., Sargent, A.I.: The Spitzer c2d legacy results: star-formation rates and efficiencies; evolution and lifetimes. ApJs **181**, 321–350 (2009). https://doi.org/10.1088/0067-0049/181/2/321
126. Faure, J., Fromang, S., Latter, H., Meheut, H.: Vortex cycles at the inner edges of dead zones in protoplanetary disks. A&A **573**, A132 (2015). https://doi.org/10.1051/0004-6361/201424162
127. Fedele, D., Carney, M., Hogerheijde, M.R., Walsh, C., Miotello, A., Klaassen, P., Bruderer, S., Henning, T., van Dishoeck, E.F.: ALMA unveils rings and gaps in the protoplanetary system <ASTROBJ>HD 169142</ASTROBJ>: signatures of two giant protoplanets. A&A **600**, A72 (2017). https://doi.org/10.1051/0004-6361/201629860
128. Federrath, C., Banerjee, S.: The density structure and star formation rate of non-isothermal polytropic turbulence. MNRAS **448**, 3297–3313 (2015). https://doi.org/10.1093/mnras/stv180
129. Flaherty, K.M., Hughes, A.M., Rose, S.C., Simon, J.B., Qi, C., Andrews, S.M., Kóspál, Á., Wilner, D.J., Chiang, E., Armitage, P.J., Bai, X.N.: A three-dimensional view of turbulence: constraints on turbulent motions in the HD 163296 protoplanetary disk using DCO^+. ApJ **843**, 150 (2017). https://doi.org/10.3847/1538-4357/aa79f9
130. Flaherty, K.M., Hughes, A.M., Rosenfeld, K.A., Andrews, S.M., Chiang, E., Simon, J.B., Kerzner, S., Wilner, D.J.: Weak turbulence in the HD 163296 protoplanetary disk revealed by ALMA CO observations. ApJ **813**, 99 (2015). https://doi.org/10.1088/0004-637X/813/2/99
131. Fleming, T., Stone, J.M.: Local magnetohydrodynamic models of layered accretion disks. ApJ **585**, 908–920 (2003). https://doi.org/10.1086/345848
132. Flock, M., Ruge, J.P., Dzyurkevich, N., Henning, T., Klahr, H., Wolf, S.: Gaps, rings, and non-axisymmetric structures in protoplanetary disks. From simulations to ALMA observations. A&A **574**, A68 (2015). https://doi.org/10.1051/0004-6361/201424693
133. Follette, K.B., Rameau, J., Dong, R., Pueyo, L., Close, L.M., Duchêne, G., Fung, J., Leonard, C., Macintosh, B., Males, J.R., Marois, C., Millar-Blanchaer, M.A., Morzinski, K.M., Mullen, W., Perrin, M., Spiro, E., Wang, J., Ammons, S.M., Bailey, V.P., Barman, T., Bulger, J., Chilcote, J., Cotten, T., De Rosa, R.J., Doyon, R., Fitzgerald, M.P., Goodsell, S.J., Graham, J.R., Greenbaum, A.Z., Hibon, P., Hung, L.W., Ingraham, P., Kalas, P., Konopacky, Q., Larkin, J.E., Maire, J., Marchis, F., Metchev, S., Nielsen, E.L., Oppenheimer, R., Palmer, D., Patience, J., Poyneer, L., Rajan, A., Rantakyrö, F.T., Savransky, D., Schneider, A.C., Sivaramakrishnan, A., Song, I., Soummer, R., Thomas, S., Vega, D., Wallace, J.K., Ward-Duong, K., Wiktorowicz, S., Wolff, S.: Complex spiral structure in the HD 100546 transitional disk as revealed by GPI and MagAO. AJ **153**, 264 (2017). https://doi.org/10.3847/1538-3881/aa6d85
134. Font, A.S., McCarthy, I.G., Johnstone, D., Ballantyne, D.R.: Photoevaporation of circumstellar disks around young stars. ApJ **607**, 890–903 (2004). https://doi.org/10.1086/383518
135. Forgan, D., Rice, K.: Stellar encounters in the context of outburst phenomena. MNRAS **402**, 1349–1356 (2010). https://doi.org/10.1111/j.1365-2966.2009.15974.x
136. Forgan, D., Rice, K., Cossins, P., Lodato, G.: The nature of angular momentum transport in radiative self-gravitating protostellar discs. MNRAS **410**, 994–1006 (2011). https://doi.org/10.1111/j.1365-2966.2010.17500.x
137. France, K., Schindhelm, E., Bergin, E.A., Roueff, E., Abgrall, H.: High-resolution ultraviolet radiation fields of classical T Tauri stars. ApJ **784**, 127 (2014). https://doi.org/10.1088/0004-637X/784/2/127
138. France, K., Schindhelm, E., Herczeg, G.J., Brown, A., Abgrall, H., Alexander, R.D., Bergin, E.A., Brown, J.M., Linsky, J.L., Roueff, E., Yang, H.: A Hubble Space telescope survey of H_2 emission in the circumstellar environments of young stars. ApJ **756**, 171 (2012). https://doi.org/10.1088/0004-637X/756/2/171

139. Frank, J., King, A., Raine, D.J.: Accretion Power in Astrophysics, 3rd ed. (2002)
140. Fricke, K.: Instabilität stationärer Rotation in Sternen. Zeitschrift für Astrophysik **68**, 317 (1968)
141. Fromang, S.: MRI-driven angular momentum transport in protoplanetary disks. In: Hennebelle, P., Charbonnel, P. (eds.) EAS Publications Series, vol. 62, pp. 95–142 (2013). https://doi.org/10.1051/eas/1362004
142. Fromang, S., Latter, H., Lesur, G., Ogilvie, G.I.: Local outflows from turbulent accretion disks. A&A **552**, A71 (2013). https://doi.org/10.1051/0004-6361/201220016
143. Fromang, S., Papaloizou, J.: Dust settling in local simulations of turbulent protoplanetary disks. A&A **452**, 751–762 (2006). https://doi.org/10.1051/0004-6361:20054612
144. Fu, R.R., Weiss, B.P., Lima, E.A., Harrison, R.J., Bai, X.N., Desch, S.J., Ebel, D.S., Suavet, C., Wang, H., Glenn, D., Le Sage, D., Kasama, T., Walsworth, R.L., Kuan, A.T.: Solar nebula magnetic fields recorded in the semarkona meteorite. Science (2014). https://doi.org/10.1126/science.1258022. http://www.sciencemag.org/content/early/2014/11/12/science.1258022.abstract
145. Fu, W., Li, H., Lubow, S., Li, S., Liang, E.: Effects of dust feedback on vortices in protoplanetary disks. ApJ **795**, L39 (2014). https://doi.org/10.1088/2041-8205/795/2/L39
146. Galicher, R., Marois, C., Macintosh, B., Zuckerman, B., Barman, T., Konopacky, Q., Song, I., Patience, J., Lafrenière, D., Doyon, R., Nielsen, E.L.: The international deep planet survey. II. The frequency of directly imaged giant exoplanets with stellar mass. A&A **594**, A63 (2016). https://doi.org/10.1051/0004-6361/201527828
147. Galvagni, M., Hayfield, T., Boley, A., Mayer, L., Roškar, R., Saha, P.: The collapse of protoplanetary clumps formed through disc instability: 3D simulations of the pre-dissociation phase. MNRAS **427**, 1725–1740 (2012). https://doi.org/10.1111/j.1365-2966.2012.22096.x
148. Gammie, C.F.: Layered accretion in T Tauri disks. ApJ **457**, 355 (1996). https://doi.org/10.1086/176735
149. Gammie, C.F.: Instabilities in circumstellar discs. In: Sellwood, J.A., Goodman, J. (eds.) Astrophysical Discs—An EC Summer School. Astronomical Society of the Pacific Conference Series, vol. 160, p. 122 (1999)
150. Gammie, C.F.: Nonlinear outcome of gravitational instability in cooling, gaseous disks. ApJ **553**, 174–183 (2001). https://doi.org/10.1086/320631
151. Garaud, P., Lin, D.N.C.: The effect of internal dissipation and surface irradiation on the structure of disks and the location of the snow line around Sun-like stars. ApJ **654**, 606–624 (2007). https://doi.org/10.1086/509041
152. Garufi, A., Quanz, S.P., Schmid, H.M., Mulders, G.D., Avenhaus, H., Boccaletti, A., Ginski, C., Langlois, M., Stolker, T., Augereau, J.C., Benisty, M., Lopez, B., Dominik, C., Gratton, R., Henning, T., Janson, M., Ménard, F., Meyer, M.R., Pinte, C., Sissa, E., Vigan, A., Zurlo, A., Bazzon, A., Buenzli, E., Bonnefoy, M., Brandner, W., Chauvin, G., Cheetham, A., Cudel, M., Desidera, S., Feldt, M., Galicher, R., Kasper, M., Lagrange, A.M., Lannier, J., Maire, A.L., Mesa, D., Mouillet, D., Peretti, S., Perrot, C., Salter, G., Wildi, F.: The SPHERE view of the planet-forming disk around HD 100546. A&A **588**, A8 (2016). https://doi.org/10.1051/0004-6361/201527940
153. Gibbons, P.G., Mamatsashvili, G.R., Rice, W.K.M.: Planetesimal formation in self-gravitating discs—the effects of particle self-gravity and back-reaction. MNRAS **442**, 361–371 (2014). https://doi.org/10.1093/mnras/stu809
154. Godon, P., Livio, M.: The formation and role of vortices in protoplanetary disks. ApJ **537**, 396–404 (2000). https://doi.org/10.1086/309019
155. Goldreich, P., Goodman, J., Narayan, R.: The stability of accretion tori. I—Long-wavelength modes of slender tori. MNRAS **221**, 339–364 (1986)
156. Goldreich, P., Schubert, G.: Differential rotation in stars. ApJ **150**, 571 (1967). https://doi.org/10.1086/149360
157. Goodman, A.A., Benson, P.J., Fuller, G.A., Myers, P.C.: Dense cores in dark clouds. VIII—Velocity gradients. ApJ **406**, 528–547 (1993). https://doi.org/10.1086/172465

158. Gorti, U., Dullemond, C.P., Hollenbach, D.: Time evolution of viscous circumstellar disks due to photoevaporation by far-ultraviolet, extreme-ultraviolet, and X-ray radiation from the central star. ApJ **705**, 1237–1251 (2009). https://doi.org/10.1088/0004-637X/705/2/1237
159. Gorti, U., Hollenbach, D.: Photoevaporation of circumstellar disks by far-ultraviolet, extreme-ultraviolet and X-ray radiation from the central star. ApJ **690**, 1539–1552 (2009). https://doi.org/10.1088/0004-637X/690/2/1539
160. Gorti, U., Hollenbach, D., Dullemond, C.P.: The impact of dust evolution and photoevaporation on disk dispersal. ApJ **804**, 29 (2015). https://doi.org/10.1088/0004-637X/804/1/29
161. Grady, C.A., Muto, T., Hashimoto, J., Fukagawa, M., Currie, T., Biller, B., Thalmann, C., Sitko, M.L., Russell, R., Wisniewski, J., Dong, R., Kwon, J., Sai, S., Hornbeck, J., Schneider, G., Hines, D., Moro Martín, A., Feldt, M., Henning, T., Pott, J.U., Bonnefoy, M., Bouwman, J., Lacour, S., Mueller, A., Juhász, A., Crida, A., Chauvin, G., Andrews, S., Wilner, D., Kraus, A., Dahm, S., Robitaille, T., Jang-Condell, H., Abe, L., Akiyama, E., Brandner, W., Brandt, T., Carson, J., Egner, S., Follette, K.B., Goto, M., Guyon, O., Hayano, Y., Hayashi, M., Hayashi, S., Hodapp, K., Ishii, M., Iye, M., Janson, M., Kandori, R., Knapp, G., Kudo, T., Kusakabe, N., Kuzuhara, M., Mayama, S., McElwain, M., Matsuo, T., Miyama, S., Morino, J.I., Nishimura, T., Pyo, T.S., Serabyn, G., Suto, H., Suzuki, R., Takami, M., Takato, N., Terada, H., Tomono, D., Turner, E., Watanabe, M., Yamada, T., Takami, H., Usuda, T., Tamura, M.: Spiral arms in the asymmetrically illuminated disk of MWC 758 and constraints on giant planets. ApJ **762**, 48 (2013). https://doi.org/10.1088/0004-637X/762/1/48
162. Gressel, O., Turner, N.J., Nelson, R.P., McNally, C.P.: Global simulations of protoplanetary disks with ohmic resistivity and ambipolar diffusion. ApJ **801**, 84 (2015). https://doi.org/10.1088/0004-637X/801/2/84
163. Güdel, M., Briggs, K.R., Arzner, K., Audard, M., Bouvier, J., Feigelson, E.D., Franciosini, E., Glauser, A., Grosso, N., Micela, G., Monin, J.L., Montmerle, T., Padgett, D.L., Palla, F., Pillitteri, I., Rebull, L., Scelsi, L., Silva, B., Skinner, S.L., Stelzer, B., Telleschi, A.: The XMM-Newton extended survey of the Taurus molecular cloud (XEST). A&A **468**, 353–377 (2007). https://doi.org/10.1051/0004-6361:20065724
164. Guilet, J., Ogilvie, G.I.: Transport of magnetic flux and the vertical structure of accretion discs—I. Uniform diffusion coefficients. MNRAS **424**, 2097–2117 (2012). https://doi.org/10.1111/j.1365-2966.2012.21361.x
165. Guilet, J., Ogilvie, G.I.: Global evolution of the magnetic field in a thin disc and its consequences for protoplanetary systems. MNRAS **441**, 852–868 (2014). https://doi.org/10.1093/mnras/stu532
166. Gullbring, E., Calvet, N., Muzerolle, J., Hartmann, L.: The structure and emission of the accretion shock in T Tauri stars. II. The ultraviolet-continuum emission. ApJ **544**, 927–932 (2000). https://doi.org/10.1086/317253
167. Gullbring, E., Hartmann, L., Briceño, C., Calvet, N.: Disk accretion rates for T Tauri stars. ApJ **492**, 323–341 (1998). https://doi.org/10.1086/305032
168. Hōshi, R.: Accretion model for outbursts of dwarf nova. Prog. Theor. Phys. **61**, 1307–1319 (1979). https://doi.org/10.1143/PTP.61.1307
169. Haghighipour, N., Boss, A.P.: On pressure gradients and rapid migration of solids in a nonuniform solar nebula. ApJ **583**, 996–1003 (2003). https://doi.org/10.1086/345472
170. Haisch Jr., K.E., Lada, E.A., Lada, C.J.: Disk frequencies and lifetimes in young clusters. ApJ **553**, L153–L156 (2001). https://doi.org/10.1086/320685
171. Hartmann, L., Calvet, N., Gullbring, E., D'Alessio, P.: Accretion and the evolution of T Tauri disks. ApJ **495**, 385–400 (1998). https://doi.org/10.1086/305277
172. Hartmann, L., Kenyon, S.J.: The FU Orionis phenomenon. ARA&A **34**, 207–240 (1996). https://doi.org/10.1146/annurev.astro.34.1.207
173. Hawley, J.F., Gammie, C.F., Balbus, S.A.: Local three-dimensional magnetohydrodynamic simulations of accretion disks. ApJ **440**, 742 (1995). https://doi.org/10.1086/175311
174. Hawley, J.F., Stone, J.M.: Nonlinear evolution of the magnetorotational instability in ion-neutral disks. ApJ **501**, 758–771 (1998). https://doi.org/10.1086/305849

175. Hayashi, C.: Structure of the solar nebula, growth and decay of magnetic fields and effects of magnetic and turbulent viscosities on the nebula. Prog. Theor. Phys. Suppl. **70**, 35–53 (1981). https://doi.org/10.1143/PTPS.70.35
176. Haynes, D.R., Tro, N.J., George, S.M.: Condensation and evaporation of H_2O on ice surfaces. J. Phys. Chem. **96**, 8502–8509 (1992)
177. Henning, T., Semenov, D.: Chemistry in protoplanetary disks. Chem. Rev. **113**, 9016–9042 (2013). https://doi.org/10.1021/cr400128p
178. Herbig, G.H.: Eruptive phenomena in early stellar evolution. ApJ **217**, 693–715 (1977). https://doi.org/10.1086/155615
179. Herbig, G.H.: History and spectroscopy of EXor candidates. AJ **135**, 637–648 (2008). https://doi.org/10.1088/0004-6256/135/2/637
180. Herczeg, G.J., Hillenbrand, L.A.: UV excess measures of accretion onto young very low mass stars and brown dwarfs. ApJ **681**, 594–625 (2008). https://doi.org/10.1086/586728
181. Hernández, J., Hartmann, L., Megeath, T., Gutermuth, R., Muzerolle, J., Calvet, N., Vivas, A.K., Briceño, C., Allen, L., Stauffer, J., Young, E., Fazio, G.: A Spitzer space telescope study of disks in the young σ Orionis cluster. ApJ **662**, 1067–1081 (2007). https://doi.org/10.1086/513735
182. Hillenbrand, L.A., Findeisen, K.P.: A simple calculation in service of constraining the rate of FU Orionis outburst events from photometric monitoring surveys. ApJ **808**, 68 (2015). https://doi.org/10.1088/0004-637X/808/1/68
183. Hirose, S., Blaes, O., Krolik, J.H., Coleman, M.S.B., Sano, T.: Convection causes enhanced magnetic turbulence in accretion disks in outburst. ApJ **787**, 1 (2014). https://doi.org/10.1088/0004-637X/787/1/1
184. Hirose, S., Turner, N.J.: Heating and cooling protostellar disks. ApJ **732**, L30 (2011). https://doi.org/10.1088/2041-8205/732/2/L30
185. Hogerheijde, M.R., Bergin, E.A., Brinch, C., Cleeves, L.I., Fogel, J.K.J., Blake, G.A., Dominik, C., Lis, D.C., Melnick, G., Neufeld, D., Panić, O., Pearson, J.C., Kristensen, L., Yıldız, U.A., van Dishoeck, E.F.: Detection of the water reservoir in a forming planetary system. Science **334**, 338 (2011). https://doi.org/10.1126/science.1208931
186. Hollenbach, D., Johnstone, D., Lizano, S., Shu, F.: Photoevaporation of disks around massive stars and application to ultracompact H II regions. ApJ **428**, 654–669 (1994). https://doi.org/10.1086/174276
187. Hopkins, P.F.: A new class of accurate, mesh-free hydrodynamic simulation methods. MNRAS **450**, 53–110 (2015). https://doi.org/10.1093/mnras/stv195
188. Ilgner, M., Nelson, R.P.: On the ionisation fraction in protoplanetary disks. I. Comparing different reaction networks. A&A **445**, 205–222 (2006). https://doi.org/10.1051/0004-6361:20053678
189. Illarionov, A.F., Sunyaev, R.A.: Why the number of galactic X-ray stars is so small? A&A **39**, 185 (1975)
190. Inaba, S., Barge, P.: Dusty vortices in protoplanetary disks. ApJ **649**, 415–427 (2006). https://doi.org/10.1086/506427
191. Ingleby, L., Calvet, N., Hernández, J., Briceño, C., Espaillat, C., Miller, J., Bergin, E., Hartmann, L.: Evolution of X-ray and far-ultraviolet disk-dispersing radiation fields. AJ **141**, 127 (2011). https://doi.org/10.1088/0004-6256/141/4/127
192. Inutsuka, S.I., Sano, T.: Self-sustained ionization and vanishing dead zones in protoplanetary disks. ApJ **628**, L155–L158 (2005). https://doi.org/10.1086/432796
193. Isella, A., Guidi, G., Testi, L., Liu, S., Li, H., Li, S., Weaver, E., Boehler, Y., Carperter, J.M., De Gregorio-Monsalvo, I., Manara, C.F., Natta, A., Pérez, L.M., Ricci, L., Sargent, A., Tazzari, M., Turner, N.: Ringed structures of the HD 163296 protoplanetary disk revealed by ALMA. Phys. Rev. Lett. **117**(25), 251101 (2016). https://doi.org/10.1103/PhysRevLett.117.251101
194. Isella, A., Pérez, L.M., Carpenter, J.M., Ricci, L., Andrews, S., Rosenfeld, K.: An Azimuthal asymmetry in the LkHα 330 disk. ApJ **775**, 30 (2013). https://doi.org/10.1088/0004-637X/775/1/30

195. Jacquet, E., Balbus, S., Latter, H.: On linear dust-gas streaming instabilities in protoplanetary discs. MNRAS **415**, 3591–3598 (2011). https://doi.org/10.1111/j.1365-2966.2011.18971.x
196. Jin, L.: Damping of the shear instability in magnetized disks by ohmic diffusion. ApJ **457**, 798 (1996). https://doi.org/10.1086/176774
197. Johansen, A., Blum, J., Tanaka, H., Ormel, C., Bizzarro, M., Rickman, H.: The Multifaceted Planetesimal Formation Process. Protostars and Planets VI, pp. 547–570 (2014)
198. Johansen, A., Klahr, H., Henning, T.: High-resolution simulations of planetesimal formation in turbulent protoplanetary discs. A&A **529**, A62 (2011). https://doi.org/10.1051/0004-6361/201015979
199. Johansen, A., Mac Low, M.M., Lacerda, P., Bizzarro, M.: Growth of asteroids, planetary embryos, and Kuiper belt objects by chondrule accretion. Sci. Adv. **1**, 1500109 (2015). https://doi.org/10.1126/sciadv.1500109
200. Johansen, A., Oishi, J.S., Mac Low, M.M., Klahr, H., Henning, T., Youdin, A.: Rapid planetesimal formation in turbulent circumstellar disks. Nature **448**, 1022–1025 (2007). https://doi.org/10.1038/nature06086
201. Johansen, A., Youdin, A.: Protoplanetary disk turbulence driven by the streaming instability: nonlinear saturation and particle concentration. ApJ **662**, 627–641 (2007). https://doi.org/10.1086/516730
202. Johansen, A., Youdin, A., Klahr, H.: Zonal flows and long-lived axisymmetric pressure bumps in magnetorotational turbulence. ApJ **697**, 1269–1289 (2009). https://doi.org/10.1088/0004-637X/697/2/1269
203. Johansen, A., Youdin, A., Mac Low, M.M.: Particle clumping and planetesimal formation depend strongly on metallicity. ApJ **704**, L75–L79 (2009). https://doi.org/10.1088/0004-637X/704/2/L75
204. Johnson, B.M., Gammie, C.F.: Nonlinear outcome of gravitational instability in disks with realistic cooling. ApJ **597**, 131–141 (2003). https://doi.org/10.1086/378392
205. Johnson, B.M., Gammie, C.F.: Vortices in thin, compressible, unmagnetized disks. ApJ **635**, 149–156 (2005). https://doi.org/10.1086/497358
206. Johnstone, C.P., Jardine, M., Gregory, S.G., Donati, J.F., Hussain, G.: Classical T Tauri stars: magnetic fields, coronae and star-disc interactions. MNRAS **437**, 3202–3220 (2014). https://doi.org/10.1093/mnras/stt2107
207. Joy, A.H.: T Tauri variable stars. ApJ **102**, 168 (1945). https://doi.org/10.1086/144749
208. Kama, M., Bruderer, S., van Dishoeck, E.F., Hogerheijde, M., Folsom, C.P., Miotello, A., Fedele, D., Belloche, A., Güsten, R., Wyrowski, F.: Volatile-carbon locking and release in protoplanetary disks. A study of TW Hya and HD 100546. A&A **592**, A83 (2016). https://doi.org/10.1051/0004-6361/201526991
209. Kamp, I., Dullemond, C.P.: The gas temperature in the surface layers of protoplanetary disks. ApJ **615**, 991–999 (2004). https://doi.org/10.1086/424703
210. Kamp, I., van Zadelhoff, G.J.: On the gas temperature in circumstellar disks around A stars. A&A **373**, 641–656 (2001). https://doi.org/10.1051/0004-6361:20010629
211. Kenyon, S.J., Hartmann, L.: Spectral energy distributions of T Tauri stars—disk flaring and limits on accretion. ApJ **323**, 714–733 (1987). https://doi.org/10.1086/165866
212. Kerswell, R.R.: Elliptical instability. Annu. Rev. Fluid Mech. **34**, 83–113 (2002). https://doi.org/10.1146/annurev.fluid.34.081701.171829
213. Kida, S.: Motion of an elliptic vortex in a uniform shear flow. J. Phys. Soc. Jpn. **50**, 3517–3520 (1981). https://doi.org/10.1143/JPSJ.50.3517
214. Kim, K.H., Watson, D.M., Manoj, P., Forrest, W.J., Furlan, E., Najita, J., Sargent, B., Hernández, J., Calvet, N., Adame, L., Espaillat, C., Megeath, S.T., Muzerolle, J., McClure, M.K.: The Spitzer infrared spectrograph survey of protoplanetary disks in Orion A. I. Disk properties. ApJs **226**, 8 (2016). https://doi.org/10.3847/0067-0049/226/1/8
215. King, A.R., Pringle, J.E., Livio, M.: Accretion disc viscosity: how big is alpha? MNRAS **376**, 1740–1746 (2007). https://doi.org/10.1111/j.1365-2966.2007.11556.x
216. Klahr, H.H., Bodenheimer, P.: Turbulence in accretion disks: vorticity generation and angular momentum transport via the global baroclinic instability. ApJ **582**, 869–892 (2003). https://doi.org/10.1086/344743

217. Kley, W., Lin, D.N.C.: The structure of the boundary layer in protostellar disks. ApJ **461**, 933 (1996). https://doi.org/10.1086/177115
218. Kley, W., Nelson, R.P.: Planet-disk interaction and orbital evolution. ARA&A **50**, 211–249 (2012). https://doi.org/10.1146/annurev-astro-081811-125523
219. Koenigl, A.: Disk accretion onto magnetic T Tauri stars. ApJ **370**, L39–L43 (1991). https://doi.org/10.1086/185972
220. Koller, J., Li, H., Lin, D.N.C.: Vortices in the co-orbital region of an embedded protoplanet. ApJ **596**, L91–L94 (2003). https://doi.org/10.1086/379032
221. Königl, A., Salmeron, R.: The Effects of Large-Scale Magnetic Fields on Disk Formation and Evolution, pp. 283–352 (2011)
222. Kounkel, M., Hartmann, L., Loinard, L., Ortiz-León, G.N., Mioduszewski, A.J., Rodríguez, L.F., Dzib, S.A., Torres, R.M., Pech, G., Galli, P.A.B., Rivera, J.L., Boden, A.F., Evans II, N.J., Briceño, C., Tobin, J.J.: The Goulds Belt Distances Survey (GOBELINS) II. Distances and structure toward the Orion molecular clouds. ApJ **834**, 142 (2017). https://doi.org/10.3847/1538-4357/834/2/142
223. Kraichnan, R.H.: Inertial ranges in two-dimensional turbulence. Phys. Fluids **10**, 1417–1423 (1967). https://doi.org/10.1063/1.1762301
224. Krasnopolsky, R., Li, Z.Y., Shang, H.: Disk formation in magnetized clouds enabled by the Hall effect. ApJ **733**, 54 (2011). https://doi.org/10.1088/0004-637X/733/1/54
225. Kratter, K., Lodato, G.: Gravitational instabilities in circumstellar disks. ARA&A **54**, 271–311 (2016). https://doi.org/10.1146/annurev-astro-081915-023307
226. Kratter, K.M., Matzner, C.D., Krumholz, M.R., Klein, R.I.: On the role of disks in the formation of stellar systems: a numerical parameter study of rapid accretion. ApJ **708**, 1585–1597 (2010). https://doi.org/10.1088/0004-637X/708/2/1585
227. Kratter, K.M., Murray-Clay, R.A., Youdin, A.N.: The runts of the litter: why planets formed through gravitational instability can only be failed binary stars. ApJ **710**, 1375–1386 (2010). https://doi.org/10.1088/0004-637X/710/2/1375
228. Kretke, K.A., Lin, D.N.C.: Grain retention and formation of planetesimals near the snow line in MRI-driven turbulent protoplanetary disks. ApJ **664**, L55–L58 (2007). https://doi.org/10.1086/520718
229. Kuiper, G.P.: On the origin of the solar system. Proc. Natl. Acad. Sci. **37**, 1–14 (1951). https://doi.org/10.1073/pnas.37.1.1
230. Kunz, M.W.: On the linear stability of weakly ionized, magnetized planar shear flows. MNRAS **385**, 1494–1510 (2008). https://doi.org/10.1111/j.1365-2966.2008.12928.x
231. Kunz, M.W., Balbus, S.A.: Ambipolar diffusion in the magnetorotational instability. MNRAS **348**, 355–360 (2004). https://doi.org/10.1111/j.1365-2966.2004.07383.x
232. Kunz, M.W., Lesur, G.: Magnetic self-organization in Hall-dominated magnetorotational turbulence. MNRAS **434**, 2295–2312 (2013). https://doi.org/10.1093/mnras/stt1171
233. Kurosawa, R., Romanova, M.M.: Spectral variability of classical T Tauri stars accreting in an unstable regime. MNRAS **431**, 2673–2689 (2013). https://doi.org/10.1093/mnras/stt365
234. Lada, C.J., Wilking, B.A.: The nature of the embedded population in the Rho Ophiuchi dark cloud—mid-infrared observations. ApJ **287**, 610–621 (1984). https://doi.org/10.1086/162719
235. Lai, D.: Magnetically driven warping, precession, and resonances in accretion disks. ApJ **524**, 1030–1047 (1999). https://doi.org/10.1086/307850
236. Larwood, J.D., Nelson, R.P., Papaloizou, J.C.B., Terquem, C.: The tidally induced warping, precession and truncation of accretion discs in binary systems: three-dimensional simulations. MNRAS **282**, 597–613 (1996)
237. Lasota, J.P.: The disc instability model of dwarf novae and low-mass X-ray binary transients. New Ast. Rev. **45**, 449–508 (2001). https://doi.org/10.1016/S1387-6473(01)00112-9
238. Latter, H.N., Papaloizou, J.: Local models of astrophysical discs. MNRAS **472**, 1432–1446 (2017). https://doi.org/10.1093/mnras/stx2038
239. Lecar, M., Podolak, M., Sasselov, D., Chiang, E.: On the location of the snow line in a protoplanetary disk. ApJ **640**, 1115–1118 (2006). https://doi.org/10.1086/500287

240. Lesur, G., Kunz, M.W., Fromang, S.: Thanatology in protoplanetary discs. The combined influence of ohmic, Hall, and ambipolar diffusion on dead zones. A&A **566**, A56 (2014). https://doi.org/10.1051/0004-6361/201423660
241. Lesur, G., Longaretti, P.Y.: On the relevance of subcritical hydrodynamic turbulence to accretion disk transport. A&A **444**, 25–44 (2005). https://doi.org/10.1051/0004-6361:20053683
242. Lesur, G., Ogilvie, G.I.: On the angular momentum transport due to vertical convection in accretion discs. MNRAS **404**, L64–L68 (2010). https://doi.org/10.1111/j.1745-3933.2010.00836.x
243. Lesur, G., Papaloizou, J.C.B.: On the stability of elliptical vortices in accretion discs. A&A **498**, 1–12 (2009). https://doi.org/10.1051/0004-6361/200811577
244. Lesur, G., Papaloizou, J.C.B.: The subcritical baroclinic instability in local accretion disc models. A&A **513**, A60 (2010). https://doi.org/10.1051/0004-6361/200913594
245. Lesur, G.R.J., Latter, H.: On the survival of zombie vortices in protoplanetary discs. MNRAS **462**, 4549–4554 (2016). https://doi.org/10.1093/mnras/stw2172
246. Levin, Y.: Starbursts near supermassive black holes: young stars in the Galactic Centre, and gravitational waves in LISA band. MNRAS **374**, 515–524 (2007). https://doi.org/10.1111/j.1365-2966.2006.11155.x
247. Li, H., Colgate, S.A., Wendroff, B., Liska, R.: Rossby wave instability of thin accretion disks. III. Nonlinear simulations. ApJ **551**, 874–896 (2001). https://doi.org/10.1086/320241
248. Li, H., Finn, J.M., Lovelace, R.V.E., Colgate, S.A.: Rossby wave instability of thin accretion disks. II. Detailed linear theory. ApJ **533**, 1023–1034 (2000). https://doi.org/10.1086/308693
249. Li, Z.Y., Banerjee, R., Pudritz, R.E., Jørgensen, J.K., Shang, H., Krasnopolsky, R., Maury, A.: The Earliest Stages of Star and Planet Formation: Core Collapse, and the Formation of Disks and Outflows. Protostars and Planets VI, pp. 173–194 (2014)
250. Liffman, K.: The gravitational radius of an irradiated disk. Publ. Aston. Soc. Aust. **20**, 337–339 (2003). https://doi.org/10.1071/AS03019
251. Lin, M.K.: Non-barotropic linear Rossby wave instability in three-dimensional disks. ApJ **765**, 84 (2013). https://doi.org/10.1088/0004-637X/765/2/84
252. Lin, M.K., Youdin, A.N.: Cooling requirements for the vertical shear instability in protoplanetary disks. ApJ **811**, 17 (2015). https://doi.org/10.1088/0004-637X/811/1/17
253. Lodato, G., Clarke, C.J.: Massive planets in FU Orionis discs: implications for thermal instability models. MNRAS **353**, 841–852 (2004). https://doi.org/10.1111/j.1365-2966.2004.08112.x
254. Lodato, G., Pringle, J.E.: Warp diffusion in accretion discs: a numerical investigation. MNRAS **381**, 1287–1300 (2007). https://doi.org/10.1111/j.1365-2966.2007.12332.x
255. Lodato, G., Rice, W.K.M.: Testing the locality of transport in self-gravitating accretion discs. MNRAS **351**, 630–642 (2004). https://doi.org/10.1111/j.1365-2966.2004.07811.x
256. Lodato, G., Rice, W.K.M.: Testing the locality of transport in self-gravitating accretion discs—II. The massive disc case. MNRAS **358**, 1489–1500 (2005). https://doi.org/10.1111/j.1365-2966.2005.08875.x
257. Lodders, K.: Solar system abundances and condensation temperatures of the elements. ApJ **591**, 1220–1247 (2003). https://doi.org/10.1086/375492
258. Loomis, R.A., Öberg, K.I., Andrews, S.M., MacGregor, M.A.: A multi-ringed, modestly inclined protoplanetary disk around AA Tau. ApJ **840**, 23 (2017). https://doi.org/10.3847/1538-4357/aa6c63
259. Lovelace, R.V.E., Li, H., Colgate, S.A., Nelson, A.F.: Rossby wave instability of Keplerian accretion disks. ApJ **513**, 805–810 (1999). https://doi.org/10.1086/306900
260. Lovelace, R.V.E., Rothstein, D.M., Bisnovatyi-Kogan, G.S.: Advection/diffusion of large-scale B field in accretion disks. ApJ **701**, 885–890 (2009). https://doi.org/10.1088/0004-637X/701/2/885
261. Lubow, S.H., Martin, R.G., Nixon, C.: Tidal torques on misaligned disks in binary systems. ApJ **800**, 96 (2015). https://doi.org/10.1088/0004-637X/800/2/96
262. Lubow, S.H., Ogilvie, G.I.: On the tilting of protostellar disks by resonant tidal effects. ApJ **538**, 326–340 (2000). https://doi.org/10.1086/309101

263. Lubow, S.H., Papaloizou, J.C.B., Pringle, J.E.: Magnetic field dragging in accretion discs. MNRAS **267**, 235–240 (1994)
264. Lubow, S.H., Papaloizou, J.C.B., Pringle, J.E.: On the stability of magnetic wind-driven accretion discs. MNRAS **268**, 1010 (1994)
265. Luhman, K.L., Allen, P.R., Espaillat, C., Hartmann, L., Calvet, N.: The disk population of the Taurus star-forming region. ApJs **186**, 111–174 (2010). https://doi.org/10.1088/0067-0049/186/1/111
266. Lynden-Bell, D.: On why discs generate magnetic towers and collimate jets. MNRAS **341**, 1360–1372 (2003). https://doi.org/10.1046/j.1365-8711.2003.06506.x
267. Lynden-Bell, D., Pringle, J.E.: The evolution of viscous discs and the origin of the nebular variables. MNRAS **168**, 603–637 (1974)
268. Lyra, W., Johansen, A., Klahr, H., Piskunov, N.: Embryos grown in the dead zone. Assembling the first protoplanetary cores in low mass self-gravitating circumstellar disks of gas and solids. A&A **491**, L41–L44 (2008). https://doi.org/10.1051/0004-6361:200810626
269. Lyra, W., Johansen, A., Zsom, A., Klahr, H., Piskunov, N.: Planet formation bursts at the borders of the dead zone in 2D numerical simulations of circumstellar disks. A&A **497**, 869–888 (2009). https://doi.org/10.1051/0004-6361/200811265
270. Lyra, W., Kuchner, M.: Formation of sharp eccentric rings in debris disks with gas but without planets. Nature **499**, 184–187 (2013). https://doi.org/10.1038/nature12281
271. Lyra, W., Mac Low, M.M.: Rossby wave instability at dead zone boundaries in three-dimensional resistive magnetohydrodynamical global models of protoplanetary disks. ApJ **756**, 62 (2012). https://doi.org/10.1088/0004-637X/756/1/62
272. Lyra, W., Turner, N.J., McNally, C.P.: Rossby wave instability does not require sharp resistivity gradients. A&A **574**, A10 (2015). https://doi.org/10.1051/0004-6361/201424919
273. Malygin, M.G., Klahr, H., Semenov, D., Henning, T., Dullemond, C.P.: Efficiency of thermal relaxation by radiative processes in protoplanetary discs: constraints on hydrodynamic turbulence. A&A **605**, A30 (2017). https://doi.org/10.1051/0004-6361/201629933
274. Manara, C.F., Fedele, D., Herczeg, G.J., Teixeira, P.S.: X-Shooter study of accretion in Chamaeleon I. A&A **585**, A136 (2016). https://doi.org/10.1051/0004-6361/201527224
275. Manara, C.F., Testi, L., Natta, A., Rosotti, G., Benisty, M., Ercolano, B., Ricci, L.: Gas content of transitional disks: a VLT/X-Shooter study of accretion and winds. A&A **568**, A18 (2014). https://doi.org/10.1051/0004-6361/201323318
276. Marcus, P.S., Pei, S., Jiang, C.H., Barranco, J.A., Hassanzadeh, P., Lecoanet, D.: Zombie vortex instability. I. A purely hydrodynamic instability to resurrect the dead zones of protoplanetary disks. ApJ **808**, 87 (2015). https://doi.org/10.1088/0004-637X/808/1/87
277. Marcus, P.S., Pei, S., Jiang, C.H., Hassanzadeh, P.: Three-dimensional vortices generated by self-replication in stably stratified rotating shear flows. Phys. Rev. Lett. **111**(8), 084501 (2013). https://doi.org/10.1103/PhysRevLett.111.084501
278. Marley, M.S., Fortney, J.J., Hubickyj, O., Bodenheimer, P., Lissauer, J.J.: On the luminosity of young Jupiters. ApJ **655**, 541–549 (2007). https://doi.org/10.1086/509759
279. Marois, C., Macintosh, B., Barman, T., Zuckerman, B., Song, I., Patience, J., Lafrenière, D., Doyon, R.: Direct imaging of multiple planets orbiting the star HR 8799. Science **322**, 1348 (2008). https://doi.org/10.1126/science.1166585
280. Marois, C., Zuckerman, B., Konopacky, Q.M., Macintosh, B., Barman, T.: Images of a fourth planet orbiting HR 8799. Nature **468**, 1080–1083 (2010). https://doi.org/10.1038/nature09684
281. Martin, R.G., Livio, M.: On the evolution of the snow line in protoplanetary discs. MNRAS **425**, L6–L9 (2012). https://doi.org/10.1111/j.1745-3933.2012.01290.x
282. Martin, R.G., Lubow, S.H.: The gravo-magneto limit cycle in accretion disks. ApJ **740**, L6 (2011). https://doi.org/10.1088/2041-8205/740/1/L6
283. Martin, R.G., Lubow, S.H.: Tidal truncation of circumplanetary discs. MNRAS **413**, 1447–1461 (2011). https://doi.org/10.1111/j.1365-2966.2011.18228.x
284. Martin, R.G., Lubow, S.H.: The gravo-magneto disc instability with a viscous dead zone. MNRAS **437**, 682–689 (2014). https://doi.org/10.1093/mnras/stt1917

285. Maslowe, S.A.: Critical layers in shear flows. Ann. Rev. Fluid Mech. **18**, 405–432 (1986). https://doi.org/10.1146/annurev.fl.18.010186.002201
286. Matzner, C.D., Levin, Y.: Protostellar disks: formation, fragmentation, and the brown dwarf desert. ApJ **628**, 817–831 (2005). https://doi.org/10.1086/430813
287. McClure, M.K., Bergin, E.A., Cleeves, L.I., van Dishoeck, E.F., Blake, G.A., Evans II, N.J., Green, J.D., Henning, T., Öberg, K.I., Pontoppidan, K.M., Salyk, C.: Mass measurements in protoplanetary disks from hydrogen deuteride. ApJ **831**, 167 (2016). https://doi.org/10.3847/0004-637X/831/2/167
288. Meheut, H., Casse, F., Varniere, P., Tagger, M.: Rossby wave instability and three-dimensional vortices in accretion disks. A&A **516**, A31 (2010). https://doi.org/10.1051/0004-6361/201014000
289. Meru, F., Bate, M.R.: Non-convergence of the critical cooling time-scale for fragmentation of self-gravitating discs. MNRAS **411**, L1–L5 (2011). https://doi.org/10.1111/j.1745-3933.2010.00978.x
290. Meyer, F., Meyer-Hofmeister, E.: On the elusive cause of cataclysmic variable outbursts. A&A **104**, L10 (1981)
291. Michael, S., Durisen, R.H., Boley, A.C.: Migration of gas giant planets in gravitationally unstable disks. ApJ **737**, L42 (2011). https://doi.org/10.1088/2041-8205/737/2/L42
292. Min, M., Dullemond, C.P., Kama, M., Dominik, C.: The thermal structure and the location of the snow line in the protosolar nebula: axisymmetric models with full 3-D radiative transfer. Icarus **212**, 416–426 (2011). https://doi.org/10.1016/j.icarus.2010.12.002
293. Miranda, R., Lai, D.: Tidal truncation of inclined circumstellar and circumbinary discs in young stellar binaries. MNRAS **452**, 2396–2409 (2015). https://doi.org/10.1093/mnras/stv1450
294. Molyarova, T., Akimkin, V., Semenov, D., Henning, T., Vasyunin, A., Wiebe, D.: Gas mass tracers in protoplanetary disks: CO is still the best. ArXiv e-prints (2017)
295. Momose, M., Morita, A., Fukagawa, M., Muto, T., Takeuchi, T., Hashimoto, J., Honda, M., Kudo, T., Okamoto, Y.K., Kanagawa, K.D., Tanaka, H., Grady, C.A., Sitko, M.L., Akiyama, E., Currie, T., Follette, K.B., Mayama, S., Kusakabe, N., Abe, L., Brandner, W., Brandt, T.D., Carson, J.C., Egner, S., Feldt, M., Goto, M., Guyon, O., Hayano, Y., Hayashi, M., Hayashi, S.S., Henning, T., Hodapp, K.W., Ishii, M., Iye, M., Janson, M., Kandori, R., Knapp, G.R., Kuzuhara, M., Kwon, J., Matsuo, T., McElwain, M.W., Miyama, S., Morino, J.I., Moro-Martin, A., Nishimura, T., Pyo, T.S., Serabyn, E., Suenaga, T., Suto, H., Suzuki, R., Takahashi, Y.H., Takami, M., Takato, N., Terada, H., Thalmann, C., Tomono, D., Turner, E.L., Watanabe, M., Wisniewski, J., Yamada, T., Takami, H., Usuda, T., Tamura, M.: Detailed structure of the outer disk around HD 169142 with polarized light in H-band. PASJ **67**, 83 (2015). https://doi.org/10.1093/pasj/psv051
296. Morbidelli, A., Chambers, J., Lunine, J.I., Petit, J.M., Robert, F., Valsecchi, G.B., Cyr, K.E.: Source regions and time scales for the delivery of water to Earth. Meteor. Planet. Sci. **35**, 1309–1320 (2000). https://doi.org/10.1111/j.1945-5100.2000.tb01518.x
297. Morfill, G.E.: Some cosmochemical consequences of a turbulent protoplanetary cloud. Icarus **53**, 41–54 (1983). https://doi.org/10.1016/0019-1035(83)90019-2
298. Muranushi, T., Okuzumi, S., Inutsuka, S.I.: Interdependence of electric discharge and magnetorotational instability in protoplanetary disks. ApJ **760**, 56 (2012). https://doi.org/10.1088/0004-637X/760/1/56
299. Muzerolle, J., Hillenbrand, L., Calvet, N., Briceño, C., Hartmann, L.: Accretion in young stellar/substellar objects. ApJ **592**, 266–281 (2003). https://doi.org/10.1086/375704
300. Nakagawa, Y., Sekiya, M., Hayashi, C.: Settling and growth of dust particles in a laminar phase of a low-mass solar nebula. Icarus **67**, 375–390 (1986). https://doi.org/10.1016/0019-1035(86)90121-1
301. Nayakshin, S., Lodato, G.: Fu Ori outbursts and the planet-disc mass exchange. MNRAS **426**, 70–90 (2012). https://doi.org/10.1111/j.1365-2966.2012.21612.x
302. Nelson, R.P., Gressel, O., Umurhan, O.M.: Linear and non-linear evolution of the vertical shear instability in accretion discs. MNRAS **435**, 2610–2632 (2013). https://doi.org/10.1093/mnras/stt1475

303. Nesvorný, D., Youdin, A.N., Richardson, D.C.: Formation of Kuiper belt binaries by gravitational collapse. AJ **140**, 785–793 (2010). https://doi.org/10.1088/0004-6256/140/3/785
304. Nixon, C., King, A.: Warp propagation in astrophysical discs. In: Haardt, F., Gorini, V., Moschella, U., Treves, Colpi, A.M. (eds.) Lecture Notes in Physics, vol. 905, p. 45. Springer, Berlin. (2016). https://doi.org/10.1007/978-3-319-19416-52
305. Nixon, C.J., Pringle, J.E.: The observable effects of tidally induced warps in protostellar discs. MNRAS **403**, 1887–1893 (2010). https://doi.org/10.1111/j.1365-2966.2010.16331.x
306. O'dell, C.R., Wen, Z., Hu, X.: Discovery of new objects in the Orion nebula on HST images—shocks, compact sources, and protoplanetary disks. ApJ **410**, 696–700 (1993). https://doi.org/10.1086/172786
307. Ogilvie, G.I.: The non-linear fluid dynamics of a warped accretion disc. MNRAS **304**, 557–578 (1999). https://doi.org/10.1046/j.1365-8711.1999.02340.x
308. Ogilvie, G.I.: Non-linear fluid dynamics of eccentric discs. MNRAS **325**, 231–248 (2001). https://doi.org/10.1046/j.1365-8711.2001.04416.x
309. Ogilvie, G.I., Latter, H.N.: Local and global dynamics of warped astrophysical discs. MNRAS **433**, 2403–2419 (2013). https://doi.org/10.1093/mnras/stt916
310. Ogilvie, G.I., Livio, M.: Launching of jets and the vertical structure of accretion disks. ApJ **553**, 158–173 (2001). https://doi.org/10.1086/320637
311. Ogilvie, G.I., Pringle, J.E.: The non-axisymmetric instability of a cylindrical shear flow containing an azimuthal magnetic field. MNRAS **279**, 152–164 (1996)
312. Olofsson, J., Augereau, J.C., van Dishoeck, E.F., Merín, B., Grosso, N., Ménard, F., Blake, G.A., Monin, J.L.: C2D Spitzer-IRS spectra of disks around T Tauri stars. V. Spectral decomposition. A&A **520**, A39 (2010). https://doi.org/10.1051/0004-6361/200913909
313. Oppenheimer, M., Dalgarno, A.: The fractional ionization in dense interstellar clouds. ApJ **192**, 29–32 (1974). https://doi.org/10.1086/153030
314. Ormel, C.W., Cuzzi, J.N.: Closed-form expressions for particle relative velocities induced by turbulence. A&A **466**, 413–420 (2007). https://doi.org/10.1051/0004-6361:20066899
315. Owen, J.E., Armitage, P.J.: Importance of thermal diffusion in the gravomagnetic limit cycle. MNRAS **445**, 2800–2809 (2014). https://doi.org/10.1093/mnras/stu1928
316. Owen, J.E., Clarke, C.J., Ercolano, B.: On the theory of disc photoevaporation. MNRAS **422**, 1880–1901 (2012). https://doi.org/10.1111/j.1365-2966.2011.20337.x
317. Owen, J.E., Ercolano, B., Clarke, C.J., Alexander, R.D.: Radiation-hydrodynamic models of X-ray and EUV photoevaporating protoplanetary discs. MNRAS **401**, 1415–1428 (2010). https://doi.org/10.1111/j.1365-2966.2009.15771.x
318. Owen, J.E., Hudoba de Badyn, M., Clarke, C.J., Robins, L.: Characterizing thermal sweeping: a rapid disc dispersal mechanism. MNRAS **436**, 1430–1438 (2013). https://doi.org/10.1093/mnras/stt1663
319. Paardekooper, S.J.: Numerical convergence in self-gravitating shearing sheet simulations and the stochastic nature of disc fragmentation. MNRAS **421**, 3286–3299 (2012). https://doi.org/10.1111/j.1365-2966.2012.20553.x
320. Paardekooper, S.J., Mellema, G.: Planets opening dust gaps in gas disks. A&A **425**, L9–L12 (2004). https://doi.org/10.1051/0004-6361:200400053
321. Paczynski, B.: A model of accretion disks in close binaries. ApJ **216**, 822–826 (1977). https://doi.org/10.1086/155526
322. Paczynski, B.: A model of selfgravitating accretion disk. Acta Astron. **28**, 91–109 (1978)
323. Papaloizou, J., Pringle, J.E.: Tidal torques on accretion discs in close binary systems. MNRAS **181**, 441–454 (1977)
324. Papaloizou, J.C.B., Pringle, J.E.: The time-dependence of non-planar accretion discs. MNRAS **202**, 1181–1194 (1983)
325. Papaloizou, J.C.B., Pringle, J.E.: The dynamical stability of differentially rotating discs with constant specific angular momentum. MNRAS **208**, 721–750 (1984)
326. Pascucci, I., Ricci, L., Gorti, U., Hollenbach, D., Hendler, N.P., Brooks, K.J., Contreras, Y.: Low extreme-ultraviolet luminosities impinging on protoplanetary disks. ApJ **795**, 1 (2014). https://doi.org/10.1088/0004-637X/795/1/1

327. Pérez, L.M., Carpenter, J.M., Andrews, S.M., Ricci, L., Isella, A., Linz, H., Sargent, A.I., Wilner, D.J., Henning, T., Deller, A.T., Chandler, C.J., Dullemond, C.P., Lazio, J., Menten, K.M., Corder, S.A., Storm, S., Testi, L., Tazzari, M., Kwon, W., Calvet, N., Greaves, J.S., Harris, R.J., Mundy, L.G.: Spiral density waves in a young protoplanetary disk. Science **353**, 1519–1521 (2016). https://doi.org/10.1126/science.aaf8296
328. Perez-Becker, D., Chiang, E.: Surface layer accretion in conventional and transitional disks driven by far-ultraviolet ionization. ApJ **735**, 8 (2011). https://doi.org/10.1088/0004-637X/735/1/8
329. Pessah, M.E., Psaltis, D.: The stability of magnetized rotating plasmas with superthermal fields. ApJ **628**, 879–901 (2005). https://doi.org/10.1086/430940
330. Petersen, M.R., Julien, K., Stewart, G.R.: Baroclinic vorticity production in protoplanetary disks. I. Vortex formation. ApJ **658**, 1236–1251 (2007). https://doi.org/10.1086/511513
331. Petersen, M.R., Stewart, G.R., Julien, K.: Baroclinic vorticity production in protoplanetary disks. II. Vortex growth and longevity. ApJ **658**, 1252–1263 (2007). https://doi.org/10.1086/511523
332. Pfalzner, S.: Encounter-driven accretion in young stellar cluster—a connection to FUors? A&A **492**, 735–741 (2008). https://doi.org/10.1051/0004-6361:200810879
333. Pinilla, P., Birnstiel, T., Ricci, L., Dullemond, C.P., Uribe, A.L., Testi, L., Natta, A.: Trapping dust particles in the outer regions of protoplanetary disks. A&A **538**, A114 (2012). https://doi.org/10.1051/0004-6361/201118204
334. Podio, L., Kamp, I., Codella, C., Cabrit, S., Nisini, B., Dougados, C., Sandell, G., Williams, J.P., Testi, L., Thi, W.F., Woitke, P., Meijerink, R., Spaans, M., Aresu, G., Ménard, F., Pinte, C.: Water vapor in the protoplanetary disk of DG Tau. ApJ **766**, L5 (2013). https://doi.org/10.1088/2041-8205/766/1/L5
335. Popham, R., Narayan, R., Hartmann, L., Kenyon, S.: Boundary layers in pre-main-sequence accretion disks. ApJ **415**, L127 (1993). https://doi.org/10.1086/187049
336. Preibisch, T., Kim, Y.C., Favata, F., Feigelson, E.D., Flaccomio, E., Getman, K., Micela, G., Sciortino, S., Stassun, K., Stelzer, B., Zinnecker, H.: The origin of T Tauri X-ray emission: new insights from the Chandra Orion Ultradeep Project. ApJs **160**, 401–422 (2005). https://doi.org/10.1086/432891
337. Pringle, J.E.: Soft X-ray emission from dwarf novae. MNRAS **178**, 195–202 (1977)
338. Pringle, J.E.: Accretion discs in astrophysics. ARA&A **19**, 137–162 (1981). https://doi.org/10.1146/annurev.aa.19.090181.001033
339. Pringle, J.E.: The properties of external accretion discs. MNRAS **248**, 754–759 (1991)
340. Pringle, J.E.: A simple approach to the evolution of twisted accretion discs. MNRAS **258**, 811–818 (1992)
341. Pringle, J.E., King, A.: Astrophysical Flows (2007)
342. Qi, C., Öberg, K.I., Wilner, D.J., D'Alessio, P., Bergin, E., Andrews, S.M., Blake, G.A., Hogerheijde, M.R., van Dishoeck, E.F.: Imaging of the CO snow line in a solar nebula analog. Science **341**, 630–632 (2013). https://doi.org/10.1126/science.1239560
343. Raettig, N., Klahr, H., Lyra, W.: Particle trapping and streaming instability in vortices in protoplanetary disks. ApJ **804**, 35 (2015). https://doi.org/10.1088/0004-637X/804/1/35
344. Raettig, N., Lyra, W., Klahr, H.: A parameter study for baroclinic vortex amplification. ApJ **765**, 115 (2013). https://doi.org/10.1088/0004-637X/765/2/115
345. Rafikov, R.R.: Properties of gravitoturbulent accretion disks. ApJ **704**, 281–291 (2009). https://doi.org/10.1088/0004-637X/704/1/281
346. Rafikov, R.R.: Viscosity prescription for gravitationally unstable accretion disks. ApJ **804**, 62 (2015). https://doi.org/10.1088/0004-637X/804/1/62
347. Railton, A.D., Papaloizou, J.C.B.: On the local stability of vortices in differentially rotating discs. MNRAS **445**, 4409–4426 (2014). https://doi.org/10.1093/mnras/stu2060
348. Rebusco, P., Umurhan, O.M., Kluźniak, W., Regev, O.: Global transient dynamics of three-dimensional hydrodynamical disturbances in a thin viscous accretion disk. Phys. Fluids **21**(7), 076,601 (2009). https://doi.org/10.1063/1.3167411

349. Reipurth, B., Clarke, C.J., Boss, A.P., Goodwin, S.P., Rodríguez, L.F., Stassun, K.G., Tokovinin, A., Zinnecker, H.: Multiplicity in Early Stellar Evolution. Protostars and Planets VI, pp. 267–290 (2014)
350. Ricci, L., Testi, L., Natta, A., Neri, R., Cabrit, S., Herczeg, G.J.: Dust properties of protoplanetary disks in the Taurus-Auriga star forming region from millimeter wavelengths. A&A **512**, A15 (2010). https://doi.org/10.1051/0004-6361/200913403
351. Rice, W.K.M., Armitage, P.J., Bate, M.R., Bonnell, I.A.: The effect of cooling on the global stability of self-gravitating protoplanetary discs. MNRAS **339**, 1025–1030 (2003). https://doi.org/10.1046/j.1365-8711.2003.06253.x
352. Rice, W.K.M., Armitage, P.J., Mamatsashvili, G.R., Lodato, G., Clarke, C.J.: Stability of self-gravitating discs under irradiation. MNRAS **418**, 1356–1362 (2011). https://doi.org/10.1111/j.1365-2966.2011.19586.x
353. Rice, W.K.M., Armitage, P.J., Wood, K., Lodato, G.: Dust filtration at gap edges: implications for the spectral energy distributions of discs with embedded planets. MNRAS **373**, 1619–1626 (2006). https://doi.org/10.1111/j.1365-2966.2006.11113.x
354. Rice, W.K.M., Lodato, G., Armitage, P.J.: Investigating fragmentation conditions in self-gravitating accretion discs. MNRAS **364**, L56–L60 (2005). https://doi.org/10.1111/j.1745-3933.2005.00105.x
355. Rice, W.K.M., Lodato, G., Pringle, J.E., Armitage, P.J., Bonnell, I.A.: Accelerated planetesimal growth in self-gravitating protoplanetary discs. MNRAS **355**, 543–552 (2004). https://doi.org/10.1111/j.1365-2966.2004.08339.x
356. Richling, S., Yorke, H.W.: Photoevaporation of protostellar disks. II. The importance of UV dust properties and ionizing flux. A&A **327**, 317–324 (1997)
357. Rigliaco, E., Natta, A., Testi, L., Randich, S., Alcalà, J.M., Covino, E., Stelzer, B.: X-shooter spectroscopy of young stellar objects. I. Mass accretion rates of low-mass T Tauri stars in σ Orionis. A&A **548**, A56 (2012). https://doi.org/10.1051/0004-6361/201219832
358. Robitaille, T.P.: HYPERION: an open-source parallelized three-dimensional dust continuum radiative transfer code. A&A **536**, A79 (2011). https://doi.org/10.1051/0004-6361/201117150
359. Rodmann, J., Henning, T., Chandler, C.J., Mundy, L.G., Wilner, D.J.: Large dust particles in disks around T Tauri stars. A&A **446**, 211–221 (2006). https://doi.org/10.1051/0004-6361:20054038
360. Romanova, M.M., Ustyugova, G.V., Koldoba, A.V., Lovelace, R.V.E.: MRI-driven accretion on to magnetized stars: global 3D MHD simulations of magnetospheric and boundary layer regimes. MNRAS **421**, 63–77 (2012). https://doi.org/10.1111/j.1365-2966.2011.20055.x
361. Ros, K., Johansen, A.: Ice condensation as a planet formation mechanism. A&A **552**, A137 (2013). https://doi.org/10.1051/0004-6361/201220536
362. Rosenfeld, K.A., Andrews, S.M., Hughes, A.M., Wilner, D.J., Qi, C.: A spatially resolved vertical temperature gradient in the HD 163296 disk. ApJ **774**, 16 (2013). https://doi.org/10.1088/0004-637X/774/1/16
363. Ruden, S.P.: Evolution of photoevaporating protoplanetary disks. ApJ **605**, 880–891 (2004). https://doi.org/10.1086/382524
364. Rybicki, G.B., Lightman, A.P.: Radiative processes in astrophysics (1979)
365. Salinas, V.N., Hogerheijde, M.R., Bergin, E.A., Cleeves, L.I., Brinch, C., Blake, G.A., Lis, D.C., Melnick, G.J., Panić, O., Pearson, J.C., Kristensen, L., Yıldız, U.A., van Dishoeck, E.F.: First detection of gas-phase ammonia in a planet-forming disk. NH_3, N_2H^+, and H_2O in the disk around TW Hydrae. A&A **591**, A122 (2016). https://doi.org/10.1051/0004-6361/201628172
366. Sano, T., Inutsuka, S.I.: Saturation and thermalization of the magnetorotational instability: recurrent channel flows and reconnections. ApJ **561**, L179–L182 (2001). https://doi.org/10.1086/324763
367. Sano, T., Stone, J.M.: The effect of the Hall term on the nonlinear evolution of the magnetorotational instability. I. Local axisymmetric simulations. ApJ **570**, 314–328 (2002). https://doi.org/10.1086/339504

368. Sano, T., Stone, J.M.: The effect of the Hall term on the nonlinear evolution of the magnetorotational instability. II. Saturation level and critical magnetic Reynolds number. ApJ **577**, 534–553 (2002). https://doi.org/10.1086/342172
369. Schäfer, U., Yang, C.C., Johansen, A.: Initial mass function of planetesimals formed by the streaming instability. A&A **597**, A69 (2017). https://doi.org/10.1051/0004-6361/201629561
370. Shakura, N.I., Sunyaev, R.A.: Black holes in binary systems. Observational appearance. A&A **24**, 337–355 (1973)
371. Shen, Y., Stone, J.M., Gardiner, T.A.: Three-dimensional compressible hydrodynamic simulations of vortices in disks. ApJ **653**, 513–524 (2006). https://doi.org/10.1086/508980
372. Shu, F., Najita, J., Ostriker, E., Wilkin, F., Ruden, S., Lizano, S.: Magnetocentrifugally driven flows from young stars and disks. 1: A generalized model. ApJ **429**, 781–796 (1994). https://doi.org/10.1086/174363
373. Simon, J.B., Armitage, P.J.: Efficiency of particle trapping in the outer regions of protoplanetary disks. ApJ **784**, 15 (2014). https://doi.org/10.1088/0004-637X/784/1/15
374. Simon, J.B., Armitage, P.J., Li, R., Youdin, A.N.: The mass and size distribution of planetesimals formed by the streaming instability. I. The role of self-gravity. ApJ **822**, 55 (2016). https://doi.org/10.3847/0004-637X/822/1/55
375. Simon, J.B., Armitage, P.J., Youdin, A.N., Li, R.: Evidence for universality in the initial planetesimal mass function. ApJ **847**, L12 (2017). https://doi.org/10.3847/2041-8213/aa8c79
376. Simon, J.B., Bai, X.N., Armitage, P.J., Stone, J.M., Beckwith, K.: Turbulence in the outer regions of protoplanetary disks. II. Strong accretion driven by a vertical magnetic field. ApJ **775**, 73 (2013). https://doi.org/10.1088/0004-637X/775/1/73
377. Simon, J.B., Bai, X.N., Stone, J.M., Armitage, P.J., Beckwith, K.: Turbulence in the outer regions of protoplanetary disks. I. Weak accretion with no vertical magnetic flux. ApJ **764**, 66 (2013). https://doi.org/10.1088/0004-637X/764/1/66
378. Simon, J.B., Beckwith, K., Armitage, P.J.: Emergent mesoscale phenomena in magnetized accretion disc turbulence. MNRAS **422**, 2685–2700 (2012). https://doi.org/10.1111/j.1365-2966.2012.20835.x
379. Simon, J.B., Hawley, J.F.: Viscous and resistive effects on the magnetorotational instability with a net toroidal field. ApJ **707**, 833–843 (2009). https://doi.org/10.1088/0004-637X/707/1/833
380. Simon, J.B., Hughes, A.M., Flaherty, K.M., Bai, X.N., Armitage, P.J.: Signatures of MRI-driven turbulence in protoplanetary disks: predictions for ALMA observations. ApJ **808**, 180 (2015). https://doi.org/10.1088/0004-637X/808/2/180
381. Simon, J.B., Lesur, G., Kunz, M.W., Armitage, P.J.: Magnetically driven accretion in protoplanetary discs. MNRAS **454**, 1117–1131 (2015). https://doi.org/10.1093/mnras/stv2070
382. Smak, J.: Eruptive binaries. XI—Disk-radius variations in U GEM. Acta Astron. **34**, 93–96 (1984)
383. Soderblom, D.R., Hillenbrand, L.A., Jeffries, R.D., Mamajek, E.E., Naylor, T.: Ages of Young Stars. Protostars and Planets VI, pp. 219–241 (2014)
384. Spruit, H.C.: Magnetohydrodynamic jets and winds from accretion disks. In: Wijers, R.A.M.J., Davies, M.B., Tout, C.A. (eds.) NATO Advanced Science Institutes (ASI) Series C, vol. 477, pp. 249–286 (1996)
385. Steiman-Cameron, T.Y., Durisen, R.H., Boley, A.C., Michael, S., McConnell, C.R.: Convergence studies of mass transport in disks with gravitational instabilities. II. The radiative cooling case. ApJ **768**, 192 (2013). https://doi.org/10.1088/0004-637X/768/2/192
386. Steinacker, J., Baes, M., Gordon, K.D.: Three-dimensional dust radiative transfer*. ARA&A **51**, 63–104 (2013). https://doi.org/10.1146/annurev-astro-082812-141042
387. Stepinski, T.F.: Generation of dynamo magnetic fields in the primordial solar nebula. Icarus **97**, 130–141 (1992). https://doi.org/10.1016/0019-1035(92)90062-C
388. Stevenson, D.J., Lunine, J.I.: Rapid formation of Jupiter by diffuse redistribution of water vapor in the solar nebula. Icarus **75**, 146–155 (1988). https://doi.org/10.1016/0019-1035(88)90133-9

389. Stoll, M.H.R., Kley, W.: Vertical shear instability in accretion disc models with radiation transport. A&A **572**, A77 (2014). https://doi.org/10.1051/0004-6361/201424114
390. Strom, K.M., Strom, S.E., Edwards, S., Cabrit, S., Skrutskie, M.F.: Circumstellar material associated with solar-type pre-main-sequence stars—a possible constraint on the timescale for planet building. AJ **97**, 1451–1470 (1989). https://doi.org/10.1086/115085
391. Supulver, K.D., Lin, D.N.C.: Formation of Icy planetesimals in a turbulent solar nebula. Icarus **146**, 525–540 (2000). https://doi.org/10.1006/icar.2000.6418
392. Suzuki, T.K., Inutsuka, S.I.: disk winds driven by magnetorotational instability and dispersal of protoplanetary disks. ApJ **691**, L49–L54 (2009). https://doi.org/10.1088/0004-637X/691/1/L49
393. Suzuki, T.K., Muto, T., Inutsuka, S.I.: Protoplanetary disk winds via magnetorotational instability: formation of an inner hole and a crucial assist for planet formation. ApJ **718**, 1289–1304 (2010). https://doi.org/10.1088/0004-637X/718/2/1289
394. Suzuki, T.K., Ogihara, M., Morbidelli, A., Crida, A., Guillot, T.: Evolution of protoplanetary discs with magnetically driven disc winds. A&A **596**, A74 (2016). https://doi.org/10.1051/0004-6361/201628955
395. Takeuchi, T., Lin, D.N.C.: Radial flow of dust particles in accretion disks. ApJ **581**, 1344–1355 (2002). https://doi.org/10.1086/344437
396. Takeuchi, T., Okuzumi, S.: Radial transport of large-scale magnetic fields in accretion disks. II. Relaxation to steady states. ApJ **797**, 132 (2014). https://doi.org/10.1088/0004-637X/797/2/132
397. Tanaka, K.E.I., Nakamoto, T., Omukai, K.: Photoevaporation of circumstellar disks revisited: the dust-free case. ApJ **773**, 155 (2013). https://doi.org/10.1088/0004-637X/773/2/155
398. Tanga, P., Babiano, A., Dubrulle, B., Provenzale, A.: Forming planetesimals in vortices. Icarus **121**, 158–170 (1996). https://doi.org/10.1006/icar.1996.0076
399. Tazzari, M., Testi, L., Ercolano, B., Natta, A., Isella, A., Chandler, C.J., Pérez, L.M., Andrews, S., Wilner, D.J., Ricci, L., Henning, T., Linz, H., Kwon, W., Corder, S.A., Dullemond, C.P., Carpenter, J.M., Sargent, A.I., Mundy, L., Storm, S., Calvet, N., Greaves, J.A., Lazio, J., Deller, A.T.: Multiwavelength analysis for interferometric (sub-)mm observations of protoplanetary disks. Radial constraints on the dust properties and the disk structure. A&A **588**, A53 (2016). https://doi.org/10.1051/0004-6361/201527423
400. Teague, R., Guilloteau, S., Semenov, D., Henning, T., Dutrey, A., Piétu, V., Birnstiel, T., Chapillon, E., Hollenbach, D., Gorti, U.: Measuring turbulence in TW Hydrae with ALMA: methods and limitations. A&A **592**, A49 (2016). https://doi.org/10.1051/0004-6361/201628550
401. Terquem, C., Papaloizou, J.C.B.: On the stability of an accretion disc containing a toroidal magnetic field. MNRAS **279**, 767–784 (1996)
402. Throop, H.B., Bally, J.: Can photoevaporation trigger planetesimal formation? ApJ **623**, L149–L152 (2005). https://doi.org/10.1086/430272
403. Throop, H.B., Bally, J.: Tail-end Bondi-Hoyle accretion in young star clusters: implications for disks, planets, and stars. AJ **135**, 2380–2397 (2008). https://doi.org/10.1088/0004-6256/135/6/2380
404. Tobin, J.J., Kratter, K.M., Persson, M.V., Looney, L.W., Dunham, M.M., Segura-Cox, D., Li, Z.Y., Chandler, C.J., Sadavoy, S.I., Harris, R.J., Melis, C., Pérez, L.M.: A triple protostar system formed via fragmentation of a gravitationally unstable disk. Nature **538**, 483–486 (2016). https://doi.org/10.1038/nature20094
405. Toomre, A.: On the gravitational stability of a disk of stars. ApJ **139**, 1217–1238 (1964). https://doi.org/10.1086/147861
406. Torres, R.M., Loinard, L., Mioduszewski, A.J., Boden, A.F., Franco-Hernández, R., Vlemmings, W.H.T., Rodríguez, L.F.: VLBA determination of the distance to nearby star-forming regions. V. Dynamical mass, distance, and radio structure of V773 Tau A. ApJ **747**, 18 (2012). https://doi.org/10.1088/0004-637X/747/1/18
407. Tout, C.A., Pringle, J.E.: Accretion disc viscosity—a simple model for a magnetic dynamo. MNRAS **259**, 604–612 (1992)

408. Trapman, L., Miotello, A., Kama, M., van Dishoeck, E.F., Bruderer, S.: Far-infrared HD emission as a measure of protoplanetary disk mass. A&A **605**, A69 (2017). https://doi.org/10.1051/0004-6361/201630308
409. Tsukamoto, Y., Iwasaki, K., Okuzumi, S., Machida, M.N., Inutsuka, S.: Bimodality of circumstellar disk evolution induced by Hall current. ArXiv e-prints (2015)
410. Tsukamoto, Y., Takahashi, S.Z., Machida, M.N., Inutsuka, S.: Effects of radiative transfer on the structure of self-gravitating discs, their fragmentation and the evolution of the fragments. MNRAS **446**, 1175–1190 (2015). https://doi.org/10.1093/mnras/stu2160
411. Turner, N.J., Benisty, M., Dullemond, C.P., Hirose, S.: Herbig stars' near-infrared excess: an origin in the protostellar disk's magnetically supported atmosphere. ApJ **780**, 42 (2014). https://doi.org/10.1088/0004-637X/780/1/42
412. Turner, N.J., Drake, J.F.: Energetic protons, radionuclides, and magnetic activity in protostellar disks. ApJ **703**, 2152–2159 (2009). https://doi.org/10.1088/0004-637X/703/2/2152
413. Turner, N.J., Sano, T.: Dead zone accretion flows in protostellar disks. ApJ **679**, L131–L134 (2008). https://doi.org/10.1086/589540
414. Turner, N.J., Sano, T., Dziourkevitch, N.: Turbulent mixing and the dead zone in protostellar disks. ApJ **659**, 729–737 (2007). https://doi.org/10.1086/512007
415. Umebayashi, T., Nakano, T.: Effects of radionuclides on the ionization state of protoplanetary disks and dense cloud cores. ApJ **690**, 69–81 (2009). https://doi.org/10.1088/0004-637X/690/1/69
416. Umurhan, O.M.: Potential vorticity dynamics in the framework of disk shallow-water theory. I. The Rossby wave instability. A&A **521**, A25 (2010). https://doi.org/10.1051/0004-6361/201015210
417. Urpin, V., Brandenburg, A.: Magnetic and vertical shear instabilities in accretion discs. MNRAS **294**, 399 (1998). https://doi.org/10.1046/j.1365-8711.1998.01118.x
418. Uyama, T., Hashimoto, J., Kuzuhara, M., Mayama, S., Akiyama, E., Currie, T., Livingston, J., Kudo, T., Kusakabe, N., Abe, L., Brandner, W., Brandt, T.D., Carson, J.C., Egner, S., Feldt, M., Goto, M., Grady, C.A., Guyon, O., Hayano, Y., Hayashi, M., Hayashi, S.S., Henning, T., Hodapp, K.W., Ishii, M., Iye, M., Janson, M., Kandori, R., Knapp, G.R., Kwon, J., Matsuo, T., Mcelwain, M.W., Miyama, S., Morino, J.I., Moro-Martin, A., Nishimura, T., Pyo, T.S., Serabyn, E., Suenaga, T., Suto, H., Suzuki, R., Takahashi, Y.H., Takami, M., Takato, N., Terada, H., Thalmann, C., Turner, E.L., Watanabe, M., Wisniewski, J., Yamada, T., Takami, H., Usuda, T., Tamura, M.: The SEEDS high-contrast imaging survey of exoplanets around young stellar objects. AJ **153**, 106 (2017). https://doi.org/10.3847/1538-3881/153/3/106
419. van Boekel, R., Henning, T., Menu, J., de Boer, J., Langlois, M., Müller, A., Avenhaus, H., Boccaletti, A., Schmid, H.M., Thalmann, C., Benisty, M., Dominik, C., Ginski, C., Girard, J.H., Gisler, D., Lobo Gomes, A., Menard, F., Min, M., Pavlov, A., Pohl, A., Quanz, S.P., Rabou, P., Roelfsema, R., Sauvage, J.F., Teague, R., Wildi, F., Zurlo, A.: Three radial gaps in the disk of TW Hydrae imaged with SPHERE. ApJ **837**, 132 (2017). https://doi.org/10.3847/1538-4357/aa5d68
420. van der Marel, N., van Dishoeck, E.F., Bruderer, S., Birnstiel, T., Pinilla, P., Dullemond, C.P., van Kempen, T.A., Schmalzl, M., Brown, J.M., Herczeg, G.J., Mathews, G.S., Geers, V.: A major asymmetric dust trap in a transition disk. Science **340**, 1199–1202 (2013). https://doi.org/10.1126/science.1236770
421. Varnière, P., Tagger, M.: Reviving dead zones in accretion disks by Rossby vortices at their boundaries. A&A **446**, L13–L16 (2006). https://doi.org/10.1051/0004-6361:200500226
422. Velikhov, E.: Stability of an ideally conducting liquid flowing between rotating cylinders in a magnetic field. Zhur. Eksptl?. i Teoret. Fiz. **36** (1959)
423. Vorobyov, E.I., Basu, S.: The origin of episodic accretion bursts in the early stages of star formation. ApJ **633**, L137–L140 (2005). https://doi.org/10.1086/498303
424. Wang, L., Goodman, J.: Hydrodynamic photoevaporation of protoplanetary disks with consistent thermochemistry. ApJ **847**, 11 (2017). https://doi.org/10.3847/1538-4357/aa8726
425. Ward, W.R.: Particle filtering by a planetary gap. In: Lunar and Planetary Science Conference. Lunar and Planetary Inst. Technical Report, vol. 40, p. 1477 (2009)

426. Wardle, M.: The Balbus-Hawley instability in weakly ionized discs. MNRAS **307**, 849–856 (1999). https://doi.org/10.1046/j.1365-8711.1999.02670.x
427. Waters, T.R., Proga, D.: Parker winds revisited: an extension to disc winds. MNRAS **426**, 2239–2265 (2012). https://doi.org/10.1111/j.1365-2966.2012.21823.x
428. Weidenschilling, S.J.: Aerodynamics of solid bodies in the solar nebula. MNRAS **180**, 57–70 (1977)
429. Weidenschilling, S.J.: The distribution of mass in the planetary system and solar nebula. Ap&SS **51**, 153–158 (1977). https://doi.org/10.1007/BF00642464
430. Weingartner, J.C., Draine, B.T.: Photoelectric emission from interstellar dust: grain charging and gas heating. ApJs **134**, 263–281 (2001). https://doi.org/10.1086/320852
431. Whipple, F.L.: On certain aerodynamic processes for asteroids and comets. In: Elvius, A. (ed.) From Plasma to Planet, p. 211 (1972)
432. Williams, J.P., Best, W.M.J.: A parametric modeling approach to measuring the gas masses of circumstellar disks. ApJ **788**, 59 (2014). https://doi.org/10.1088/0004-637X/788/1/59
433. Williams, J.P., Cieza, L.A.: Protoplanetary disks and their evolution. ARA&A **49**, 67–117 (2011). https://doi.org/10.1146/annurev-astro-081710-102548
434. Yang, C.C., Johansen, A.: On the feeding zone of planetesimal formation by the streaming instability. ApJ **792**, 86 (2014). https://doi.org/10.1088/0004-637X/792/2/86
435. Yang, C.C., Johansen, A., Carrera, D.: Concentrating small particles in protoplanetary disks through the streaming instability. A&A **606**, A80 (2017). https://doi.org/10.1051/0004-6361/201630106
436. Youdin, A.N., Chiang, E.I.: Particle pileups and planetesimal formation. ApJ **601**, 1109–1119 (2004). https://doi.org/10.1086/379368
437. Youdin, A.N., Goodman, J.: Streaming instabilities in protoplanetary disks. ApJ **620**, 459–469 (2005). https://doi.org/10.1086/426895
438. Youdin, A.N., Lithwick, Y.: Particle stirring in turbulent gas disks: including orbital oscillations. Icarus **192**, 588–604 (2007). https://doi.org/10.1016/j.icarus.2007.07.012
439. Youdin, A.N., Shu, F.H.: Planetesimal formation by gravitational instability. ApJ **580**, 494–505 (2002). https://doi.org/10.1086/343109
440. Zhang, K., Blake, G.A., Bergin, E.A.: Evidence of fast pebble growth near condensation fronts in the HL Tau protoplanetary disk. ApJ **806**, L7 (2015). https://doi.org/10.1088/2041-8205/806/1/L7
441. Zhu, Z., Hartmann, L., Calvet, N., Hernandez, J., Muzerolle, J., Tannirkulam, A.K.: The hot inner disk of FU Orionis. ApJ **669**, 483–492 (2007). https://doi.org/10.1086/521345
442. Zhu, Z., Hartmann, L., Gammie, C.: Long-term evolution of protostellar and protoplanetary disks. II. Layered accretion with infall. ApJ **713**, 1143–1158 (2010). https://doi.org/10.1088/0004-637X/713/2/1143
443. Zhu, Z., Hartmann, L., Gammie, C., McKinney, J.C.: Two-dimensional simulations of FU Orionis disk outbursts. ApJ **701**, 620–634 (2009). https://doi.org/10.1088/0004-637X/701/1/620
444. Zhu, Z., Hartmann, L., Gammie, C.F., Book, L.G., Simon, J.B., Engelhard, E.: Long-term evolution of protostellar and protoplanetary disks. I. Outbursts. ApJ **713**, 1134–1142 (2010). https://doi.org/10.1088/0004-637X/713/2/1134
445. Zhu, Z., Nelson, R.P., Dong, R., Espaillat, C., Hartmann, L.: Dust filtration by planet-induced gap edges: implications for transitional disks. ApJ **755**, 6 (2012). https://doi.org/10.1088/0004-637X/755/1/6
446. Zhu, Z., Stone, J.M.: Dust trapping by vortices in transitional disks: evidence for non-ideal magnetohydrodynamic effects in protoplanetary disks. ApJ **795**, 53 (2014). https://doi.org/10.1088/0004-637X/795/1/53
447. Zhu, Z., Stone, J.M., Bai, X.N.: Dust transport in MRI turbulent disks: ideal and non-ideal MHD with ambipolar diffusion. ApJ **801**, 81 (2015). https://doi.org/10.1088/0004-637X/801/2/81

448. Zhu, Z., Stone, J.M., Rafikov, R.R., Bai, X.N.: Particle concentration at planet-induced gap edges and vortices. I. Inviscid three-dimensional hydro disks. ApJ **785**, 122 (2014). https://doi.org/10.1088/0004-637X/785/2/122
449. Zweibel, E.G.: Ambipolar diffusion. In: Astrophysics and Space Science Library, vol. 407, 285 (2015)

Chapter 2
Planet Formation and Disk-Planet Interactions

Wilhelm Kley

Abstract This review is based on lectures given at the 45th Saas-Fee Advanced Course "From Protoplanetary Disks to Planet Formation" held in March 2015 in Les Diablerets, Switzerland. Starting with an overview of the main characterictics of the Solar System and the extrasolar planets, we describe the planet formation process in terms of the sequential accretion scenario. First the growth processes of dust particles to planetesimals and subsequently to terrestrial planets or planetary cores are presented. This is followed by the formation process of the giant planets either by core accretion or gravitational instability. Finally, the dynamical evolution of the orbital elements as driven by disk-planet interaction and the overall evolution of multi-object systems is presented.

2.1 Introduction

The problem of the formation of the Earth and the Solar System has a very long tradition in the human scientific exploration, and has caught the attention of many philosophers and astronomers. Often it is referred to as *one of the most fundamental problems of science. Together with the origin of the Universe, galaxy formation, and the origin and evolution of life, it forms a crucial piece in understanding, were we, as a species, come from.* This statement was made in 1993 by J. Lissauer in his excellent review about the planet formation process [159], just before the discovery of the first extrasolar planet orbiting a solar type star. Today, as about 20 years have passed since the discovery of the first extrasolar planet orbiting a solar type star in 1995, the understanding of the origin of planets and planetary systems has indeed become a major focus of research in modern astrophysics.

Applying different observational strategies the number of confirmed detections of exoplanets has nearly reached 2000 as of today. While already the very first discoveries of hot Jupiter planets such as 51 Peg [182] and very eccentric planets,

W. Kley (✉)
Institute of Astronomy & Astrophysics, Universität Tübingen,
Morgenstelle 10, 72070 Tübingen, Germany
e-mail: wilhelm.kley@uni-tuebingen.de

M. Audard et al. (eds.), *From Protoplanetary Disks to Planet Formation*,
Saas-Fee Advanced Course 45, https://doi.org/10.1007/978-3-662-58687-7_2

such as 16 Cyg B [54], have hinted at the differences to our own Solar System, later the numerous detections by the Kepler Space Telescope and others have given us full insight as to the extraordinary diversity of the exoplanetary systems in our Milkyway. Planets come in very different masses and sizes and show interesting dynamics in their orbits. Full planetary systems with up to 7 planets have been found as well as planets in binary stars systems, making science fiction become a reality.

At the same time, it has become possible to study so called protoplanetary disks in unprecedented detail. These are flattened, disk-like structures that orbit young stars, as seen for example clearly in the famous silhouette disks in the Orion nebula, observed by the Hubble Space telescope, also in 1995 [183]. Being composed by a mixture of about 99% gas and 1% dust, these disks hold the reservoir of material from which planets may form. Indeed protoplanetary disks are considered to be the birthplaces of planets as anticipated already long time ago by Kant and Laplace [124, 145] in their thoughts about the origin of the Solar System. Following the close connection between planets and disks, particular structures (gaps, rings, and non-axisymmetries) observed in these disks are often connected to the possible presence of young protoplanets. The most famous recent example is the ALMA-observation of the disk around the star HL Tau that shows a systems of ring-like structures which may have been carved by a planetary system forming in this relatively young disk [4].

As it is well established that planets form in protoplanetary disks, many aspects of this formation process are still uncertain and depend on details of the gas disk structure and the embedded solid (dust) particles. At the same time the evolution of planets and planetary systems is driven by the evolving disk, and we can understand the exoplanet sample and its architecture only by studying both topics (disks and planets) simultaneously. In this lecture we will summarize the current understanding of the planet formation and evolution process, while aspects of the physics of disks have been presented in the chapter by P. Armitage in this volume. The presentation will focus more on the basic physical concepts while for the specific aspects we will refer to the recent review articles and other literature.

2.1.1 The Solar System

Any theory on planet formation has to start by analyzing the physical properties of the observed planetary systems. Here, we start out with a brief summary of the most relevant facts of the Solar System, with respect to its formation process, for a more detailed list see the review by J. Lissauer [159]. The Solar System is composed of 8 planets, 5 dwarf planets, probably thousands of minor bodies such as trans-Neptunian objects (TNOs), asteroids, and comets, and finally millions of small dust particles. The planets come in two basic flavors, terrestrial and giant planets. The first group (Mercury, Venus, Earth and Mars) are very compact, low mass planets that occupy the inner region of the Solar System, from 0.4 to 2.5 AU. Separated by the asteroid-belt the larger outer planets (Jupiter, Saturn, Uranus and Neptune) occupy the region from about 5–30 AU. Sometimes the giant planets are sub-divided into the *gas-giants* (Jupiter, Saturn) that have a mean density similar or even below that of water (1 g/cm^3) and are composed primarily of Hydrogen and Helium, and the *ice-giants* (Uranus, Neptune) with a mean density of about 1.5 g/cm^3. All giants are believed to have a solid core (rocks) in their centers, while the atmospheres of the ice-giants are much less massive than those of the gas-giants and contain more ices of water, ammonia and methane. The dwarf planets, asteroids and TNOs are primarily composed of solid material.

The most important *dynamical* property of the Solar System is its *flatness*. The maximum inclination (i.e. the angle of the planetary orbit with the ecliptic plane) a planet has is about 7° for Mercury, while all the other, larger planets have $i < 3.4°$. All planets orbit the Sun in the same direction (prograde), their angular momentum vectors are roughly aligned with that of the Sun, and their orbits are nearly circular. The giant planets have an eccentricity $e < 0.055$, only the smallest planets have a significant eccentricity, $e = 0.2$ for Mercury and in particular $e = 0.1$ for Mars, which allowed Johannes Kepler to infer the elliptic nature of the planetary orbits. The spin-axis of the planets is also approximately aligned with the orbital angular momentum, only Venus and Uranus (and Pluto, even though not a planet anymore) represent exceptions. From meteoric dating the age of the planets and asteroids in the Solar System has been estimated to be about 4.56 billion years, i.e. the planets have about the same age as the Sun itself which implies a coeval origin of the Solar System as a whole. The orbital spacings of the planets are such that their mutual separation increases with semi-major axis, in a way that they can be ordered into a geometric series, the Titius-Bode law. It has often been suggested that this orbital sequence must be a direct consequence of the formation process, which has become very doubtful after the discovery of the extrasolar planets and noticing the importance of physical processes like migration and scattering. Instead, the simple requirement of long-term stable planetary orbits implies a sort of geometric spacing between planets [111].

The prevalence of solid material suggests that the main formation process has started from the accumulation of small bodies via *sequential accretion* where bodies grow through a sequence of trillions of collisions from small dust particles to full fledged planet. Additionally, the flat structure of the Solar System indicates that this process has taken place within a protoplanetary disk, the *Solar Nebula*, that orbited the early Sun. These findings are supported by the observational fact that many protostars are surrounded by a flat disk consisting of gas and dust, with extensions similar to that of the Solar System. This nebular hypothesis of the Solar System's origin was already the basis of the formation theories of Kant and Laplace. While Kant focused on the evolution of the solid dust material in the Solar System [124], Laplace focused more on the hydrodynamical aspects [145]. Much later, von Weizsäcker has taken up these ideas to develop his hydrodynamical theory of planet formation in a disk containing several vortices [284]. An overview of these and many subsequent ideas are contained in [291] or [85]. A modern (pre-extrasolar planet) review of the formation of the Solar System from an astrophysicists perspective is given by [159], while the chronological aspect is emphasized by [193].

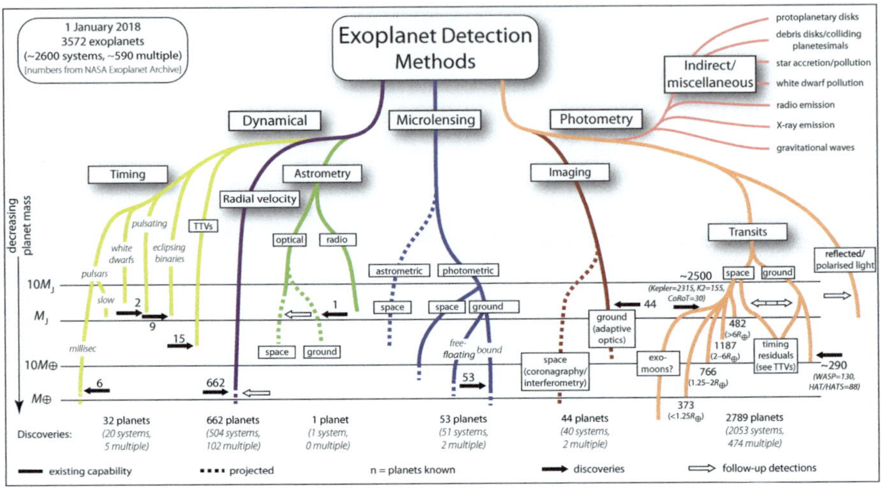

Fig. 2.1 Graphics to indicate the number of exoplanet detections using a particular detection method. *Courtesy* M. Perryman, in [219] © Cambridge University Press, reproduced with permission

2.1.2 Properties of the Extrasolar Planets

During the past 20 years numerous extrasolar planets orbiting Sun-like stars have been detected. Their physical and dynamical properties provide us with the opportunity to examine the generality of our ideas about the planet formation process, and modify those chapters of the whole story that are not compliant with the new observational data. We will not go in any detail into the detection methods that are used to discover new planets, an overview is given in [219], see also the lecture notes by A. Quirrenbach in the Saas-Fee Advanced Course 31 [228]. In Fig. 2.1 we display graphically the number of planets discovered using a particular detection method as of July, 2015. As shown, in absolute numbers the transit method has been by far the most successful method since over half of all detections have been achieved using transits, most of them with the Kepler space telescope. From the ground the radial velocity method is still the most successful. As shown, there are nearly 500 multiple systems. After the end of the regular Kepler-mission the rate of detections has somewhat slowed, but new space missions such as TESS [243] or PLATO [231] will surely enhance it again. A status of the actual numbers can be found in several online catalogues such as http://exoplanet.eu/, http://exoplanets.org, or http://www.openexoplanetcatalogue.com/.

The amount of data collected so far allows for extensive statistical analyses of the exoplanet properties, which serves in particular to understand the similarities, as well as the differences with respect to our Solar System. A very important diagram is shown in Fig. 2.2 where we display the mass of the planet against semi-major axis. Unfortunately, for the Kepler planets their mass is known only for a small fraction

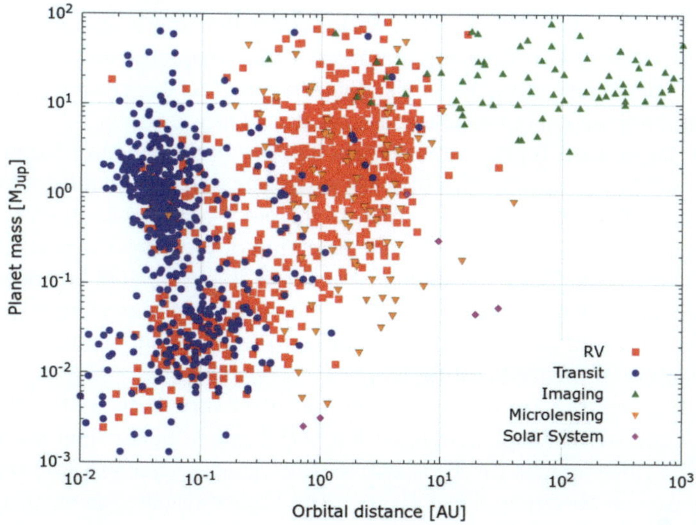

Fig. 2.2 Mass of the extrasolar planets with respect to the distance from their host star. Based on data from http://exoplanet.eu

of all discovered planets because the low apparent magnitude of the host stars does not allow for an easy radial velocity follow-up, but see [168], where this has been successful. Hence, the plotted data in Fig. 2.2 refer primarily to the radial velocity measurements. Here, one has to keep in mind that for this detection method the quoted masses refer to the minimum masses because of the unknown inclination to the systems. In the plot the different detection techniques have been indicated and it is clear that they are sensitive to different regions in the $M - a$ diagramme, but we will not discuss here any detection limits or biases. The planets populate different regions in the plot which have been used to separate the whole planet population into various planetary species. The top left corner, for semi-major axis smaller below about 0.1 AU and masses above 100 $M_\oplus$, is inhabited by the so-called *Hot Jupiters*, as these planets are comparable in mass to Jupiter in the Solar System but are located very close to the star such that the stellar irradiation leads to surface temperatures often well above 1000 K. Already the very first exoplanet discovered, 51 Peg [182], with a 4 day period and a separation from the Sun of about 1/20 AU, belongs to this category. It is very difficult to form such a planet at that location because the high disk temperatures do not allow for simple condensation of material. Hence, already this first discovery of an exoplanet required a modification to the standard planet formation theory by allowing some migration process [155]. Within the same mass range, or even a bit higher, up to 10 $M_{\rm Jup}$, and with a larger distance between 1 and 10 AU we find the *classical giant* planets, i.e. planets of Jupiter mass or higher within a similar distance from the star. Below these, there is a whole group of planets with masses between 2 and about 30 $M_\oplus$, and distances from 0.03 up to about 1 AU. These are the so-called *Super Earths*,[1] as they are located in the same distance regime as our home planet but with a somewhat larger mass. Even though not directly apparent in this limited sample plot, from the Kepler data it is clear that the smaller planets are by far the largest population of the detected exoplanets. Only less than 10% of all Kepler planets are larger than Jupiter while nearly half are in the range of Neptune, with (2–6) $R_\oplus$, and the rest smaller (for the most up to date statistics on the distribution of planetary radii of Kepler-planets we refer to the NASA-Homepage). Observational limitations do presently not allow for the detection of the very low mass and small objects in this regime, but from statistical analyses it is estimated that at least *11% of Sun-like stars harbor an Earth-size planet receiving between one and four times the stellar intensity as Earth* [220], a statement with strong implication with respect to the possibility of life in the Milkyway.

Concerning the shape of the orbits it became clear very early on that the average orbital eccentricity of exoplanets is with $\bar{e} \sim 0.3$ much higher than for the Solar System. This value applies for the radial velocity planets, which show a weak trend of an increase in e for larger planet masses. The Kepler systems show, on average, a smaller eccentricity and have lower planet masses. The main orbital and dynamical properties of exoplanets have recently been summarized in [290]. From a formation point of view it is noteworthy that the multi-planet systems show only a very small mutual inclinations between the planetary orbits, i.e. planetary systems are born flat. Interestingly, this applies to the circumbinary planets as well which indicates that the protoplanetary disks from which the planets formed were closely aligned with the central binary star.

2.1.3 Pathways to Planets

After having summarized briefly the observational aspects of planetary systems including our own, we will in the following sections outline the process of formation, based on the modern day view. Historically, the Greek philosophers have already theorized a long time ago about the origin of the Earth and possible other *Worlds*. In the 5th century BCE Leukippos suggested that *the worlds form in such a way, that the bodies sink into the empty space and connect to each other*, as translated in [107]. In a very broad sense this is still the view today, because with *World* the whole Solar System is implied, Sun and all planets. As we know today, stars form within a collapsing molecular cloud that turns into a highly flatted configuration due to angular momentum conservation. In the center the proto-sun forms and in the disk the planets. The problem we are facing is, that we need to illuminate a process that for the Solar System took place 4.5 billion years ago. Nevertheless, the observational data from the Solar System, the data from protoplanetary disks and last not least the large and growing sample of extrasolar planetary systems allow us to draw a coherent

[1]Even though some of the planets lie more in the size range of Neptune (4 $R_\oplus$) and are sometimes termed Sub-Neptunes, we refer in this text to the whole group as Super-Earths.

framework in which many details are left to be worked out but the main processes have probably been understood.

Two main pathways to make planets have been discussed in the literature. In the first scenario, planets are believed to have formed directly from the protoplanetary disk by gravitational instability (GI). In this *top-down* process spiral arms become gravitationally unstable and fragment directly to form large protoplanets. The advantage of such a process may be the possibly very fast formation timescale ($\sim$1000 years) but for typical protoplanetary disks the required fast cooling of the disk material is probably not satisfied expect possibly for very large distances (several 10s of AU) from the star. In addition, the Solar System contains a multitude of small solid objects and the giant planets have most likely massive solid cores in their centers. These problems of the GI-scenario, and the properties of the Solar System have led to the current view that the planet formation has occurred predominantly via a *bottom-up* process in which small dust particles have grown through millions and millions of sticking collisions to form eventually large protoplanets, terrestrial planets, and the cores of the giant planets. The giants collect by their strong gravitational force in a final step a large amount gas.

Hence, in the following we describe the planet formation process on the basis of the sequential accretion scenario, where we start out in Sect. 2.2 with the growth from small dust to km-sized planetesimals, study the formation of terrestrial planets in Sect. 2.3. The growth to massive planets via core-accretion will be described in Sect. 2.4 followed a description of the GI pathway in Sect. 2.5. An important part of the planet formation process is the occurrence of dynamical evolution due to disk-planet interaction. We will describe the most important results in Sect. 2.6 of this lecture followed by a study of the dynamical behaviour of multi-object planetary systems in Sect. 2.7.

The physical and dynamical properties of the Solar System indicate that the planets formed in a flat disk orbiting the young proto-sun. The information drawn from the solid bodies implies that the growth occurred via a bottom-up process where planets were formed in a sequential accretion process starting from tiny interstellar dust grains, all the way to full grown planets. The additional information drawn from the large sample of extrasolar planetary systems supports this basic scenario but requires to take into account dynamical effects in multi-planet systems and the gravitational interaction between the disk and the growing planets.

2.2 From Dust to Planetesimals

Protoplanetary disks typically consist of a mixture of about 99% gas and a small amount (~1%) of solid, dust particles [286]. As mentioned above in the introduction, in the Solar System and many of the observed extrasolar planetary systems, the formation of planets is believed to be accomplished primarily through a sequential growth process starting from small interstellar dust particles. Hence, to study this early planet formation phase we need two ingredients, an initial ensemble of dust particles and a suitable disk model.

Concerning the dust, the now classical measurements of interstellar extinction [179] have shown that interstellar dust grains have a size range of about 0.1–1.0 μm with a size distribution of $n(a) \propto a^{-3.5}$, where $n(a)$ denotes the number of particles with a given size a. New results indicate some deviation from this MRN-profile (after Mathis et al. [179]), but the typical expected initial sizes of interstellar dust particles is probably within the indicated size range [70]. Through a sequence of trillions and trillions of collisions, starting from these tiny dust grains full grown planets are eventually assembled. In this section we deal with the first phase of this process and consider the initial growth up to about km-sized planetesimals.

Concerning the disk structure, one often refers to simple disk models where the density and temperature distribution can be described by suitable power laws, in order to simplify the analyses. Starting from the observed locations and masses of the planets in the Solar System, Hayashi and Weidenschilling [110, 279] constructed a simple relation for the variation of the surface density of the material with distance from the Sun. To this purpose they spread out the observed mass of the planets in a number of radial bins and added the amount of gas to match the observed dust/gas ratio in observed protoplanetary disk, typically 1/100 [286]. This mass distribution is now referred to as the *Minimum Mass Solar Nebula* (MMSN) as it contains just the right total amount of matter and density slope to make the Solar System, as we observe it today. The resulting gas density distribution is then given by [110]

$$\Sigma_{\rm g}(r) \approx 1700 \left(\frac{r}{1\,\rm AU}\right)^{-3/2} \ {\rm g/cm^2}\,. \tag{2.1}$$

Such a model yields a typical total disk mass of about 0.01–0.05 $M_\odot$. Of course, initially the mass could have been substantially larger and the density distribution different, because planets tend to move (migrate) in disks and do not remain at their birth locations. Consequently, the simple distribution (2.1) has been criticized frequently and different slopes have been suggested [5, 64], but is nevertheless still often used, let it be only as a suitable reference model. Observed disks typically tend to have a flatter profile and show at some radius a cutoff, as seen also in time-dependent viscous disk models [163]. Such distributions can be described in parameterized form as an exponentially tapered power law

$$\Sigma(r) \propto \left(\frac{r}{R_c}\right)^{-\gamma} \exp\left[-\left(\frac{r}{R_c}\right)^{2-\gamma}\right]. \tag{2.2}$$

Here the exponent γ and the characteristic radius, R_c, can be chosen to match the observations. For γ mean values of about 0.9 have been determined while the value of R_c ranges from 20 to 200 AU [286]. For the temperature in the disk one can assume as a first approximation a passive disk, where internal heat is solely generated by the illumination of the central star. At larger radii (beyond a few AU) this is a good approximation because there the internally generated heat is lower than stellar irradiation. In this case the radial temperature of the disk is given by [110]

$$T(r) \approx 280 \left(\frac{r}{1\ \mathrm{AU}}\right)^{-1/2} \mathrm{K}. \tag{2.3}$$

This distribution assumes a vertically isothermal stratification, hence Eq. (2.3) describes the midplane as well as the surface temperature of the disk. Assuming now a constant temperature in the z-direction, the vertical hydrostatic equation can be solved for the density, and a Gaussian distribution is obtained, with

$$\rho(r, z) = \rho_0(r) \exp(-z^2/H^2). \tag{2.4}$$

Here $\rho_0(r)$ is the midplane density, $H(r) = c_s/\Omega_K$ the vertical scale height (often termed the *disk half-thickness*), $\Omega_K(r)$ is the Keplerian rotational velocity in the disk, and $c_s(r)$ the local sound speed. By integrating $\rho(r, z)$ over z one obtains the radial surface density distribution

$$\Sigma(r) = \int_{-\infty}^{\infty} \rho(r, z) dz. \tag{2.5}$$

By varying the exponent γ in Eq. (2.2) different disk models can be constructed, that can be used as a basis for analyzing the dust motion within the disk. More detailed analytical disk models including internal as well as external heating have been calculated by [53].

These disk models can be used as a basis to study the motion and growth of embedded dust particles. As shown in the accompanying chapter by P. Armitage in this volume the particles have a different velocity as the gas because they do not feel the effects of gas pressure. As a consequence they experience a drag force which is proportional to the velocity difference between the gas and dust particles and depends on the size of the particle and the gas density. This frictional force leads to a drift of the particles which is directed towards the pressure maximum of the gas. As can easily be inferred from the gas and temperature distribution the pressure, that is $\sim \rho T$, drops rapidly with radius in a protoplanetary nebula. Consequently, the particles experience a rapid inward radial drift which is for typical disk parameter about $v_{\mathrm{drift}} \sim 50$ m/s. This leads to a rapid loss of particles into the central star, and any successful growth process has to be sufficiently fast to overcome this *drift-barrier*. For more details on the motion of particles in gas disks see [278], or the review article by [296]. The overall growth phase from dust to planetesimals has been reviewed more recently in [52] or [119].

2.2.1 *Study the Initial Growth Phase*

Starting from the initially μm sized solid particles that are embedded in the accretion disk the growth process can only proceed through a sequence of collisions where two partners hit each other and stick together due to some adhesive forces. The outcome of these mutual collisions and the important sticking probability depend on the friction coefficients, the compactification of the material (i.e. how much kinetic energy can be dissipated upon collisions) and the relative velocity of the components in the collision process.

The actual growth of small particles has been studied in the laboratory and via numerical simulations, where for the latter the experimental results have been used to calibrate the simulations. On the experimental side the most complete set of studies have been performed using spherical, mono-disperse SiO_2 particles (silica) that have the additional advantage of an easy direct usage in numerical studies [35]. Although these particles seem to be rather special, experiments have indicated that the variations in material properties between silica and silicates (that are often detected in protoplanetary disks) do not seem to play a major role in comparison to morphological and size differences [226]. Hence, the usage of this material in the laboratory studies seems to be well justified. In the following we describe the basic results of the laboratory work and the numerical studies.

2.2.1.1 Experiments

The experimental studies have been performed under a variety of different laboratory setups, partly under vacuum and zero gravity conditions. For the latter the following facilities have been utilized: the fall-tower in Bremen, parabola flights as offered for example by ESA, and the international space station. The outcome of these series of experimental studies have been summarized exhaustively in the review article [35], and we will focus here on the main results.

For the very small μm sized particles the initial growth is very clearly fractal, i.e. the mass-size relation is given by

$$m \propto s^{D_\mathrm{f}} \tag{2.6}$$

with a fractal dimension D_f smaller than 3. Here s denotes the size of the particle (effective radius) with mass m. The fractal growth can be understood by the fact that small particles are well coupled to the gas and do show only very small relative velocities with respect to each other. For the experiments relative velocities $v_\mathrm{rel} \approx 10^{-4}$–$10^{-2}$ m/s between the individual collision partners have been used.

Hence, upon collisions the direct sticking of individual particles/aggregates dominates over restructuring effects, and the aggregates display a strongly elongated shape, with a fractal dimension typically such that $1.4 \le D_{\mathrm{f}} \le 1.8$. Images of the fractal agglomerates and their mass growth in time are shown in [36, 141].

The subsequent growth after the initial fractal phase has been studied through experiments using multiple collisions between the aggregates. In the experimental setup of [283] particles were enclosed in a plexiglass box whose floor could vibrate with a frequency of about 100 Hz. Using a high speed camera snapshots of the aggregates were taken and their mutual velocities and changes in mass/volume during collisions were measured. The observed typical collision velocities for the particles have been in the range of 0.1–0.3 m/s. An important quantity to describe the compactness of an agglomerate is the filling factor ϕ that gives the mass (density) ratio of a porous aggregate to a solid object of the same base material and same volume (i.e. ϕ is also referred to as the volume fraction of the material)

$$\phi = \rho_{\mathrm{agg}}/\rho_{\mathrm{mat}} \,. \tag{2.7}$$

Initially, the prepared 'dust-cakes' are extremely fluffy and have a small filling factor of $\phi \approx 15\%$ which refers to a mean density of only $\rho_{\mathrm{agg}} = 0.3$ g/cm^3 given that the solid matter density of SiO_2 is about 2.0 g/cm^3. Often the *porosity* is quoted which is in a sense complementary to ϕ and given by $1 - \phi$. The main outcome of these experiments (using mm sized particles) is a restructuring and compactification of the aggregates such that the fractal dimension D_{f} is increasing [283]. In terms of filling factor, starting from $\phi_0 = 0.15$ it increases with the number of collisions, n, according to

$$\phi(n) = \phi_{\mathrm{max}} - \Delta\phi \cdot e^{-n/\nu} \,, \tag{2.8}$$

where ϕ_{max} is the final value of the filling factor, $\Delta\phi = \phi_{\mathrm{max}} - \phi_0$, and ν a constant determined to about 700. The final filling factor reached is about $\phi_{\mathrm{max}} = 0.365$. This compaction of the aggregates changes the surface to mass ratio, which is crucial for the dynamical behaviour of the particles in the protosolar nebula [280].[2] At the same time this restructuring modifies the macroscopic material parameter that determine the tensile, shear and compressive strength of the material (see below), which is turn alters their behaviour in subsequent collisions.

In addition to the described results, a multitude of additional collision experiments have been performed and analyzed, see e.g. [35] for an overview. In [104] the physical outcomes are classified with respect to different collision channels corresponding to sticking, bouncing, fragmentation, or mass transfer from one aggregate to the other, see also [92] for an alternative classification scheme. A summary of the experimental results is presented in Fig. 2.3 where on the y-axis the mass of collision partners is plotted and on the x-axis the relative collision velocity. In all analyzed collisions the two colliding objects have the same size. The areas enclosed by the

[2]This is known very well from daily experience where a feather with the same weight as a small pebble falls much slower to the ground than the pebble due to the air resistance.

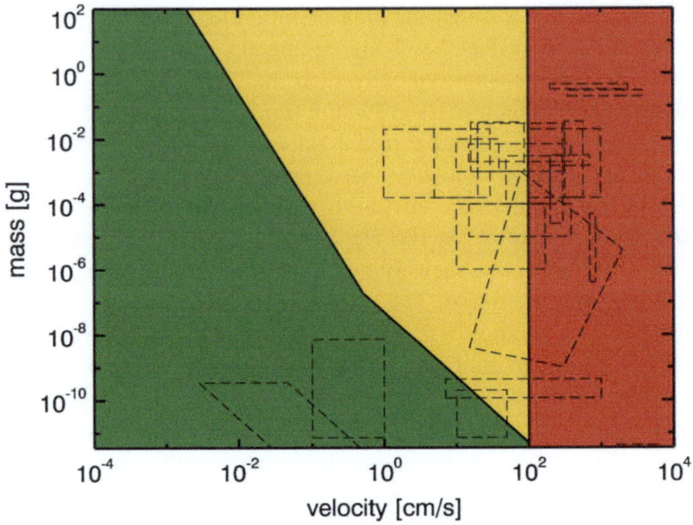

Fig. 2.3 The results of laboratory experiments with respect to the outcome of mutual collisions of equal sized dust aggregates. Green areas indicate sticking collisions, yellow bouncing ones, and red corresponds to fragmentation of the aggregates. The collision experiments correspond to the areas within the dashed lines. *Courtesy* Jürgen Blum

dashed lines have been covered by the experiments of the research groups in Braunschweig (around Jürgen Blum) and Duisburg (around Gerhard Wurm). The colored regions indicate different growth or destruction regimes. In the green area the two aggregates stick together (growth regime), in the yellow one they bounce off each other (neutral regime), and in the red areas they are shattered into pieces and fragment (destruction regime). For net growth, one clearly has to be in the green regime, i.e. sufficiently small relative velocities. The diagram makes it very clear that there are different obstacles to the successful growth from small sized grains to planetesimals, these are basically the *bouncing barrier* [299] (the transition from green to yellow) and the *fragmentation barrier* (the transition from yellow to red). Obviously, above $v_{rel} \approx 1$ m/s growth is extremely difficult to achieve. Together with the *drift barrier*, mentioned above, these obstacles to successful growth are sometimes referred to as the *meter-sized barrier* to planetesimal growth [35], as it is difficult to achieve growth beyond 1 m in size. We will comment on possible pathways to overcome these obstacles to further growth in Sect. 2.2.2 below.

2.2.1.2 Numerical Studies

A very useful alternative to real laboratory experiments is performing *numerical experiments* and study collisions in the computer using appropriate models to simulate reality. In the case of planetesimal growth two approaches have proven to be

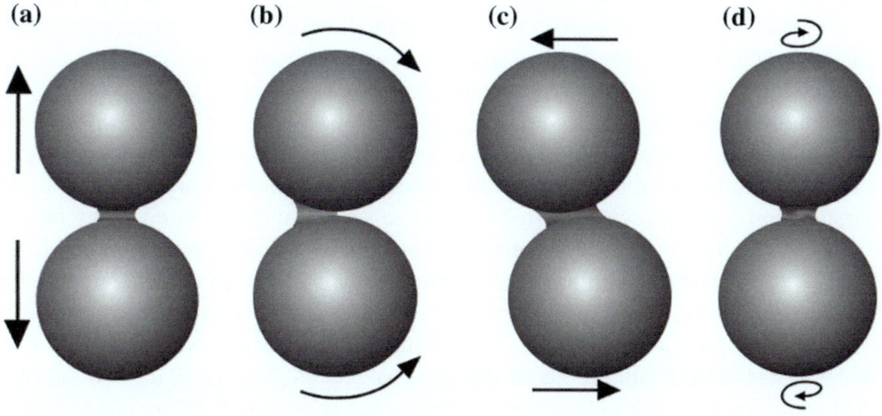

Fig. 2.4 Schematic view of the different type of interactions between two individual spherical monomers. These correspond to **a** compression/adhesion, **b** rolling, **c** sliding, and **d** twisting. *Courtesy* Alexander Seizinger. Taken from Fig. 1 in [251], © ESO, reproduced with permission

useful. In the first approach the behaviour of aggregate collisions is simulated on a microscopic level using a direct molecular dynamics (MD) approach with suitable forces between the individual μm-sized monomers that the whole aggregate is composed off. In a second approach, suitable for larger particles, a continuum model is constructed using averaged material parameters (such a elasticity, strengths, sound speed, etc.). We will discuss both approaches in this section.

The MD approach relies on modeling in detail the microscopic interaction between two individual monomers, using nanoscale molecular forces, such as the van der Waals force or other electrostatic forces. In the case of dust aggregates the forces can be divided into 4 categories as depicted in Fig. 2.4, see [271] and references therein. The formulation of the forces is based on a microscopic approach, and the implementation and application to the physics of dust coagulation under astrophysical conditions has been described in [68]. For the standard normal relative motions between individual monomers (panel **a** in Fig. 2.4) the force is based on the JKR-theory (after Johnson et al. [121]), that constitutes an improvement to the original model which is due to to Hertz [114]. The rolling (**b**), sliding (**c**) and twisting (**d**) frictional forces have been calculated by Dominik and Tielens, for example in [67], see summary in [68]. As hinted in Fig. 2.4 by the 'necks' between the two spheres, the intermolecular forces show a hysteretic behaviour. The forces between two monomers do not only depend on the actual distance but on the past history. The forces between approaching particles, that have not been in contact yet, is different from the force (at the same separation) if they just had been in contact. The way to imagine this is to assume that the particles are surrounded by a thin layer of honey or glue. Before coming into contact there is no attractive fore, but after separation a neck is formed that pulls the particles back together. The physical model by Dominik and Tielens does take account for these hysteretic, dissipative effects.

Later the model was improved by deriving the forces (torques) from appropriate potentials which allows for detailed tracking of the different energy channels during the collisions [271]. Before applying the numerical model to physical collisions several numerical parameter have to be adjusted properly. This is done by comparing the results of numerical simulations directly with laboratory experiments. In these well constructed experiments a small dust aggregate (*dust-cake*) is prepared by simple ballistic aggregation procedure which leads to a sample with an initial filling factor, $\phi = 0.15$. The dust-cake is then contained between two walls and compressed from above [105]. Upon the compression the filling factor ϕ increases and the pressure on the adjacent walls as well. The important quantity to calibrate is the pressure-porosity relation, that shows a characteristic behaviour, which depends on the geometry of the particular setup. Calibrations of this type have been done by [215] and later by [251] who found for example a modification of the rolling and sliding coefficients.

After a successful calibration process the numerical simulations allow to determine continuum properties of the aggregate, such as sound speed or the shear modulus. These are sometimes difficult or impossible to determine experimentally. In numerical simulations special treatments, such as adding a sort of glue between the particles and the wall can be introduced to make these measurements numerically possible. Having successfully determined these parameter a new type of simulations can be performed where the aggregates are treated as a smooth continuum average over the local details. In the astrophysical community a very important, successful and widespread method is *Smooth Particle Hydrodynamics* (SPH). Here, the continuum is modeled again by individual 'particles' but now they do not represent individual monomers or real physical particles but they can rather be considered as a sort of Lagrangian nodes that move in space and time. Quantities such as density or pressure are obtained from a suitable averaging procedure over neighboring particles. The method was initially developed for simulating purely hydrodynamic flow (see review by Monaghan [192]) but was later extended to model the dynamics of solid objects, where time dependent equations to follow the stresses within the body are added and solved simultaneously. In addition, the method has been augmented by an elasto-plastic model, and special treatments for handling cracks and fragmentation have been implemented [26]. Hence, these type of simulations allow to model the collisions of larger objects, starting from dm-size up to very large objects where internal gravity may play a role. Indeed, using the SPH method collisions between objects that range in size from m-sized bodies up to the Moon forming impact between the proto-Earth and a Mars-sized object have been simulated [27, 28].

Similar to the collisions between tiny particles those between macroscopic objects will lead to a fragmentation and destruction of the (larger) target if the relative collision speed becomes too large. To analyse the outcome of these object-object 'encounters' the specific collision energy between target and projectile is a useful quantity. It is defined as

$$Q_{\rm D} = \frac{1}{2} \frac{m_{\rm proj} v_{\rm rel}^2}{M_{\rm target} + m_{\rm proj}}, \tag{2.9}$$

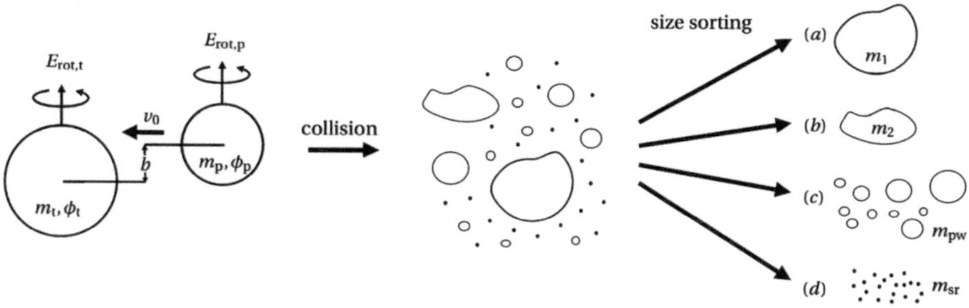

Fig. 2.5 Schematic representation of the outcome of a collision between two objects. The left side displays the situation before the collision: the target (t) and projectile (p) collide with impact velocity v_0, and impact parameter b. Before the collision they are described by their mass (m), filling factor (ϕ) and rotational energy ($E_{\rm rot}$). The outcome consists of different groups: **a** the largest fragment; **b** the second largest fragment; **c** the power law population; and **d** the sub-resolution population. This classification scheme is based on the four-population model by [92]. Taken from Fig. 2 in [92], © ESO, reproduced with permission

where $m_{\rm proj}$ is the mass of the projectile, $M_{\rm target}$ the mass of the target and $v_{\rm rel}$ the relative velocity between the two objects.

The general outcome of collisions is well described in Fig. 2.5. It consists typically of a limited number of bigger objects (here cases **a**, **b** on the right hand side), a large number of smaller particles (**c**) that follow approximately a power law size distribution [152], and a 'sea' of very small particles (**d**) that are too small to be resolved (at least numerically). The four-population model of [92] assumes that there are only two major objects after the collision. The catastrophic destruction threshold, $Q^*_{\rm D}$, is now defined as that specific collision energy at which the largest remaining fragment has 1/2 of the target mass. As a function of target size $Q^*_{\rm D}$ has a V-shaped behaviour, in the strength dominated regime (for target radii smaller than about 1 km) $Q^*_{\rm D}$ is decreasing with increasing target size. For larger objects gravitational re-accumulation becomes important and the mass of the largest post-collision object increases, hence $Q^*_{\rm D}$ increases again with target size in this gravity dominated regime. The minimum occurs at the transition between the two regimes. Using the SPH-method numerical experiments of collisions between two basalt spheres with radii between 100 m and 10 km using different relative velocities have been performed [27]. The results show that for a typical velocity, $v_{\rm rel} = 20$ km/s, the weakest bodies (with minimum $Q^*_{\rm D}$) are those with radii of about 300 m, here $Q^*_{\rm D}$ has values of about 10^2 (J/kg). Later [152] constructed fit formulae to calculate $Q^*_{\rm D}$ as a function of target radii which can be used in statistical simulations for an ensemble of objects (e.g. asteroids or Kuiper belt objects) to follow their time evolution [248].

2.2.2 How to Overcome Growth Barriers

As sketched out for example in Fig. 2.3, from laboratory experiments it is known that the mutual collisions of mm to cm-sized aggregates result frequently in bouncing, e.g. [112, 118, 282, 283]. Incorporating the data from the various experimental setups, [104] constructed an algorithm to to describe analytically the results of mutual collisions of aggregates as a function of their relative collision speed, their mass and their porosity. This model was then used to simulate the evolution of a swarm of dust aggregates in a protoplanetary disk in a statistical manner [299]. Initially the aggregates grow by the above described process, but for larger sizes the relative velocities increase. Due to the higher kinetic impact energy the aggregates become more and more compacted during successive collisions. If the aggregates get too compact, i.e. the filling factor reaches values up to about $\phi \sim 0.4$, then mutual collisions do not result in sticking anymore, rather they bounce off each other and the growth process is terminated [104]. In agreement with the experiments this occurs within a size regime of centimeters to decimeters, and was termed the *bouncing barrier* [299].

In early numerical studies to understand the origin of the bouncing it was found that for the aggregate parameters used in the experiments typically sticking occurs, and bouncing only for much larger filling factors above $\phi = 0.5$ [272]. In subsequent numerical simulations the condition for rebound (bouncing) has been studied in more detail by Wada et al. [273] who showed that it depends on the *coordination number*, n_c, of the individual monomers, which describes the number of contact points (neighbors) an individual monomer has. Clearly, the more contacts an aggregate has the stiffer it reacts to collisions promoting bouncing rather than sticking, and a value of $n_c = 6.0$ above which bouncing occurs has been found in the simulations [273]. Obviously, the higher the filling factor the larger the coordination number has to be. However, as shown recently there is not a one-to-one relation between ϕ and n_c [250]. The number of contacts that a specific aggregate with a given porosity has, depends on the process by which it has been constructed. In laboratory experiments using aggregates composed of micron-sized dust grains, it is usually only possible to determine the global filling factor (via the mean density) but not the local coordination number which is a microscopic quantity. Thus, one has to be very careful when comparing results from numerical simulations for a given filling factor directly to results from laboratory experiments that use the same ϕ.

Even though the numerical studies are still not in full agreement with the experiments in the bouncing regime, there are indications that non head-on collisions lead to increased bouncing [250]. Additionally, it was shown that for larger aggregate sizes the fragmentation velocity becomes higher, about 10 m/s. Despite these purely geometric characteristics of the collision the sticking probability could be enhanced by special material properties, such as sticky organic materials, magnetic or charged particles. However, as discussed in [35] the effects probably do not change the growth efficiency significantly. Another option is the aerodynamic re-accretion as suggested by [293]. Here, the idea is that after a nearly destructive collision the small fragments that surround the largest fragment fall back on it due to aerodynamic drag exerted on them. By this mechanism it is possible that the majority of particles become re-accreted leading to a net growth. The process depends on the properties of the growing body in particular its porosity with respect to the gas flow through it and on the speed of the ejecta, its importance for a successful growth is not fully settled [35].

In a process related to aerodynamic re-accretion the bouncing barrier can be overcome by a sweep-up process. Using the standard results from the collision experiments in a coagulation code one finds for the general dust population that bouncing collisions prevent any growth above millimeter-sizes. However, adding in the models a few cm-sized particles to a sea of smaller ones, which could happen in a real disk for example by vertical mixing or radial drift, these can act as a catalyst by starting to sweep up the smaller particles, which results in very rapid growth. As shown in [287] using this mechanism, 100-m-sized bodies can be formed on a timescale of 1 Myr at a distance of 3 AU from the central star. In this process the existence of the bouncing barrier is highly beneficial for promoting growth, as it prevents the formation of too many larger particles that would otherwise destroy each other via fragmenting collisions. Hence, a reservoir of small particles is maintained that can be swept up by larger bodies. The single requirement for this greatly enhanced growth process is the creation of a few *lucky* particles of cm-size or larger that can then sweep-up the smaller ones [287].

A possibility to overcome the fragmentation barrier is through collisions of particle with very different sizes. As shown in some experimental studies [294], collisions between millimeter-sized dust projectiles and centimeter-sized dust targets lead to net mass growth of the target up to collision velocities of 25 m/s. The authors suggest that for even higher velocities growth can be achieved which supports the idea that planetesimal formation via collisional growth is a viable mechanism at higher impact velocities. Following up on this this idea of high velocities and different collision partners, [288] and [91] demonstrated that it is the combination of a statistical relative velocity distribution function in the coagulation models and high-mass-ratio collisions that allows for successful growth of larger bodies. The velocity distribution allows to overcome the bouncing barrier, and the different masses to cross the fragmentation barrier [294]. Garaud et al. [91] even suggest that via this mechanism the problem of planetesimal formation close to the central star, the presence of millimeter- to centimeter-sized particles far out in the disk, and the persistence of μ-sized grains for millions of years, can be solved.

In the outer regions of the protoplanetary-disk the growth of particles is made easier by two effects. First, water has condensed to ice beyond the so-called *snow line* at temperatures below about 170 K, which increases the local surface density of solids by about a factor up to 4 [110]. In addition to the enhanced surface density of the solid particles the icy particles stick together up to much higher relative velocities of about 50 km/s [273]. The higher density and improved stickiness clearly will enhance the growth of small bodies. For disk parameter according to the MMSN the snow line lies around 2.7 AU in good agreement with the location of the planets in the Solar System [110]. In addition, the sudden change in the opacity of the disk at the snow line can lead to a reduction in turbulent activity in the disk and subsequently to a pressure maximum of the gas, which in turn will further increase the local surface density of solid particles because they drift towards the local pressure maxima [278]. Hence, the inward drift of particles is prevented at the snow line and the growth of planetesimals enhanced [142]. Finally, considering the low catastrophic destruction threshold of 10 m sized objects made of basalt, recent SPH simulations have indicated that the inclusion of a porous equation of state allows for non-fragmenting collisions to higher velocities [92].

2.2.3 *Dust Concentration*

As mentioned above there are several new suggestions available now how to overcome the growth barriers on the way from dust to planetesimals. The suggested mechanism rely on a better understanding of individual collisions and a statistical treatment of a whole ensemble of growing objects. Another, often discussed mechanism to cross the growth barriers is the pre-concentration of particles in inhomogeneities of the protoplanetary disk. Often discussed has been the collection of particles in turbulent eddies or in vortices. When the dust concentration has reached a critical value a streaming instability can set in [297], and gravity can lead to rapid growth to larger objects [120]. Similarly, gravitational instability in the condensed dust layer in the midplane of the disk dust layer could enhance planetesimal growth [97]. These mechanisms have been presented in detail in the contribution by P. Armitage and will not be discussed here further.

> The initial growth from dust to planetesimals suffers a growth crisis when particles reach a typical size of about decimeter to meter. In this regime, the particles experience the fastest radial drift and possible loss into the star and collisions often lead either to destruction or bouncing and no net growth. While this obstacle is presently not fully mastered, several possible new paths have been investigated recently to solve the problem. These include a better understanding of individual collisions, improvements in the statistical treatment as well as large scale collective effects by interaction with the ambient disk.

2.3 Terrestrial Planet Formation

In this chapter we describe the growth from planetesimals to full terrestrial type planets. Following [43] we define a planetesimal to be an object large enough that the motion of nearby particles are significantly perturbed through gravitational interaction. This implies that the velocity perturbations induced by planetesimals are larger than the typical drift velocity of particles which is about 10 m/s. Hence, the term planetesimal refers then to objects of a km in size or larger, while below this size the term pre-planetesimal is sometimes used [92]. As we shall see, this growth from km-sized planetesimals to a full grown terrestrial planet several 1000 km in size proceeds in two major steps that are quite different in duration. In the first phase the planetesimals grow in a fast, runaway process to produce a relatively small number of Moon to Mars sized *planetary embryos* (also called *protoplanets*). After this phase, the inner parts of the protoplanetary disk contains very little amounts of gas and the embryos grow via a much longer collisional phase to a set of terrestrial type planets. Excellent overviews on this era in the history of the planet formation process have been presented elsewhere [47, 194].

2.3.1 Concepts

Before going into the actual growth scenario from planetesimals to protoplanets we explain a few important physical concepts or processes that are of relevance with respect to this phase in the planet formation process. The problem we are facing when growing in size from few km-sized planetesimals to Moon-sized planetary embryos, is the fact that for this mass range the aerodynamic drag forces become negligible and cannot serve anymore to provide for any change in the relative velocity between particles necessary to have collisions. Additionally, the induced inhomogeneities in the disk by the growing object are not strong enough yet to provide significant drag and subsequent migration. Finally, with initial 1 km-sized planetesimals over 10^{11} particles are required to make an embryo. This is numerically very demanding as it requires lots of particles and very long evolution times (a few 10 million years), see [43]. Consequently, this part of planetary growth is described best by a combination of statistical and numerical methods.

2.3.1.1 Gravitational Focusing

The solid bodies can only growth via physical collisions, i.e. the shortest approach of two objects must be smaller than the sum of their radii. Here, we assume that the growing planetesimal has a radius $R_{\rm p}$ and that the accreted particle with radius R is much smaller than this, $R \ll R_{\rm p}$. The geometrical cross section for such a collision is then given simply by $\sigma_0 = \pi R_{\rm p}^2$. If there were no other means of increasing σ_0 then planetary growth would finish soon. Fortunately, when the objects become a few km in size the situation becomes much more favorable because the gravitational interaction between the two bodies will start to play an important role in bringing the particles closer together, a process called *gravitational focusing*. Let us consider the situation of two objects that have masses $m_{\rm p}$ (the large body) and m (the small body to be accreted) which have initially the relative velocity $v_{\rm rel}$ with the impact parameter b, as shown in Fig. 2.6. To have a physical collision between the two objects the

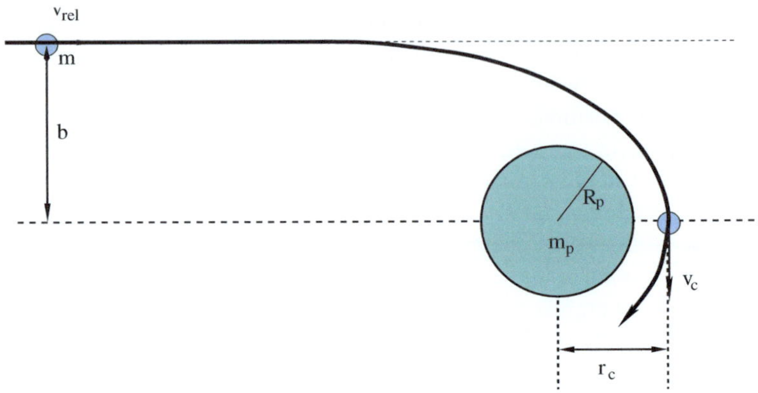

Fig. 2.6 Principle of the gravitational focusing process. A small body with mass m approaches a larger planetesimal of mass m_p with impact parameter b and relative velocity v_rel. Gravitational attraction results in a close encounter with smallest distance r_c, and velocity v_c at the time of closest approach

distance of closest approach, r_c, must be smaller than R_p, i.e. $r_\mathrm{c} < R_\mathrm{p}$. As the sketch in Fig. 2.6 implies the number of particles that possibly can directly hit the growing planetesimal is greatly enhanced over the purely geometrical cross section σ_0. The enhancement of the cross section through gravitational focusing can be obtained from the conservation of angular momentum and energy, by considering the initial state when the objects are very far away from each other and the situation of closest approach. We neglect the presence of the very distant central star and consider only these two objects. To have a physical collision we require that $r_\mathrm{c} = R_\mathrm{p}$, and angular momentum conservation yields

$$b\, v_\mathrm{rel} = R_\mathrm{p}\, v_\mathrm{c}\,, \tag{2.10}$$

where v_c is the velocity at the closest approach (Fig. 2.6). Energy conservation (in the center of mass frame) gives

$$\frac{1}{2}\mu\, v_\mathrm{rel}^2 = \frac{1}{2}\mu\, v_\mathrm{c}^2 - \frac{G(\mu M_\mathrm{tot})}{R_\mathrm{p}}\,, \tag{2.11}$$

with the reduced mass $\mu = m m_\mathrm{p}/(m + m_\mathrm{p})$ and total mass $M_\mathrm{tot} = m + m_\mathrm{p}$. Here we assumed that initially the two particles are sufficiently far apart such hat we can neglect the potential energy on the left side of Eq. (2.11), and on the right-hand side we evaluated the energy again at the point of closest approach. Combining these two equations results in the following expression for the *effective cross section*, σ, of the interaction

$$\sigma \equiv \pi b^2 = \pi R_\mathrm{p}^2\, F_\mathrm{grav} = \sigma_0 \left[1 + \left(\frac{v_\mathrm{esc}}{v_\mathrm{rel}}\right)^2\right], \tag{2.12}$$

where we introduced the joint *escape velocity*

$$v_{\rm esc} = \left(\frac{2GM_{\rm tot}}{R_{\rm p}}\right)^{1/2}, \qquad (2.13)$$

and the gravitative *focussing factor* over the purely geometrical cross section

$$F_{\rm grav} = \left[1 + \left(\frac{v_{\rm esc}}{v_{\rm rel}}\right)^2\right]. \qquad (2.14)$$

From Eq. (2.12) and Fig. 2.6 we can see that for large relative velocities only particles that arrive directly from the front of the object will collide with it, i.e. the effective cross section is just the geometric one. On the other hand for small relative speed, i.e. in a cold disk of planetesimals with $v_{\rm rel} \ll v_{\rm esc}$, particles feel their mutual gravitational attraction for a longer time and hence particles from further away will be the drawn in. The effective cross section is then *much* higher than without gravity, i.e. $\sigma \gg \sigma_0$ or $F_{\rm grav} \gg 1$. The quantity $\theta = (v_{\rm esc}/v_{\rm rel})^2$ in Eqs. (2.12) or (2.14) is sometimes called the *Safronov number* after an early pioneer in studying the formation of the Solar System [245]. If the approaching body is of size similar to the accreting planetesimal, then we can set $R_{\rm p} \to R_{\rm p} + R$ in the above formula for $v_{\rm esc}$.

2.3.1.2 The Hill-Radius

Another important concept in this context is the definition of the Hill-radius. Here one considers the motion of a small (massless) particle in the presence of two larger gravitating bodies, that orbit each other on circular Keplerian motion. This is the so-called circular *restricted three-body problem*. In today's Solar System the primary (largest) object is the Sun, the secondary for example Jupiter and the third massless particle a small asteroid. In our planet formation context the following objects are considered: a protostar with mass M_*, the growing planetesimal (mass $m_{\rm p}$) and a small particle (much smaller mass) to be accreted onto the planetesimal. As the motion of the star and the planetesimal is not affected by the small particle, these orbit each other with Keplerian motion, i.e. with an orbital speed

$$\Omega_{\rm p} = \sqrt{\frac{G(M_* + m_{\rm p})}{a_{\rm p}^3}} \approx \sqrt{\frac{GM_*}{a_{\rm p}^3}}, \qquad (2.15)$$

where $a_{\rm p}$ is the semi-major axis of the planetesimal.

The type of motion of a massless (low mass) particle in the presence the other two can be most easily derived from the effective gravitational potential written in a coordinate frame that rotates with the orbital speed, $\Omega_{\rm p}$, of the growing planetesimal see [198]. In this corotating frame it reads

$$\Phi(\mathbf{r}) = -\frac{GM_*}{|\mathbf{r} - \mathbf{r}_*|} - \frac{Gm_\mathrm{p}}{|\mathbf{r} - \mathbf{r}_\mathrm{p}|} - \frac{1}{2}\,\Omega_\mathrm{p}^2\,r^2\,, \tag{2.16}$$

where $\mathbf{r}_*$ and $\mathbf{r}_\mathrm{p}$ denote the positions of the star and planetesimal, respectively. The first two terms of the right hand side refer to the individual potentials of the star and planetesimal while the last term denotes the *centrifugal potential* as we consider the motion in the corotating frame. The equations of motion of the massless particle follow, as usual, from the gradient of the potential. The extrema of this potential are given as the roots of a 5th order polynomial, and one obtains the classical 5 Lagrange points, L_1 to L_5, as depicted in Fig. 2.7 that shows selected equipotential lines for a secondary to primary mass ratio of $m_1/M_* = 0.01$. The Lagrange points mark possible equilibrium positions of the particle but only two of those are stable, L_4 and L_5, which lie 60° in front and behind the secondary. Typically one finds that the orbits are around the primary object, either inside the distance of the secondary (e.g. the main belt asteroids in the Solar System), or outside of it as indicated by the inner and outer yellow lines in Fig. 2.7. For particles having the same semi-major axis as the secondary (here the planetesimal) the motion is either around one of the stable Lagrange points (L_4 or L_5) in so-called tadpole (dark blue and purple lines), or around both in horse shoe (light blue line) orbits [198]. The prime example from the Solar System are the Trojan asteroids that have the same semi-major axis and hence period as Jupiter and orbit around L_4 and/or L_5. For stability reasons these particles come never too close to the secondary object as can be seen by the equipotential lines in Fig. 2.7. If the object is in the very close vicinity of the secondary it is physically

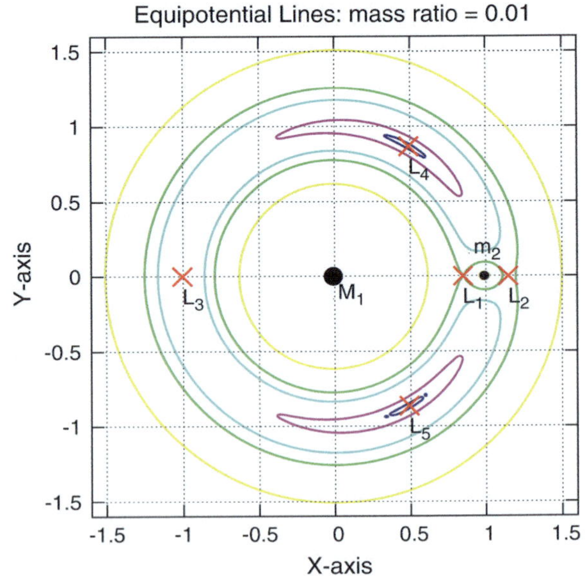

Fig. 2.7 The dynamical structure of the restricted circular three-body problem. Shown are equipotential lines of the corotating potential (Eq. 2.16) of a central star (M_1) which is orbited by secondary object (m_2). The classical 5 Lagrange points, L_1 to L_5 (marked by the red crosses), correspond to extrema of the potential

bound to it and orbits the secondary (e.g. the Galilean moons orbiting Jupiter). For our planetesimal this is the case for particle distances within a sphere of radius

$$R_{\rm H} = \left(\frac{m_{\rm p}}{3M_*} \right)^{1/3} a_{\rm p} \tag{2.17}$$

from the planetesimal. Here the index p refers to the planetesimal. This sphere is enclosed between the Lagrange point L_1 and L_2 of the corotating potential, see Fig. 2.7. Using direct 3-body simulations of a star, a planetesimal and small particles it was shown [100] that the motion of the particle to be accreted onto the planetesimal is highly chaotic in the vicinity of the planetesimal, in particular inside of the Hill-radius $R_{\rm H}$. These 3-body effects lead eventually to a limitation of the gravitational focusing factor $F_{\rm grav}$, that would otherwise diverge for small $v_{\rm ref}$, to maximum values of about 10^4 [159].

2.3.1.3 Modes of Growth

The accretion of small particles leads to an increase in mass of the growing planetesimal. In general one can distinguish two modes of mass growth, ordered or runaway, as is schematically displayed in Fig. 2.8. In the ordered growth phase all objects grow at roughly the same rate and all the planetesimals in the whole ensemble have approximately the same size, hence this mode is sometimes called *oligarchic* growth [136]. On the other hand in the runaway case, one (or few) objects grow very rapidly at the expense of the smaller ones. What type of growth mode operates can be analysed for example by considering the relative growth of two particles with mass m_1 and m_2 which can be obtained by expanding the time derivative of the mass ratio

$$\frac{d}{dt}\left(\frac{m_1}{m_2}\right) = \frac{m_1}{m_2}\left(\frac{1}{m_1}\frac{dm_1}{dt} - \frac{1}{m_2}\frac{dm_2}{dt}\right) . \tag{2.18}$$

From Eq. (2.18) we can infer that the *relative mass growth*, $1/m(dm/dt)$, for each object is important. Let us assume we start out with $m_1 > m_2$, then we see that this ratio is increasing if the relative growth increases with m, because then the right-hand side in (2.18) is positive. This is exactly the situation for runaway-growth. On the other hand, if relative growth decreases with m we have an ordered growth because then the mass ratio m_1/m_2 tends to unity. As we shall see in the following both growth modes occur during the assembly of protoplanets, the first growth is via a runaway process which is followed later by an orderly growth.

Using the cross section from Eq. (2.12) we find for the mass growth of a planetesimal with mass $m_{\rm p}$

$$\frac{dm_{\rm p}}{dt} = \rho_{\rm part}\, v_{\rm rel}\, \sigma = \rho_{\rm part}\, v_{\rm rel}\, \pi R_{\rm p}^2\, F_{\rm grav} \tag{2.19}$$

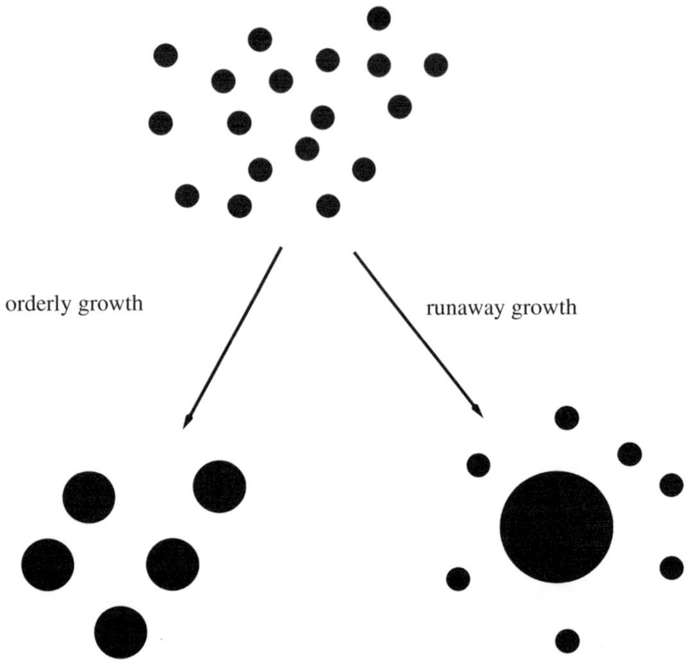

Fig. 2.8 The two modes of mass growth. In the case of orderly growth the whole ensemble has always similar particle sizes, while in the case of runaway growth one particle grows rapidly in a swarm of smaller ones. Adapted from Fig. 1 in [135]

where $\rho_{\rm part}$ is the density of the incoming particle stream. In deriving Eq. (2.19) we have assumed that each collision will results in growth (100% sticking efficiency). The outcome of collisions of km-sized objects can obviously not be studied in the laboratory (where the maximum size is a fraction of a meter) and one has to rely on numerical simulations. Here, the results indicate that typically the collisions lead to net accretion [27, 151] unless the relative speeds are very high or the collisions are only grazing, but details depend on the internal strength of the colliding objects [257].

Before we evaluate (2.19) for specific particle densities let us look at two illustrative examples (see [6]) that illustrate the different growth modes. Assuming a constant focusing factor, $F_{\rm grav} = const.$, we have for the mass growth of the planetesimal

$$\frac{1}{m_{\rm p}} \frac{dm_{\rm p}}{dt} \propto m_{\rm p}^{-1/3} . \tag{2.20}$$

For objects with approximately constant density during the growth the mass scales as $m_{\rm p} \propto R_{\rm p}^3$, and substituting this relation into Eq. (2.20) one finds $\dot{R}_{\rm p} = const.$, which implies a linear growth of the particle with radius, $R_{\rm p} \propto t$. Assuming now constant relative velocities between growing planetesimal and incoming particles,

$v_{\rm rel} = const.$ in Eq. (2.19), and using the definition for the escape velocity then one obtains

$$\frac{1}{m_{\rm p}} \frac{dm_{\rm p}}{dt} \propto R_{\rm p} \propto m_{\rm p}^{1/3}\,, \tag{2.21}$$

which implies a growth of the particle to infinite mass, $m_{\rm p} \to \infty$, in a finite time, corresponding to strong runaway. Of course, before this happens, the dynamics and space density of the ambient swarm of planetesimals will be changed which leads to a modification of relation (2.21).

To obtain estimates of the actual growthrates of planetesimals within the protoplanetary disk we assume that the incoming particle density is given by

$$\rho_{\rm part} \approx \frac{\Sigma_{\rm part}}{2H_{\rm part}} = \frac{\Sigma_{\rm part}\,\Omega_{\rm K}}{2v_{\rm rel}}\,.$$

Here $\Sigma_{\rm part}$ is the surface density of the particles, obtained by vertical integration over $\rho_{\rm part}$. To obtain the vertical thickness of the particle layer, $H_{\rm part}$, we assume that it is comparable to the thickness of the gas density in the accretion disk, i.e. $H = c_{\rm s}/\Omega_{\rm K}$, where $c_{\rm s}$ is the local sound speed and $\Omega_{\rm K}$ the Keplerian rotational angular velocity (see Chapter by P. Armitage). In the case of the particle disk, we replace $c_{\rm s}$ by the 'velocity dispersion', which is given here by the relative velocity $v_{\rm rel}$. Using this in Eq. (2.19) we obtain for the mass growth

$$\frac{dm_{\rm p}}{dt} = \frac{1}{2}\Sigma_{\rm part}\,\Omega_{\rm K}\,\pi R_{\rm p}^2 \left[1 + \left(\frac{v_{\rm esc}}{v_{\rm rel}}\right)^2\right]. \tag{2.22}$$

As can be noticed, for the mass increase, $dm_{\rm p}/dt$, of the planetesimal the following conditions hold:

- growth is proportional to $\Sigma_{\rm part}$
- growth is proportional to $\Omega_{\rm K}$, i.e. slower at larger distances
- $v_{\rm rel}$ enters only through the focusing factor

During its growth the planetary embryo begins to influence and eventually alter its environment through its increasing gravitational force which increases the velocity dispersion $v_{\rm rel}$. At the same time the particle density in its environment will be depleted, either due to accretion or scattering. This will finally lead to a slow down of the runaway and the growth terminates. For the relative velocity one can use here $v_{\rm rel} \approx \sqrt{e^2 + i^2}\, v_{\rm K}$ where e and i are the (mean) eccentricity and inclination of the particle distribution and $v_{\rm K}$ the Keplerian velocity.

2.3.2 Growth to Protoplanets

Using the above estimate for the mass growth, Eq. (2.22), one can construct models that simulate the growth of a whole ensemble of planetesimals to larger objects. Numerically, this growth process to protoplanets can be described by different methods which combine statistical and direct methods, a topic that has been nicely reviewed by J. Lissauer [159]. Here, we present briefly two types of approaches, the direct N-body method and a statistical method, based on solving a Boltzmann type of equation.

(a) Direct N-Body methods

In this method the equations of motion for N planetesimals are solved by direct integration of Newton's equations of motion. For the i-th planetesimal, which has the position $\mathbf{x}_i$, the velocity $\mathbf{v}_i$ and mass m_i the equation then reads

$$\frac{d\mathbf{v}_i}{dt} = -GM_* \frac{\mathbf{x}_i}{|\mathbf{x}_i|^3} - \sum_{j \neq i}^{N} Gm_j \frac{\mathbf{x}_i - \mathbf{x}_j}{|\mathbf{x}_i - \mathbf{x}_j|^3} + \mathbf{f}_{\rm gas} + \mathbf{f}_{\rm coll} \,. \tag{2.23}$$

In addition the positions need to be updated via

$$\frac{d\mathbf{x}_i}{dt} = \mathbf{v}_i \,. \tag{2.24}$$

In Eq. (2.23) the first term on the right-hand side is the gravitational force of the central star, the second refers to the gravitational attraction of the $N-1$ other planetesimals, $\mathbf{f}_{\rm gas}$ is the frictional force exerted on the planetesimals by the gas in the protoplanetary disk, and $\mathbf{f}_{\rm coll}$ is the velocity change upon collisions between the individual planetesimals. For details how to model these forces see for example [135] and references therein. The velocity dispersion of the growing planetesimals, $v_{\rm disp}$, is damped by the gas drag which enhances their growth because of the reduced relative velocity between them, see Eq. (2.19). The advantage of this direct method is its accuracy because the growth of each individual particle is modeled, and this is also its disadvantage because it requires to follow the evolution of very many particles. Since it is impossible to include all planetesimals in such a simulation, the numerical computations follow the evolution of so called super-particles that represent a collection of many planetesimals [136]. In treating the outcome of physical collisions, the total momentum has to be conserved, while energy will have to be dissipated in the growth processes. Special numerical methods have been developed to integrate the N gravitationally interacting bodies accurately over long times [1].

(b) Statistical method

In this method the mean density of particles in phase space, i.e. the probability distribution function $f(\mathbf{r}, \mathbf{v})$ is evolved in time. The function f gives the density of particles per space and velocity interval, such that the particle density is given by integration over all velocities, $n(\mathbf{r}) = \int f d^3v$. The evolution of the distribution is described by the collisional Boltzmann-equation

$$\frac{\partial f}{\partial t} + \dot{\mathbf{r}}\frac{\partial f}{\partial \mathbf{r}} + \dot{\mathbf{v}}\frac{\partial f}{\partial \mathbf{v}} = \left.\frac{\partial f}{\partial t}\right|_{\mathrm{coll}} + \left.\frac{\partial f}{\partial t}\right|_{\mathrm{grav}}, \tag{2.25}$$

where f_{coll} describes the changes by individual collisions and f_{grav} the gravitational scattering by the other particles. It is typically assumed that the motion of individual particles is given approximately by Keplerian orbits with eccentricity e and inclination i with randomly oriented orbits. Then, the distribution function f can be simplified to follow a Rayleigh distribution

$$f_{\mathrm{R}}(x) \propto \frac{x}{\sigma^2} \exp\left[-\frac{x^2}{2\sigma^2}\right], \tag{2.26}$$

where σ is related to the mean value of the distribution. Here f_{R} has two arguments, e and i, given by $f_{\mathrm{R}}(e, i)$ (see [159] for details).

The actual growth of the particles in this method is described by the coagulation equation

$$\frac{dn_k}{dt} = \frac{1}{2}\sum_{i+j=k} A_{ij}n_i n_j - n_k \sum_{i=1}^{\infty} A_{ik}n_i\,, \tag{2.27}$$

where n_k is proportional to the number of particles with a given mass m_k, and A_{ij} represent the outcome of physical collisions between two planetesimals. The first term on right-hand side describes a gain in the number of objects with mass m_k and the second one a loss. The outcome of individual physical collisions depends in a complicated way on the velocities and masses of the collision partners and requires extensive parameter studies [150, 257]. The advantage of the Boltzmann method is that the complete ensemble is modeled, the disadvantage is its statistical nature. How these kind of equations are actually solved numerically and the application to planetesimal growth have been described in detail elsewhere, see [281, 285] and references therein.

2.3.2.1 An Illustrative Example

To be specific, we present in the following an example for the typical outcome of the planetesimal growth process. In [43] representative results of the second (statistical) type of approach have been described in more detail [281, 285] and we refer the reader to that excellent summary. For a complementary point of view, we summarize here results of the N-body approach, that has been used for example by E. Kokubo [136, 137] to describe planetesimal growth. In [135] the main results of such simulations are presented, and we give here a short summary. Initially a number of planetesimals are spread over certain region around the central star, here the Sun, typically centered

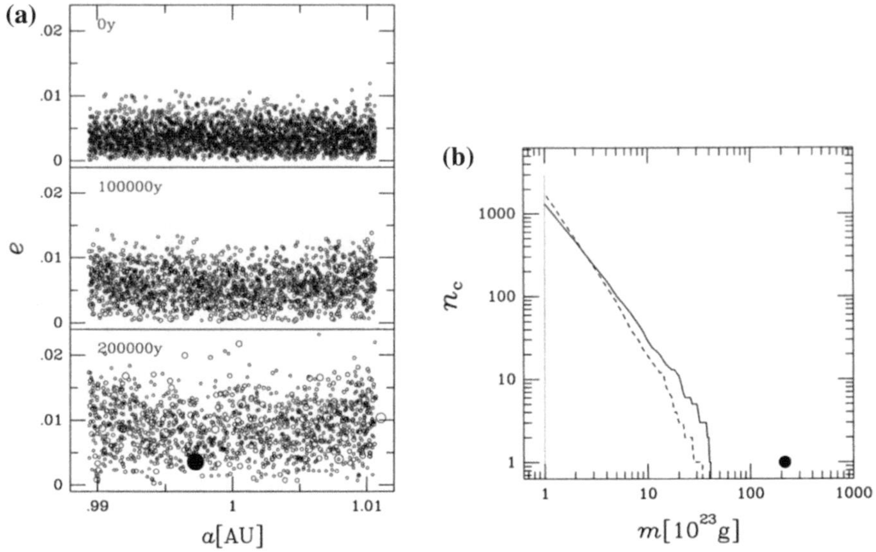

Fig. 2.9 Example of the runaway growth process from planetesimals to planetary embryos using the N-body method as described by Kokubo [135]. The basic setup and initial conditions are given in the main text in Sect. 2.3.2.1. The left panel shows the eccentricity and distance distribution of the bodies for the initial setup (top panel) and at times 100,000 and 200,000 yrs (middle and bottom). The right panel shows the cumulative mass distribution of the formed bodies where $n_c(m)$ = number of particles with mass larger than m. The results are shown at the same times as in the left panel: $t = 0$ (vertical thin line), $t = 100{,}000$ (dashed) and $t = 200{,}000$ (solid). The large bullet in both panels denotes the single outstanding large object that has formed through a runaway process. Taken from Figs. 3 and 4 in [135], © German Astronomical Society (Astronomische Gesellschaft, AG), reproduced with permission

at 1 AU with certain radial width. In [135] a radial extent of 0.02 AU, centered a $a = 1$ AU, was chosen. The simulations used 3000 bodies, each with an initial mass of $m = 10^{23}$ g. The mean material density of the growing planetesimals was assumed to be $\rho_p = 2\ \mathrm{g\,cm^{-3}}$. The whole ensemble was evolved in time for several 100,000 yrs.

As shown in Fig. 2.9, starting from the set of equal mass particles, at time $t =$ 200,000 yrs the distribution has evolved towards the situation where one large body (the •) has formed which has about 200 times the initial mass. It is embedded in a sea of smaller particles that have a continuous mass distribution, see right panel in Fig. 2.9. The low eccentricity (and inclination) of (•) comes through dynamical friction with the small objects. Through (distant) gravitational interactions the smaller particles are dynamically excited, and their mean e and i increase in time. Using equipartition of energy between e and i yields on average the following relation $<e^2> = 4<i^2>$ where $<x>$ denote mean values averaged over the ensemble of small particles. The one large object orbits the star on a nearly circular orbit.

For the same N-body simulation, the right panel in Fig. 2.9 shows the cumulative particle distribution, after 10^5 yrs (dashed), and after 2×10^5 yrs (solid). The objects between 10^{23} and 10^{24} g contain the majority of mass of the whole sample. The distribution follows a power-law

$$\frac{dn_c}{dm} \propto m^q \,. \tag{2.28}$$

Here, q describes the exponent of the power-law mass distribution, where a value of $q = -2$ is equivalent to equal mass in each logarithmic mass bin. A steeper distribution, $q < -2$, is characteristic for a runaway process [160], where only very few particle reach larger masses. Indeed, in the simulation shown in Fig. 2.9 the slope is about $q \approx -2.7$ and one very massive particle (•) is separated from the continuous distribution, i.e. it serves as a sink of particles. These results indicate very clearly that in the early phase the planetesimal growth proceeds through a runaway phase.

Very similar results to those shown in Fig. 2.9 are obtained for example by [281, 285] using the statistical method, see summary in [43]. Their distribution of particle sizes follows a very similar slopes to that shown in the right panel of Fig. 2.9, again indicative of runaway growth.

2.3.2.2 The End of the Growth

Obviously a runaway process, as just described, cannot continue forever. It is slowed down and eventually stopped mainly by two processes. Upon growing to larger objects the planetesimals stir up their environment such that the velocity dispersion is increasing and hence the relative velocity between them rises which leads, according to Eq. (2.12) to a reduction in the collisional cross section. Secondly, the accretion process reduces the local density of particles, $\rho_{\rm part}$, and the body isolates itself from further growth, because fewer and fewer collision partners are available within the *feeding zone*. One can get an estimate on the final size an embryo can reach, the *isolation mass* $M_{\rm iso}$, by assuming that the volume of accretion, the feeding zone, has a radial extent given by the width of the horse shoe region, which is approximately given by the Hill-radius (2.17). The total mass of all the particles within a region Δa inside and outside of semi-major axis a is given by $m = 2\pi\, 2a\, \Delta a\, \Sigma_{\rm part}$. Using now $\Delta a = C R_{\rm H}$ we obtain with $m = M_{\rm iso}$

$$M_{\rm iso} = 4\pi a\, C \left(\frac{M_{\rm iso}}{3\, M_\odot}\right)^{1/3} a\, \Sigma_{\rm part} \,, \tag{2.29}$$

where C is a factor of order unity [6].

We consider now for example the growth of the Earth in the Solar System, and assume that there were initially $2\, M_\oplus$ located between 0.5 und 1.5 AU. Using the standard condition of the MMSN with $\Sigma_{\rm part} \sim a^{-3/2}$ and $\Sigma_{\rm part} = 8$ g cm^{-3} at 1 AU and $C = 1$, then we obtain for the isolation mass

$$M_{\rm iso} \approx 0.05\,M_\oplus\,, \tag{2.30}$$

i.e. a mass of a few lunar masses. This runaway process is essentially a local phenomenon, because the embryos accrete primarily from their immediate neighborhood, their feeding zone. The limited extent of each embryo's feeding zone implies that several objects in the protoplanetary nebula will experience runaway growth and grow at a similar rate. This is the *oligarchic phase* of terrestrial planet formation which results eventually in about 40 planetary embryos which have a mean spatial separation of about $\approx$0.01−0.025 AU [47, 158]. The timescale for this growth of the oligarchs is about 0.1–1.0 Myr [262, 281]. Due to the locality of the runaway growth, any radial compositional gradient present in the protoplanetary disk should be reflected in the embryos' chemical compositions [194]. Starting from these protoplanets the final assembly of the terrestrial planets can ensue.

2.3.3 Assembly of the Terrestrial Planets

After the oligarchic phase there are only a few objects, the *embryos*, left over with masses of about Moon to Mars size. The sea of planetesimals has mostly been depleted and only the gravitational interaction between these planetary embryos remains, i.e. in contrast to the previous growth phases the problem is physically relatively clean. To model this final assemblage, classical N-body simulations are the standard choice. In principle this is a straight-forward exercise because there are only very few particles ($\approx$100) left over whose motion needs to be integrated, but this process occurs over a very long time scale, about 10^7-10^8 yrs. Hence, the longterm integration of the equations of motion (similar to Eq. 2.23 with vanishing gas and collision terms) requires good symplectic integrators that conserve automatically the total energy of the system. A well known, and often used example is the MERCURY-code developed by J. Chambers [45] which is publicly available. Other codes are for example the SWIFT-package [76] or the REBOUND-code [239], a modern N-body code, with the capability to treat collisions.

As an example we discuss briefly the results of Chambers et al. [46, 48]. The authors performed a series of N-body simulations, where as starting conditions they used about 50 embryos in the first paper [48] and about 155 embryos in the second paper [46]. In total about 2 $M_\oplus$ were distributed between 0.3 and 2.0 AU with different types of initial mass distributions: all equal, bimodal or with a radial mass profile. In all simulations Jupiter and Saturn were included on their present day orbits. The collisions were treated as 100% sticking (perfectly inelastic) and the angular momentum of the coalesced bodies went into their spin. The presence of Jupiter and Saturn may be surprising at this still early phase during the growth of the terrestrial planets but, as we shall see in the following chapter, the formation timescale for these massive planets is indeed shorter than the time necessary for the final assembly of the terrestrial planets.

The results of those type of simulations show that indeed planetary system containing several terrestrial planets are produced. The formation timescale is a few 10^7 years, which is very long compared to the previous phases. The reason is the fact that in this final growth phase only very few objects are remaining which reduces the frequency of mutual collisions considerably and it takes a long time to produce full grown terrestrial planets. The presence of the massive planets Jupiter and Saturn is required as they dynamically stir up the sample of embryos and prevent the formation of an additional planet within the region of the main belt asteroids. The whole evolution is a highly *chaotic process* because objects from different regions are scattered around and lead to a variety of collisions starting from head-on to near misses. Many objects may be lost as they fall into the Sun, and it is estimated [46] that about 1/3 of the initial objects within 2 AU may have to fear this fate. The typical outcome of these N-body simulations is a system with 3–4 planets on stable well separated orbits, with the tendency for more planets in those runs that have more initial embryos. Hence, the final systems in these simulations resemble roughly the situation in the Solar System where the most massive planets are in the Venus-Earth region, while the innermost planets and those in the Mars region are on average much smaller. The smallness of the Mercury type objects can be understood in terms of the high collision speeds for the innermost orbits that often lead to fragmentation rather than growth. Indeed the smallness and high density of Mercury can be attributed to a high speed impact during this chaotic phase of terrestrial planet formation [25]. Most of the objects initially residing within the main asteroid region are scattered out due to resonant action by Jupiter and Saturn [203].

Nevertheless there are important differences when compared directly to the Solar System: the mass concentration in the planets is not as high as for Venus and Earth, the planets have on average too high e and i, and the spin-orientations are arbitrary. Specifically, the typical mass of a 'Mars'-object turns out to be too large when compared to the Solar System, by a factor of about 5. The presence of residual gas from the protoplanetary disk will reduce the eccentricities of the growing objects and shorten the formation time but too much gas will lower the collision rates such that massive planets like Venus and Earth will not form at all [138]. More recently, new evolutionary N-body simulations have been performed, some using a much larger number of initial bodies, a few thousand, that were spread over a wider radial domain ranging from 0.5 to 5.0 AU [203, 235, 236]. The influence of several input parameters to the simulations, such as the disk mass and radial density profile, the particle distribution in space and in mass, the orbits of the giant planets, and the treatment of collisions have been analysed in detail in more elaborate simulations, see overview in [194]. From these one can infer that for the reduction of the orbital excitement of the terrestrial planets the dynamical friction of the remaining population of planetesimals plays an important role. Concerning the orbital architecture of the giant planets it was shown that for more eccentric giants the terrestrial planets grow faster and have more circular orbits [234]. While the overall architecture of the formed planetary systems resembles approximately the Solar System, the 'Mars-problem' still remains. One scenario to overcome the problem of the too small Mars is the *Grand Tack* scenario where during the early Solar System the giant planets Jupiter and Saturn migrated first far inward and then turned around to move out to their present locations [274]. We will not discuss this scenario any further in contribution and refer the reader to excellent reviews [194, 233].

The formation of terrestrial planets from planetesimals proceeds in different steps. In the first phase the gravitational attraction between the growing planetesimals leads to a fast runaway growth which is followed by a slower oligarchic growth phase, at the end of which an ensemble of about 50 Moon to Mars sized objects has formed, spatially well separated. On timescales of tens of millions of years these planetary embryos evolve under their mutual gravitational force and form through a sequence of collisions and impacts terrestrial type planets and the cores of giant planets.

2.4 The Formation of Massive Planets by Core Accretion

In this section we describe how the growth of massive, gaseous planets is believed to proceed. The massive planets of the Solar System are about 100 times (Saturn) or 300 times (Jupiter) more massive than the Earth, which is the most massive object in the terrestrial planet region. Their mean densities are roughly comparable to that of water or even lower in the case of Saturn which implies that they consist of a huge amount of gas in addition to a possible central solid core. The question arises, how it is possible to collect large amounts of Hydrogen and Helium to form those planets under the conditions in the protoplanetary nebula. These lightest elements are highly volatile and difficult to condense. Two main pathways for the formation of massive planets have been discussed [39].

In the first scenario, the *core accretion* model, an initial seed object forms onto which the gas can later accumulate. It is a bottom-up process, where initially a solid core forms along a similar evolutionary path as for the embryos in the formation of terrestrial planets. Upon reaching a certain critical mass the gravity of the core becomes high enough that a rapid, runaway gas accretion onto the core is possible, leading eventually to the formation of a gaseous planet. In the second scenario, the *gravitational instability* (GI) model, the formation pathway is similar to that of star formation and it is a top-down process. The scenario is believed to operate if the initial gas density of the protoplanetary disk is so high that a dynamical gravitational instability ensues that leads directly to the collapse of a local patch of the disk.

The physical structure of the gaseous planets in the Solar System has been studied extensively from the ground and through space missions. For the Solar System, the favored scenario is the core accretion model because it explains straightforward the existence of cores in the centers of the massive planets and the large amount of solids, with Jupiter enriched about 1.5–6 times above solar composition and Saturn about

6–14 times [247]. Additionally, the atmospheres of Saturn and Jupiter are enriched in metals and in noble gases, in particular Argon [102]. The formation via the GI process may have operated in the creation of the observed distant, directly imaged extrasolar planets. In this section we first focus on the core accretion pathway, while the GI model will be discussed below in the next Sect. 2.5. The status of knowledge about the internal structure of the Solar and the extrasolar giant planets are reviewed in [87] and recently in [102]. A classic review of this phase of planet formation is presented in [159], while modern reviews are given in the relevant chapters of the *Protostars and Planets* series, here in PPVI [44, 113].

2.4.1 Background

Planetary growth in the core accretion model consists of three phases. In the first phase a solid object forms in a manner similar to the processes that have led to the planetary embryos in the case of the terrestrial planets. In the previous section we have analysed the isolation mass of a growing body which is given by the amount of solid material that can be accreted directly from the *feeding zone* of an embryo. In contrast to the terrestrial region of the protoplanetary disk the abundance of solids is much higher at the location of Jupiter because of the low temperatures in the disk which allows for the condensation of many additional molecules, most importantly water. H_2O is the most abundant molecule in the Universe and in the disk it freezes out to ice beyond the so-called *snowline*. The exact condensation temperature depends on the ambient gas pressure but for conditions in the protosolar nebula T has to be smaller than (150−170) K for ice to condense. For a passively heated disk where the stellar irradiation dominates the heat input to the disk, one finds for a Solar type star $r_{\rm cond} \approx 2.7\,\mathrm{AU}$ for the condensation radius. Hence, in the standard Hayashi-model of the protosolar nebula, for distances to the star larger than $r_{\rm cond}$, the amount of solids available for embryo formation is typically assumed to be 4 times higher than in the inner regions [110]. While newer calculations indicate possibly a lower value [166], we still use the standard value here, then the surface density for solids $\Sigma_{\rm s}$ is given by

$$\Sigma_{\rm s}(\text{rock/ice}) = 30\,(r/1\,\mathrm{AU})^{-3/2}\ \mathrm{g/cm^2} \quad \text{for} \quad r > 2.7\,\mathrm{AU}\,. \tag{2.31}$$

Using the formula given in the previous section for the *isolation mass* we obtain at the current distance of Jupiter ($a_{\rm Jup} = 5.2\,\mathrm{AU}$) the following estimate

$$M_{\rm iso} \approx (5-9)\,M_\oplus\,,$$

which is considerably higher than in the terrestrial region. For the subsequent phases in the formation of giant planets the core mass is an important quantity, as it determines whether the embryo can capture a sufficient amount of gas in a short timescale to become a giant.

The first phase of gas accretion proceeds in a slow hydrostatic manner. The question arises: How much of an atmosphere can a growing planet hold? To answer this question, we assume that the minimum requirement for an atmosphere is that the escape velocity from the planet is larger than the sound speed in the ambient disk $v_{esc} > c_s$. In the discussion below we follow P. Armitage [6] and use

$$m_p = \frac{4\pi}{3}\rho_m R_p^3, \qquad v_{esc} = \sqrt{\frac{2Gm_p}{R_p}}, \quad \text{and} \quad c_s = \frac{H}{r}u_K\,,$$

where m_p, R_p denote the mass and radius of the growing planet (embryo) with density ρ_m, v_{esc} is the escape velocity from its surface, c_s is the sound speed and H the vertical half-thickness of the disk. For an icy body at 5 AU around a Solar mass star we will have some atmosphere (where v_{esc} begins to be larger than c_s) for $m_p \gtrsim 5 \cdot 10^{-4}\, M_{Earth}$. This is a very small mass planet, but such an atmosphere has no dynamical importance.

Let us consider the situation where the atmosphere or rather envelope takes up a sizable fraction, f_{env}, of the planet such that $M_{env} = f_{env}\, m_p$. For an isothermal atmosphere with $c_s(\text{atm}) = c_s(\text{disk})$, and where the outer density matches that of the ambient disk (ρ_0) one obtains at $r = 5\,\text{AU}$ with $\rho_0 = 2 \cdot 10^{-11}\text{g/cm}^3$, $c_s = 7 \cdot 10^4\text{cm/s}$ for the minimum mass of a planet to hold an atmosphere of about 10% of its mass ($f_{env} = 0.1$) [6]

$$m_p \gtrsim 0.2\, M_\oplus\,. \tag{2.32}$$

As a comparison, at the location of the Earth, one obtains for $f_{env} = 0.1$ at $r = 1$ AU using $\rho_0 = 6 \cdot 10^{-10}$ g/cm^3, $c_s = 1.5 \cdot 10^5$ cm/s, and $\rho_m = 3$ g/cm^3

$$m_p \sim M_\oplus\,, \tag{2.33}$$

in very rough agreement with the actual conditions in the Solar System. Planets are assembled over the life time of the disk and the mass of the embryo will be given by the isolation mass as defined in Eq. (2.29). To hold an atmosphere, M_{iso} must be larger than the critical mass to acquire an atmosphere. An analysis [6] shows that inside r_{cond} the isolation mass is always smaller than the mass to hold an atmosphere, while outside r_{cond} the isolation mass is larger. This relation explains qualitatively very well the fact that the inner planets in the Solar System consist of terrestrial planets with very little atmosphere while outside we find the gas or gas/ice giants with a very extended envelope. Further details will depend on the actual disk model, and the final location where gas giants are found eventually will depend on the amount of migration that occurred. To understand the final phase of the assembly of massive planets, we will have to study in more detail the internal structure of the growing giants.

2.4.2 The Growth to a Giant

To calculate the structure of a growing star or planet one assumes typically that the overall evolution of the mass growth proceeds on a timescale that is very long in comparison to the adjustment timescale of the interior. The latter can be estimated on the basis of the sound crossing time of an object that is of the order $\tau_{\text{cross}} \sim (G\bar{\rho})^{-1/2}$, where $\bar{\rho}$ denotes the mean density of the body. For a growing planet τ_{cross} is only a few minutes. Indeed, this is much smaller than typical accretion time scales, and the evolution of the growing planet can be well described by a sequence of *hydrostatic models*, similar to stellar evolution.

Hence, the basic equations that describe the inner structure of a growing planet are given by the standard stellar structure equations [125], adapted to the planet formation process [101]

$$\text{mass conservation:} \qquad \frac{dm}{dr} = 4\pi r^2 \rho \tag{2.34}$$

$$\text{hydrostatics:} \qquad \frac{dp}{dr} = -\frac{Gm(r)}{r^2}\rho \tag{2.35}$$

$$\text{radiative diffusion:} \qquad \frac{L(r)}{4\pi r^2} = -\frac{4\sigma T^3}{3\kappa_R \rho}\frac{dT}{dr} \tag{2.36}$$

$$\text{energy generation:} \qquad \frac{dL}{dr} = -4\pi r^2 \rho \left(\varepsilon - T\frac{\partial S}{\partial t}\right) \tag{2.37}$$

with: $m(r)$ mass interior to the radius r; $L(r)$ luminosity at radius r; ε internal energy generation, σ Stefan-Boltzmann constant, and κ_R Rosseland-opacity. Equations (2.34–2.37) are essentially the same equations as for stellar interiors with the difference that the luminosity is not due to nuclear burning of hydrogen but due to the contraction and cooling of the gas that changes the entropy S, as given by the last term in the energy equation. Impacting planetesimals give an additional contribution. In case of convection (for a super-adiabatic stratification) the adiabatic temperature gradient is used.

To calculate the structure of the whole planet, from the very center to the surface, equations of states for matter under extreme conditions are required. A simple estimate for the *minimum* central pressure, p_c, can be obtained from the hydrostatic equation using the assumption of a constant density inside the body, $\rho(r) = \bar{\rho}$. Integrating Eq. (2.35) then yields

$$p_c = \frac{1}{2}\bar{\rho}\frac{GM}{R}, \tag{2.38}$$

where M and R are the total mass and radius of the planet. For the observed mass and radius of Jupiter ($M_{\text{Jup}} = 1.9 \cdot 10^{27}$ kg $= 318\, M_{\text{Earth}}$ and $R_{\text{Jup}} = 7 \cdot 10^7$ m $= 11.2\, R_{\text{Earth}}$) the mean density is given by $\bar{\rho} = 1.33$ g/cm^3, and Eq. (2.38) gives $p_c \approx 2.5 \cdot 10^{12}$ Pa $= 2.5 \cdot 10^7$ bar. This result shows that to describe the interior structure of (massive) planets the conditions of matter under extreme conditions has to be known.

The obtained result for Jupiter's p_c is slightly beyond that what present day experimental setups can reach (about $0.6 \cdot 10^{12}$ Pa in diamond anvil cells [71]), and due to the assumption of constant density the estimated value is the minimum p_c that Jupiter can have. Hence, the equations of state (EOS) that can be used to describe the inner cores of giant planets are constructed through a combination of theoretical calculations and experimental results. A summary of frequently used EOS is presented in [87], see also [13].

A simple equation of state used for the outer regions (envelope and atmosphere) is given by the ideal gas law

$$p = \frac{R_{\rm gas}\, \rho T}{\mu}, \tag{2.39}$$

with the gas constant, $R_{\rm gas}$, and the mean molecular weight, μ, in units of $m_{\rm H}$.

To solve the structure equations (2.34–2.37) for a growing planet suitable *boundary conditions* have to be chosen. These are now different from the standard conditions of stellar growth because the planet is still embedded in the protoplanetary disk that acts as a mass and heat reservoir.

The inner boundary is given in this situation not by the central values but rather by the size and luminosity of the core. The size, $R_{\rm core}$, is given by solving the structure equations for the core for a given composition and core mass, $M_{\rm core}$. As a first approximation a constant density can be chosen for the core. The luminosity, $L_{\rm core}$, has different contributions. The main part comes from the residual kinetic energy of the accreted planetesimals after they have fallen through the envelope and land on the core. Additional core luminosity comes from radioactive decay in the core, core contraction and cooling of the core.

The outer boundary conditions at the planet's radius, R, depend on the evolutionary status of the planet and the ambient disk. Mainly 3 different options have been considered [196]

(1) *Attached or nebular phase*:
For small masses (M lower than about 10–20 $M_{\rm Earth}$) the protoplanet is still deeply embedded in the disk such that its radius is much smaller than the disk scale height, H. Hence, its envelope is smoothly attached to the nebula. The planet's radius in this case is given by the smaller of the accretion radius, $R_{\rm A} = GM_{\rm p}/c_{\rm s}^2$, and the Hill radius, $R_{\rm H} = (M_{\rm p}/3M_*)^{1/3} a_{\rm p}$ (Eq. 2.17), where a smooth transition is used between these two values. The accretion radius is half of the Bondi-radius which is often used in spherical accretion problems. The temperature and pressure at the outer radius, R, are then identical to the disk temperature at that location.

(2) *Detached or transition phase*:
For larger masses, no solution satisfying the attached state exists, and the proto-planet contracts to a radius much smaller than $R_{\rm H}$. The luminosity follows from radial infall, via a spherical accretion shock, or through accretion from a circumplanetary disk [212]. The gas accretion is not anymore determined by planet, but by nebula conditions. For the spherical case the pressure at the surface is determined by the

accretion shock (gas in free-fall) at the photospheric pressure. In reality the accretion onto the planet will not be completely spherical symmetric anymore and mass will be accreted through an accretion disk around the planet, the circumplanetary disk. Numerical simulations [126, 162] show that accretion occurs via 2 'streams' that are the extensions of the spiral shock waves generated by the planet that extend all the way into the Hill sphere of the planet, see Fig. 2.12. High resolution three dimensional simulations of the flow near growing planets show that a substantial amount of material is accreted through the polar regions [60, 259].
(3) *Isolated or evolution phase*:
The planet evolves at constant mass and receives irradiation from the central star that is spread over the whole planetary surface. The albedo of the planet is required for an accurate determination of the absorbed radiation. The photosphere is given, as in stellar models at an optical depth, $\tau = 2/3$.

For the case where the luminosity is provided by the accretion of planetesimals the planetesimal-envelope interaction during the infall is crucial for the energy and mass deposition profile in the envelope. To determine the trajectory of infalling particles one has to consider first the gravity and gas drag (i.e. the envelope increases the capture radius of the growing planet), and secondly thermal ablation and aerodynamical disruption. A modern example of this process is given by the infall of comet Shoemaker-Levy 9, where temperatures over 30,000 K were reached in the bow shock around the infalling body. This interaction with the envelope will heat up the planetesimal that re-radiates this energy and will loose some of its mass, and may eventually disintegrate. The accumulation of energy by the envelope or core during these processes will determine the time evolution of the planet.

When performing evolutionary simulations of growing planets one extremely important finding was the fact that beyond a critical value of the core mass, $M_{\rm crit}$, the envelope cannot be in hydrostatic equilibrium anymore and will collapse onto the core [190, 191, 218]. This finding has given rise to the name *core-instability* model, sometimes also called *core-accretion* scenario. In the simulations, the hydrostatic equations stated as above were solved, where it was assumed that all planetesimals reach the core and provide the luminosity, $L_{\rm core}$, hence in the envelope $L(r) = const. = L_{\rm core}$. The critical mass, $M_{\rm crit}$, is a function of the opacity of the accreted material and it lies around 10 $M_{\rm Earth}$ for typical ISM composition. Subsequent numerical simulations confirmed exactly these early results and the runaway mass accretion is now *the* cornerstone in the core-accretion scenario. Noteworthy classic simulations for the in-situ formation of Jupiter in the Solar System are presented [38, 225]. More elaborate models on the formation of massive planets that include the effects of accretion through a circumplanetary disk have been formulated [195, 212], which have been incorporated into population synthesis models of planets. The latter article [195] contains a very readable introduction to that field.

Initially difficult to understand the physical mechanism of the core-instability phenomenon, Stevenson [256] presented a simple analytical model of the process, that helped to uncover the underlying principles, and that is summarized nicely in [6]. For typical parameters in the protosolar disk one obtains [256]

$$M_{\text{crit}} \simeq 20\kappa_R^{3/7}\, M_\oplus\,, \tag{2.40}$$

where κ_R denotes the Rosseland mean opacity in units of (cm^2/g). For $M_{\text{core}} > M_{\text{crit}}$ the envelope contracts rapidly and runaway gas accretion sets in. Full numerical simulations are required to obtain the exact evolution of this phase that includes the details on the boundaries as given above, in particular the contact with the ambient disk and the solid particle accretion process.

An example of such a calculation for the in-situ formation of Jupiter in the Solar nebula [195] is shown in Fig. 2.10. The vertical lines and roman numbers indicate different evolutionary phases: (I–III) refer to the attached phase with the protoplanet in contact with the disk, where (I) denotes the assembly of the core, (II) refers to the the continued slow core accretion and hydrostatic gas accumulation, isolation of solid particles, and in (III) the core reaches the critical mass with the onset of rapid gas accretion. (D) is the detached phase, and (E) the long term isolation, or evolution phase. The authors [195] point out that the depicted evolution serves as a test case that was described very similarly in early simulations [38, 225]. The overall evolution indicates that the typical time scale to form Jupiter is about a million years. However, there are many factors that have an impact on the overall evolutionary timescale. Details will depend on the:

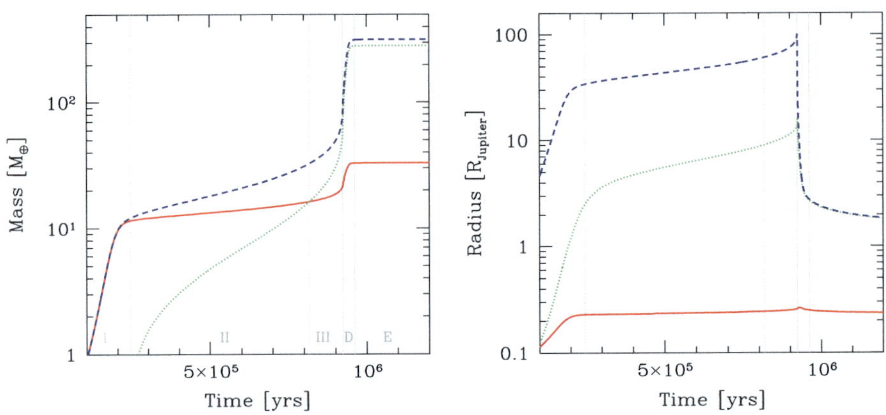

Fig. 2.10 The evolution of the mass and radius of Jupiter in the Solar Nebula. During the whole growth the planet was fixed at its present position (in-situ formation). The core mass (made of solids) is given by the red line, the gas mass by the green line, and the total mass by the dashed blue line. The different phases of the accretion process have been marked by the vertical lines and additional roman numbers in the left panel, see text for more details. In the right panel, the red line refers to the core radius, the blue dashed line to the total radius, and the green line to the capture radius of planetesimals. *Courtesy* Chr. Mordasini. Taken from Fig. 2 (top and bottom left panels) in [195], © ESO, reproduced with permission

- Opacity
  Low values of κ_R will allow faster envelope growth. It is determined essentially by the amount of dust present in the envelope because the gas opacities alone are much too low.
- Convection
  The onset of convection in the envelope will enhance the efficiency of energy transport and hence change the time scale for accretion.
- Chemical composition
  The chemical composition influences the opacity and the mean molecular weight, μ. The latter has through the equation of state (2.39) a direct impact on the pressure and hence the radial stratification through the requirement of hydrostatics.
- Accretion rate
  The accretion onto the core, $\dot{M}_{\rm env}$, is often estimated from one-dimensional models. However, only through two-dimensional (circumplanetary disk) and three-dimensional simulations (polar accretion) can the actual accretion rate been obtained.
- Migration
  The migration of the growing planet through the disk has an influence on the mass accretion rate. For example, new reservoirs for the solid particle as well as gas accretion will be opened.

One very important ingredient in determining the overall growth time scale is the formation time of the core, which is determined by the accretion of solids from the disk. Obviously this is directly proportional to the solid particle density in the disk, $\Sigma_{\rm solid}$. The simulations indicate that for $\Sigma_{\rm solid} < 10$ g/cm^2 the time scale for core formation is too long such that the gas reservoir of the disk will also be depleted given the typical disk lifetime of a few million years. On the other hand, for $\Sigma_{\rm solid} > 10$ g/cm^2 there are too many heavy elements in comparison to today's Jupiter composition. Given that the surface density of solids for the MMSN is only 5 g/cm^2 (at the location of Jupiter), there appears to be a timescale problem in that the growth time for the core formation is too long for Jupiter, and the problem becomes even more severe in case of Uranus and Neptune formation.

A new option that is presently discussed in more detail is the idea to enhance the growth of planetary cores by rapid accretion of small pebbles [144]. In contrast to the larger planetesimals, these typically cm to dm-sized particles feel the gas drag. When pebbles approach a growing core which has some atmosphere already collected within its environment, the gas drag will slow them down and they will spiral deeper into the Hill sphere towards the planet and become eventually accreted by it. Depending on the mass of the growing core two different types of accretion regime can be distinguished [144]. First we define a Bondi-radius for the growing object as $R_{\rm B} = GM_{\rm p}/\Delta v_{\rm K}$, where $\Delta v_{\rm K}$ is the velocity difference between the Keplerian rotation and particle speed due to the gas drag. By equating the Bondi-radius and the Hill radius one can define a transition mass, $M_{\rm t}$, where both are equal. For the MMSN one finds

$$M_t \approx 0.016\, M_\oplus \left(\frac{r}{30\,\mathrm{AU}}\right)^{3/4} \left(\frac{\Delta v_K}{0.1 c_s}\right)^3 . \tag{2.41}$$

For $m < M_t$ ($R_B < R_H$) accretion occurs in the so-called drift regime, while for $m > M_t$ (i.e. $R_B > R_H$) it occurs in the Hill regime. For the growth of massive planets the more interesting phase will be the Hill regime with $m > M_t$.

Classically, the accretion rate of planetesimals onto a planetary core (100% sticking) is given by $\dot{M}_p = \Sigma \Omega R_p^2 F_{grav}$ [158] (see also Eq. 2.22), where R_p is the planetary radius and F_{grav} the gravitational focusing factor (Eq. 2.14). Since only a fraction $\alpha_p = R_p/R_H$ of the whole Hill sphere is captured for a growing core [93, 144], the mass accretion will be changed to $\dot{M}_p(\text{planetesimals}) = \Sigma \Omega \alpha_p R_H^2 F_{grav}$. On the other hand, since pebbles feel the gas drag, all objects that reach the Hill sphere will be accreted and the accretion rate becomes $\dot{M}_p(\text{pebbles}) = \Sigma \Omega R_H^2 F_{grav}$. Clearly, for a growing core with $m > M_t$, the Hill radius will be much larger than its actual, physical radius, such that $\alpha_p \ll 1$. Then the ratio $\dot{M}_p(\text{pebbles})/\dot{M}_p(\text{planetesimals})$ will be much larger than unity, reducing the accretion time scale for the core considerably. Indeed, the simulations of Lambrechts and Johansen [144] show that under the assumption of a steady influx of new material (from outside) and pebbles well settled to the midplane of the disk, the growth time for a 10 M_{Earth} core is less than 10^5 yrs at 5 AU, and less than 10^6 yrs at 50 AU, see Fig. 2.11. Clearly, these growth times are much shorter than those expected by pure growth of planetesimals and would ease the timescale problems for massive planet formation considerably. The

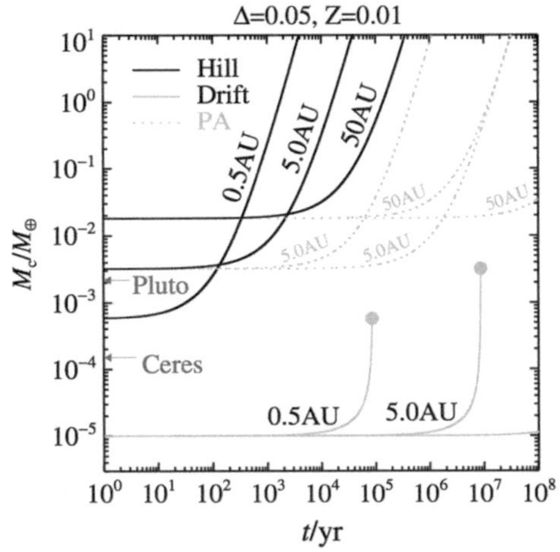

Fig. 2.11 The growth of a planetary core as a function of time at different radii in the disk. Curves are shown for the Hill regime ($m_p > M_t$, with the transition mass M_t, see Eq. 2.41), the drift regime ($m_p < M_t$) and the standard pebble accretion (PA). The result shown are for a headwind parameter $\Delta \equiv \Delta v_K/c_s = 0.05$ (see Eq. 2.42) and a dust to gas mass ratio $Z \equiv \Sigma_p/\Sigma = 0.01$. Taken from Fig. 11 in [144], © ESO, reproduced with permission

exact efficiency of this problem will depend on several physical mechanisms, such as the so-called headwind parameter

$$\Delta = \Delta v_\phi / c_s \,, \tag{2.42}$$

the vertical pebble concentration, $Z = \Sigma_p/\Sigma$, a possible migration of the core and of course the structure of the envelope within the Hill sphere of the planet. In Eq. (2.42) Δv_ϕ is the velocity difference between the pebbles, which have a Keplerian velocity, and the gas in disk that moves slower due to the radial pressure gradient. The details need to be worked out, and also the question how the formation of lower (Neptune) mass planets fit into this new model.

2.4.3 *The Final Mass*

In the previous section we have seen that in the final stages the planet grows in a runaway fashion, and the question arises what determines the final mass of a planet. The simplest answer to this might be fact that the amount of gas that is available in the disk is necessarily finite which sets eventually a natural mass limit. While the limited mass reservoir in the disk certainly plays a role, there is an important additional factor that does inhibit mass growth, and this is the opening of a gap. We might expect that when the Hill radius of the planet exceeds the vertical scale height of the disk, $R_H \gtrsim H$, this will have a significant impact on the disk structure, and alter the mass accretion rate. We will analyse this from two perspectives, a particle based approach and a hydrodynamical one.

In the first *particle approach*, the velocity change experienced by an individual (gas) particle that passes by a growing planet in the disk can be calculated most easily in the *impulse approximation* [156]. Here, it is assumed that the motion of the particle is primarily Keplerian in most part of the orbit and only perturbed in the vicinity of the planet. This creates an additional gravitational force (impulse) on the particle that leads to a deflection of its trajectory and a slight change in its azimuthal velocity, and hence angular momentum. Due to the collisions with other nearby disk particles this angular momentum change is immediately 'shared' between them. Integrating the angular momentum exchange over all particles, i.e. over the whole disk, one can calculate the total rate of angular momentum input to the disk mediated by the planet

$$\dot{j}_{\rm grav} = -\frac{8}{27} \left(\frac{r_p}{\Delta r_0}\right)^3 \left(\frac{m_p}{M_*}\right)^2 \Omega_p^2 \, \Sigma(r_p) \, r_p^4 \,, \tag{2.43}$$

where r_p denotes the distance of the planet from the star, Δr_0 is the closest approach of a disk particle to the planet, and the index p denotes that all quantities are evaluated at r_p. The minus sign indicates that the inner disk loses angular momentum while the outer gains it, i.e. in the planet region the disk will be 'pushed' away from the planet, see below. In spite of the simple approximation this result is nearly exact. A nice pedagogical treatment of the derivation is given in the lecture notes of the *Saas-Fee Advanced Course 31* by P. Cassen [43]. From Eq. (2.43) it is clear that in the vicinity of planet ($\Delta r_0 \to 0$) there is a strong increase of $\dot{J}$, and the total amount deposited will depend on the choice of Δr_0.

In the second *hydrodynamic approach*, the continuum behaviour of the gas in the disk is analyzed and the angular momentum deposition follows from more complex wave phenomena. A planet embedded in a disk produces disturbances in the disk's density distribution. These are sound waves that spread out from the planet's position because its presence impacts the dynamics of the ambient gas. The Keplerian shear flow in the disk turns these sound waves into a spiral wave pattern with an outer trailing arm and an inner leading one, as shown in Fig. 2.19. The spiral wave pattern is stationary in a frame corotating with the planet and hence, at a certain radial distance from the planet they reach a supersonic speed with respect to the Keplerian disk flow. At this point the waves turn into shock waves, dissipate energy and deposit angular momentum to the disk. In the outer disk the spiral arm is faster than the disk material and it deposits positive angular momentum, i.e. the outer disk material gains angular momentum and hence speed, and moves away from the planet, because in a Keplerian disk the angular momentum increases as $\propto r^{1/2}$. Inside of the planet the situation is reversed, the spiral arm is slower than the disk and negative angular momentum is deposited which leads to an inward motion of the disk material. In total, the disk matter is receding from the location of the planet lowering the density in its vicinity. The effect increases with planet mass and eventually an annular gap in the density is formed at the location of the planet, which happens even without considering direct mass accretion onto the planet. This lowering of the ambient density decreases the available mass reservoir and the gas mass accretion onto the planet will be reduced accordingly.

From Eq. (2.43) it is clear that the deposition rate of angular momentum will increase with the mass of the growing planet, and in the absence of other effects there will be no mass left at the planet location. However, there are two main competitors, viscosity and pressure, working against this continuing gap deepening. The viscous disk torques at the location of the planet are given by

$$\dot{J}_{\rm visc} = \dot{M}_{\rm disk}\, j_{\rm p} \; = \; 3\pi\, \Sigma(r_{\rm p})\nu\, r_{\rm p}^2 \Omega_{\rm p} \tag{2.44}$$

with the kinematic disk viscosity ν and the specific angular momentum of the planet, $j_{\rm p} = r_{\rm p}^2 \Omega_{\rm p}$. In writing Eq. (2.44) we have assumed a stationary accretion disk with a globally constant mass accretion rate [227]. The *viscous criterion* for gap formation is then $\dot{J}_{\rm grav} \geq \dot{J}_{\rm visc}$. If one assumes for the smallest distance the Hill radius (2.17) of the planet, $\Delta r_0 = R_{\rm H}$, then

$$q \geq q_{\rm visc} \simeq \frac{10\nu}{\Omega_{\rm p}\, r_{\rm p}^2}\,, \tag{2.45}$$

where q is the planet to star mass ratio, $q = m_\mathrm{p}/M_*$. A second, *pressure criterion* of gap formation is obtained from the condition that the Hill sphere of the growing planet is larger than the disk thickness, i.e. $R_\mathrm{H} \geq H$ which gives

$$q \geq q_\mathrm{Hill} \simeq 3 \left(\frac{H}{r}\right)_\mathrm{p}^3 . \tag{2.46}$$

For typical parameter of the protoplanetary disk both criteria yield similarly a limiting mass for gap formation between Saturn and Jupiter. This explains the mass of Jupiter in the Solar System quite well. Using two-dimensional hydrodynamical simulations a more general criterion for gap formation has been derived [58]

$$\frac{3}{4}\frac{H}{R_\mathrm{H}} + \frac{50}{q\,Re} \leq 1 , \tag{2.47}$$

with the Reynolds number $Re = r_\mathrm{p}^2 \Omega_\mathrm{p}/\nu$. This last criterion assumes that the density of the disk at the location of the planet has been reduced to 10% in comparison to the unperturbed disk density. Additional, new estimates on the depth and width of the gap have been developed more recently, they will be discussed in more detail in Sect. 2.6 below.

The finding that the mass of Jupiter coincides with the mass required to open a significant gap in the disk has often been taken as an indication that it is the gap creation that will eventually limit the final mass of a planet. This is after all in agreement with the fact that Jupiter is the most massive planet in the Solar System. However, the discovery of several extrasolar planets with masses much above Jupiter's indicated that there must be a way to increase the mass even further in spite of the gap formation. Indeed, hydrodynamical simulations showed that the gap is not as impermeable as thought, because mass can enter the horseshoe region and either be accreted onto the planet or move from outer disk (beyond the planet) to inner disk or vice versa. The detailed flow field of the gas in the close vicinity of the planet is depicted in Fig. 2.12 for a Jupiter mass planet. Clearly, even though a clear gap has formed material can still be accreted onto the planet—the mass entering within the critical while lines on the right panel. For a Jupiter mass planet the mass accretion rate onto the planet is of the order of the equilibrium disk accretion rate, $\dot{M}_\mathrm{disk} = 3\pi\Sigma\nu$ which leads to a doubling time of a few 10^5 yrs for typical disk masses [126]. Increasing the mass from 1 to 6 M_Jup the accretion rate onto the planet drops by nearly one order of magnitude [162]. The results indicate that one Jupiter mass is not the limiting mass for planets in agreement with the observations, but beyond 5–6 M_Jup growth times become very long. Despite this apparent limitation, it turns out that further mass accretion is nevertheless possible because massive planets will induce a significant eccentricity in the outer disk such that periodically the planet will enter into the disk allowing for more mass accumulation onto the planet [129].

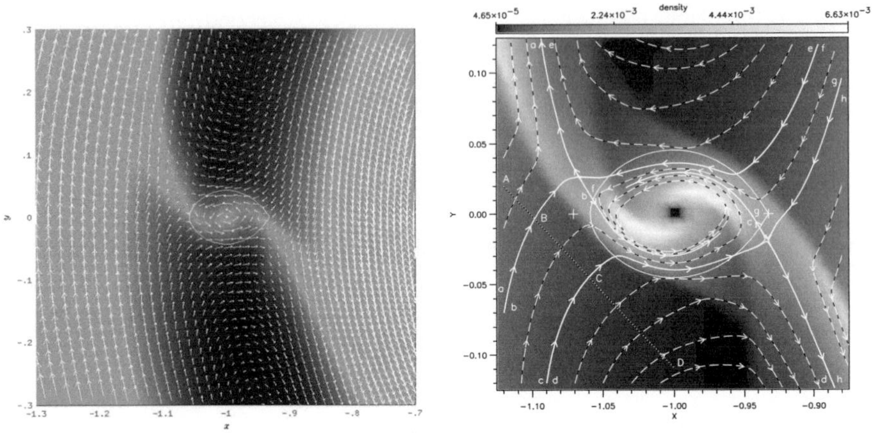

Fig. 2.12 The gas flow around a Jupiter type planet embedded in a protoplanetary disk. The planet is at location $x = -1$, $y = 0$ (in units of 5.2 AU), and the star is located at the origin. Both panels show a density image (in Cartesian coordinates) of the Roche lobe region near the planet. The motion of the planet around the star would be counter clockwise in an inertial frame of reference. (**Left**) The flow field around the planet, displayed in a reference frame corotating with the planet. The solid white line indicates the Roche lobe of the planet. Taken from Fig. 18 in [126], © Royal Astronomical Society, by permission of Oxford University Press. (**Right**) Density contours with sample streamlines given by the dashed lines. The left (right) plus sign marks the L2 (L1) point. Critical streamlines that separate distinct regions are the solid white lines. Material approaching the planet within these critical lines (i.e. between 'b' and 'c' on the outside, and between 'f' and 'g' on the inside) can become accreted onto the planet, while material at the outside (or inside) either circulates or enters into the horseshoe region and crosses it. Taken from Fig. 4 in [162], © American Astronomical Society, reproduced with permission

2.4.4 *Interior Structure of Planets*

After having studied the formation of planets let us very briefly comment on the information on the internal composition that can be drawn from the sample of observed exoplanets. The transit method allows for a determination of the planet's radius, or at least the ratio of planetary to stellar radius, R_p/R_*, because the reduction in flux during the transit is directly proportional to the square of this quantity. The Kepler mission allowed the determination of planetary radii for over thousand exoplanets and even from the ground there have been over 200 transit detections. To know in addition the mass of the planet, the radial velocity signal is required. Unfortunately, this has only been possible for a small fraction of the Kepler planets but for many of the ground based transit detections. Knowing the mass and radius of the planet, the mean density of the planet is determined and rough estimates as to its composition can be made. The important mass-radius (M-R) diagram can be constructed and compared to the theoretical models for the planet interiors. To calculate the interior models the structure equations as written above (2.34–2.37) need to be solved for long evolutionary times, after the disk has dispersed. What determines the final radius of a planet with a given mass is primarily its *composition*. Different constituents directly alter the mean density and will determine the important equation of state (EOS), that relates pressure to density (and temperature). Hence, the EOS defines the compressibility of the material and has direct influence on the planetary radius. The second factor is the *age* of the planet, as this determines the remnant heat that is still incorporated within and has not been lost by some cooling process. A third factor is the *distance from the star* as this determines the amount of external heating that is received by the planet for example by direct irradiation from the central star or the strength of the tidal interaction and dissipation.

In order to obtain accurate models, details of additional physical processes have to be considered that are often not known very well. The equation of state has to be known within a regime up to about 20,000 K and 70 Mbar which is only partly accessible by experiments. Shock-wave and compression experiments give only data up to a few Mbar, and theoretical calculations based on Quantum Molecular Dynamics simulations have to be performed in addition. Knowledge about possible phase transitions has to be acquired. For the cooling rate of planets the amount of radioactive elements need to be known, the efficiency of convection or plate tectonics (viscosity of the material) to be determined. For a recent summary of the present status of the field see the PPVI review [13], where Fig. 2.13 is taken from. In the figure, the M-R diagram is shown for a sample of detected transiting exoplanets and some Solar System planets together with calculated theoretical curves. Because the transit probability increases strongly with shorter periods, i.e. shorter distances from the star, the exoplanets sample refers basically to 'hot' planets.

As seen in the diagram, these can be divided in two major groups: hot Neptunes (Super-Earths) and hot Jupiters. The first group which is the actually most abundant in absolute numbers, as discovered by the Kepler-mission, is under-represented here due to the mentioned lack of radial velocity data of the Kepler planets. Clearly, as can be inferred from the reference objects in the Solar System, the locations of the points give indeed a good first indication of the composition of the planets, indicating that many of the Super-Earths may consist of rock and iron material while the Jupiter mass objects are gaseous planets that will consist primarily of Hydrogen and Helium. Sometimes very exotic options make it into the public media. One example is the so called 'diamond' planet 55 Cnc e, for which some radius estimates indicated a value that matched exactly that of a planet with a composition of 100% diamond material. However, a mixture of carbon, silicates and iron material will match the observations equally well [165]. An additional feature that can be inferred from the Fig. 2.13 is that the majority of the hot Jupiter planets are much larger than predicted even if purely solar composition is assumed, i.e. they are inflated. Assuming irradiation from the star does indeed increase the radius somewhat but is by no means sufficient to explain the observed large radii. Observationally, it has been found that the amount of inflation decreases clearly with distance from the central star [63], implying that the central star is responsible for the effect. Possible suggested mechanisms are irradiation, tidal friction between the orbiting planet and the star during the circularization process of eccentric planets, or electrical currents generated through the interaction of ionized

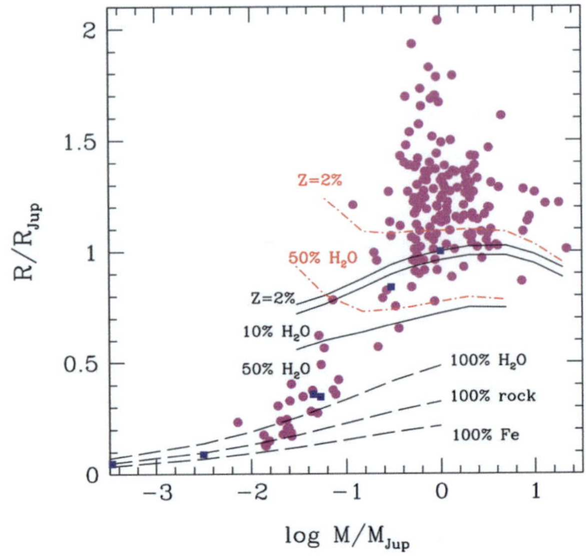

Fig. 2.13 Mass versus radius of known exoplanets, including Solar System planets (blue squares, Mars to Neptune) and transiting exoplanets (magenta dots). The curves correspond to interior structure and evolution models at 4.5 Gyr with various internal compositions, and for a mass range in 0.1 M_{Earth} to 20 M_{Jup}. The solid curves refer to a mixture of H, He and heavy elements, as indicated by the labels. The long dashed lines correspond to models composed of pure water, rock or iron from The 'rock' composition here is olivine (forsterite Mg_2 SiO_4) or dunite. Solid and long-dashed lines (in black) refer for non-irradiated models while dash-dotted (red) curves correspond to irradiated models at 0.045 AU from a Sun. Taken from Fig. 1 in [13], *Protostars and Planets VI*, ed. by Henrik Beuther. © 2014 The Arizona Board of Regents. Reprinted by permission of the University of Arizona Press

particles with the planetary magnetic field [63]. However, the main cause is not known as of today.

Information drawn on the inner structure of the massive planets in the Solar System has given rise to the core accretion scenario of giant planets. Within this model, first solid cores of a few Earth masses are forming in a manner similar to the assembly of terrestrial planets. Once grown big enough, the ambient gas will be accreted onto the core, hence the terminology core accretion scenario. This evolutionary phase can be described by the classical stellar evolution equations augmented by suitable boundary conditions accounting for the fact that the planet is still embedded into the disk and the luminosity is created by the infall of solid material. As it turns out, once the core has grown to a critical mass, the gas accumulation proceeds in a runaway fashion such that on a timescale of a few million years a gas giant can be created at the location of Jupiter's orbit. Going to larger distances from the star the evolutionary timescales (for core formation and gas accretion) become longer than the typical lifetimes of the disk. A solution to this timescale problem may be given by the pebble accretion scenario where solid, cm-sized particles are continuously accreted such that the critical core mass can be reached very fast, reducing the formation time significantly. Information on the interior composition can be obtained for transiting extrasolar planets if additional radial velocity data allow for a mass determination. The observations show that, due to star-planet interactions, most of the Jupiter type planets are significantly inflated in their radii.

2.5 Planets Formed by Gravitational Instability

Having discussed the formation of planets via the core accretion (CA) scenario in the preceding section we will now turn to the alternative scenario of planet formation, the gravitational instability (GI) of the disk. While CA is the preferred scenario for the planets in the Solar System, GI is a possible a pathway considered for planets at large distances from their host stars. First, we will present observational examples of directly imaged planets that are indeed located at large distances from their host stars. Then we will consider the question under what conditions a disk can fragment directly to form planets. For such an analyses primarily two methods have been used. First a *linear stability analyses* and secondly full nonlinear numerical simulations of self-gravitating disks. We will describe the main findings below.

2.5.1 *Background*

The most prominent example of directly imaged planets is the system HR 8799 in the constellation Pegasus. For this system, in 2008 the detection of 3 planets was announced, discovered using adaptive optics at the Keck telescopes in Hawaii [169]. The system was observed at different epochs such that the motion of the individual planets across the plane of the sky was detected. Hence, this discovery marks clearly a breakthrough in exoplanetary science because for the very first time the actual motion of planets around another star was directly detected, following the classical terminology of a planet being a 'wandering star'. Coincidentally, the position of this first directly imaged 'real' planetary system in the sky lies very close to the first planet discovered by the RV method, 51 Peg. The system is observed nearly face on, and the observed motion of the planets agrees very well with the Keplerian motion about the host star that has a mass of 1.5 $M_\odot$. The 3 planets are 24, 38, and 68 AU away from the star and have estimated masses of 10, 10 and 7 $M_{\rm Jup}$. These masses are upper limits set by dynamical stability arguments where it must be assumed that they

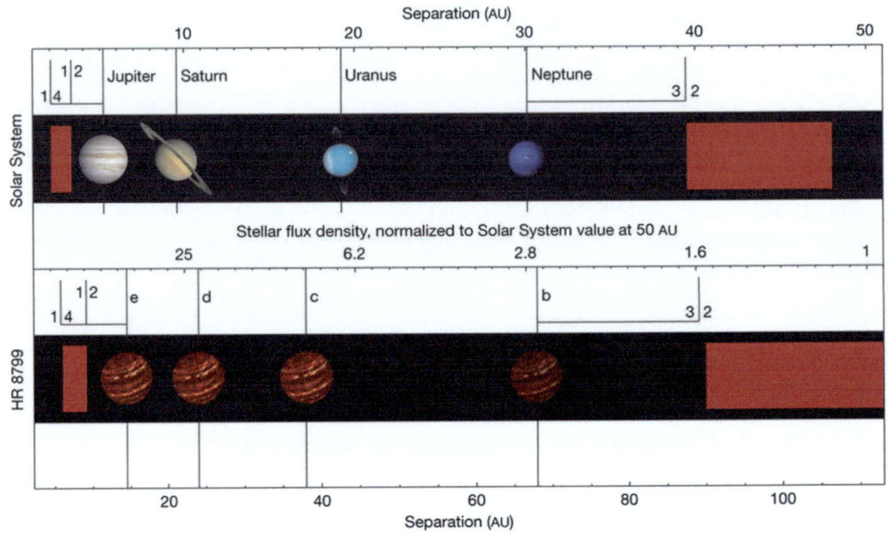

Fig. 2.14 The structure of the planetary system HR 8799, that contains 4 massive planets all discovered by direct imaging, in comparison to the outer Solar System. The x-axis has been compressed according to the luminosities of the host stars by the factor $\sqrt{L_{\rm HR8799}/L_\odot}$, with $L_{\rm HR8799} = 4.9\,L_\odot$. This means that the planets in HR 8799 are about two times farther away from the star but have the same equilibrium temperature as the Solar System planets because of the higher luminosity of the host star in HR 8799. The red rectangles indicate the regions where debris material is orbiting the stars. The lines marked with 1:4, 1:2 and 3:2 indicate the locations mean-motion resonances with respect to the planets. Taken from Fig. 4 in [170], © Springer Nature, reproduced with permission

are engaged in a resonant configuration [84]. Later, in 2010 a fourth planet located at only 14 AU distance from the star was discovered [170]. In Fig. 2.14 the layout of the whole planetary system HR 8799 is displayed. It is shown in a special scaling such that the similarities to the outer Solar System planets become apparent. Even the spatial distribution of the debris material is similar. So, one may speculate about the existence of terrestrial planets in that system. Otherwise there have only been a few systems where directly imaged planets in the few Jupiter mass range have been found, for example Fomalhaut b, β Pictoris b or Gliese 504 b, even though the status of Fomalhaut b has been debated.

Despite these interesting similarities the large absolute distances of the planets in HR 8799 pose a challenge for their formation, due to severe timescale and stability problems. Comparing the effectiveness of various formation scenarios to form planets at large distances such as core accretion (with or without migration), outward scattering from the inner disk, or gravitational instability it was suggested that the last scenario is the most likely [66]. For example, in the classical core accretion scenario the growth time of Neptune at its present orbit will be of the order [6]

$$\tau_{\rm grow} = \frac{m_{\rm p}}{dm_{\rm p}/dt} \approx 5 \cdot 10^{10}\, F_{\rm grav}^{-1} \quad {\rm yr}\,, \tag{2.48}$$

where we assumed $\Sigma_{\mathrm{part}} = 1\ \mathrm{g/cm^3}$ in Eq. (2.22). This is a very long timescale unless the gravitational focusing factor, F, is very large ($\sim 10^4$). Considering the larger distances of the planets in HR 8799 and the youth (60 million yrs) of the host star it appears unlikely that those planets have formed at these large distances by core accretion, although pebble accretion may help in this case (see above). And, since the formation via scattering often led to unstable systems it was concluded that formation via gravitational instability was the most viable mechanism to form the HR 8799 system [66], possibly related to later periods of mass fall-in from the envelope [270]. In models explaining HR 8799, the resonant structure of the planets places further constraints as it seems to require some migration process.

2.5.2 *Linear Stability Analyses*

To study the linear stability of a thin disk rotating around a central object we first assume that the disk is infinitesimally thin and follow the evolution in a two-dimensional setup. In the context of galactic dynamics, the classical studies are given by Toomre [263] and Lin and Shu [154], and a local shearing sheet analysis is presented in [29]. To give an idea on how such a stability analyses is performed we sketch briefly the procedure. The set of hydrodynamical equations in cylindrical coordinates (r, φ) for a disk confined in the $z = 0$ plane are given for example in [146] and read

$$\frac{\partial \Sigma}{\partial t} + \nabla \cdot (\Sigma \mathbf{u}) = 0 \tag{2.49}$$

$$\frac{\partial (\Sigma v)}{\partial t} + \nabla \cdot (\Sigma v \mathbf{u}) = \Sigma\, r \Omega^2 - \frac{\partial P}{\partial r} - \Sigma \frac{\partial \psi}{\partial r} \tag{2.50}$$

$$\frac{\partial (\Sigma r^2 \Omega)}{\partial t} + \nabla \cdot (\Sigma r^2 \Omega \mathbf{u}) = -\frac{\partial P}{\partial \varphi} - \Sigma \frac{\partial \psi}{\partial \varphi} \tag{2.51}$$

where P is the 2D vertically integrated pressure, $\mathbf{u} = (u_r, u_\varphi) = (v, r\Omega)$ the 2D velocity, and ψ the gravitational potential

$$\psi = \psi_* + \psi_{\mathrm{d}} + \psi_{\mathrm{p}}\,, \tag{2.52}$$

that is given here as the sum of the stellar potential, $\psi_* = -GM_*/r$, the disk contribution

$$\psi_{\mathrm{d}}(r, \varphi) = -G \int_{\mathrm{disk}} \frac{\Sigma(r') r' dr' d\varphi'}{\sqrt{r^2 + r'^2 - 2rr' \cos \varphi'}}\,, \tag{2.53}$$

and possibly a planetary contribution, ψ_{p}, that is neglected in this discussion. In Eq. (2.53) the integration has to be performed over the whole disk. For the pressure we assume that it is given as a function of the surface density, $P = P(\Sigma)$, as is the case for the isothermal or adiabatic equations of state. To study the stability we start from an axisymmetric equilibrium state

$$r\Omega_0^2 - \frac{1}{\Sigma_0}\frac{\partial P_0}{\partial r} - \frac{\partial \psi_0}{\partial r} = 0\,, \tag{2.54}$$

where the subscript 0 refers to the unperturbed basic state which is a function of the radius alone. Now the system is perturbed by adding a small perturbation

$$f(r,\varphi,t) = f_0(r) + f_1(r,\varphi,t)\,, \tag{2.55}$$

with $f \in \{\Sigma, v, \Omega, \psi\}$. Here, we only consider perturbations within the plane of the disk. The ansatz (2.55) is substituted into the full time dependent hydrodynamical equations (2.49)–(2.51) which are then *linearized*, i.e. two main assumptions are applied: (a) The functions f_1 are assumed to be small compared to their equilibrium counterparts, i.e. $f_1 \ll f_0$. This implies that terms that are quadratic in the perturbations f_1 can be neglected with respect their linear counterparts. (b) The stratification of the background varies only slowly, which implies that the radial derivatives of the basic functions are assumed to be small in comparison to those of the perturbed functions, i.e. $\partial f_0/\partial r \ll \partial f_1/\partial r$. After these simplifications the non-linear hydrodynamic equations have been transformed into a set of linear equations for the perturbed quantities, that can in principle be integrated numerically. However, better insight is obtained by further analysis.

Because the basic state has neither a time nor an azimuthal dependence, one can quite generally expand the perturbations in a Fourier series

$$f_1 = \tilde{f}_1(r) e^{i(m\varphi - \sigma t)}\,, \tag{2.56}$$

where m denotes the azimuthal wave number of the disturbances and σ the frequency of the temporal variations. As written in Eq. (2.56), the perturbation functions $\tilde{f}_1$ depend now only on the radius, and the time and azimuthal derivatives become $\frac{\partial}{\partial t} \Rightarrow -i\sigma$ and $\frac{\partial}{\partial \varphi} \Rightarrow im$, respectively. With the expansion (2.56) the linearized equations become as set of ordinary differential equations in radius

$$\tilde{\Sigma}_1(\sigma - m\Omega_0) = -i\Sigma_0 \tilde{u}_1' + \Sigma_0 m \tilde{\Omega}_1 \tag{2.57}$$

$$\tilde{u}_1(\sigma - m\Omega_0) = i2r\Omega_0\tilde{\Omega}_1 - i\frac{c_{s_0}^2}{\Sigma_0}\tilde{\Sigma}_1' - i\tilde{\psi}_1' \tag{2.58}$$

$$\tilde{\Omega}_1(\sigma - m\Omega_0) = -i\frac{\kappa_0^2}{2r\Omega_0}\tilde{u}_1 - \frac{c_{s_0}^2}{\Sigma_0} im\tilde{\Sigma}_1 + \frac{1}{r^2} im\tilde{\psi}_1\,, \tag{2.59}$$

with the radial derivative $f' = \partial f/\partial r$. Here, κ_0 denotes the *epicyclic frequency*

$$\kappa_0^2 \equiv \frac{1}{r^3}\frac{\partial}{\partial r}\left[(r^2\Omega_0)^2\right] = 4\Omega_0^2 + 2\Omega_0 r\frac{\partial \Omega_0}{\partial r}\,. \tag{2.60}$$

For a disk in pure Keplerian rotation, $\kappa_0 = \Omega_K$, a relation which is also approximately fulfilled in self-gravitating disks. The sound speed in the disk is denoted with c_s. Now, to simplify matters, we expand also the radial direction in a Fourier-series, i.e. the radial dependence is given by $\propto e^{ikr}$, which implies that $\frac{\partial}{\partial r} \Rightarrow ik$.

Finally, we make the so-called *tight winding approximation*, which means $kr \gg m$, i.e. the radial wavelength $(1/k)$ is small against the azimuthal (r/m). This implies that all terms containing the azimuthal wavenumber m on the right hand side in Eqs. (2.57)–(2.59) can be neglected. The last problem relates to the evaluation of the perturbation of the disk potential, $\tilde{\psi}_1$. This is obtained not from the Poisson-integral (2.53) but rather from the linearized Poisson-equation

$$\nabla^2 \psi_1 = 4\pi G \Sigma_1 \delta(z)\,, \tag{2.61}$$

where we assume a matter distribution that is only non-zero within the plane of disk ($z = 0$), i.e. $\delta(z)$ denotes the δ-function. Integrating Eq. (2.61) over a small volume around the disk one obtains [6]

$$\tilde{\psi}_1 = -\frac{2\pi G \tilde{\Sigma}_1}{|k|}\,, \tag{2.62}$$

and Eqs. (2.57)–(2.59) turn into the *dispersion relation* [154]

$$(\sigma - m\Omega_0)^2 = \kappa_0^2 + c_{s_0}^2 k^2 - 2\pi G |k| \Sigma_0\,. \tag{2.63}$$

Remembering the time dependence of the perturbations, $\propto e^{i\sigma t}$, it is clear that in general perturbations are: (a) Stable, for $\sigma^2 > 0$ because in this case σ is real and the disturbances oscillate in time with a frequency σ, or they are (b) Unstable, for $\sigma^2 < 0$ because in this case σ is imaginary and the disturbances can grow exponentially, leading to an unlimited growth, i.e. instability. The point of marginal stability is given by $\sigma = 0$. From our relation (2.63) one can see that the epicyclic oscillations (κ_0-term) are stabilizing at all spatial scales, this is the classical *Rayleigh stability criterion*. Indeed, for individual particles orbiting a central object that are slightly perturbed, κ_0 is the oscillation frequency of the particle around its equilibrium position. In the case of a spread out gas the propagation of sound waves and self-gravity come into play. For the sound waves (c_s-term) the stabilizing effect is larger for larger k, i.e. for smaller spatial scales. On the other hand, the last term in Eq. (2.63) refers to the effect of the self-gravity of the disk which is always destabilizing due to the minus sign and proportional to the local surface density, Σ_0.

Let us now consider axisymmetric disturbances with $m = 0$. Since for stability the frequencies must be real, $\sigma^2 \geq 0$, the most unstable oscillations are those where σ^2 is minimal. Because σ is a function of k the most unstable *critical* wavelength, $k_{\rm crit}$, can be calculated from $d\sigma^2/dk = 0$, and we obtain

$$k_{\rm crit} = \frac{\pi G \Sigma_0}{c_s^2}\,, \tag{2.64}$$

where we drop for simplicity the index 0 at the sound speed. Substituting this into the dispersion relation (2.63) and investigating the point of marginal stability by setting $\sigma^2(k_{\rm crit}) = 0$ we obtain after rearranging

$$Q \equiv \frac{c_{\rm s}\kappa_0}{\pi G \Sigma_0} = 1\,, \tag{2.65}$$

where we defined the *Toomre parameter* Q. This relation implies that $Q = 1$ defines the borderline between stable and unstable configurations, the marginal state. Indeed, from the dispersion equation (2.63) that is quadratic in k one can show that in the case of axisymmetric disturbances the following inequality must be satisfied for stability [253]

$$\text{For stability:} \qquad Q > 1 \qquad \text{(Toomre-Criterion)}\,. \tag{2.66}$$

As just discussed above and directly seen from (2.66), for a given $\kappa_0 \approx \Omega_{\rm K}$ the disk will be stabilized by higher temperatures (increase in $c_{\rm s}$) implying thicker disks, while a larger surface density will lead to destabilization. Hence, whenever Q is of the order unity the disk is prone to instability. Using these definitions for $k_{\rm crit}$ and Q the dispersion relation (2.63) can be rewritten as

$$\left(\frac{\omega - m\Omega_0}{\kappa_0}\right)^2 = 1 + \frac{1}{Q^2}\left(\frac{k^2}{k_{\rm crit}^2} - 2\frac{|k|}{k_{\rm crit}}\right)\,. \tag{2.67}$$

This function is displayed graphically in Fig. 2.15 for different Q values. Obviously, the minima of the parabolas occur always at $k = k_{\rm min}$ and for $Q = 1$ it coincides with marginal stability.

After having obtained now a useful criterion for disk instability, it remains to be seen what happens actually to an unstable disk. Before doing so, let us consider the situation in the Solar System. For the protosolar nebula at 10 AU with $H/r \approx 0.05$ (i.e. $c_{\rm s} \approx 0.33$ km/s), a value $Q = 1$ requires $\Sigma_0 \simeq 10^3$ g/cm^2, which is much larger than the MMSN value ($\approx$50 g/cm^2). This implies that for the Solar System the gravitational instability could only have worked in an early evolutionary phase, when the mass of the protosolar nebula was still high. Using the critical wavelength $\lambda_{\rm crit} = 2\pi/k_{\rm crit}$ the mass of such a fragment can be estimated to be around

$$M_{\rm p} \sim \pi\,\Sigma_0 \lambda_{\rm crit}^2 = \frac{4\pi c_{\rm s}^4}{G^2\,\Sigma_0} \sim 2\,M_{\rm Jup} \tag{2.68}$$

which lies in the range of the most massive gas giant in the Solar System. The idea of Solar System planet formation via gravitational instability goes back to Kuiper [143] or Cameron [42].

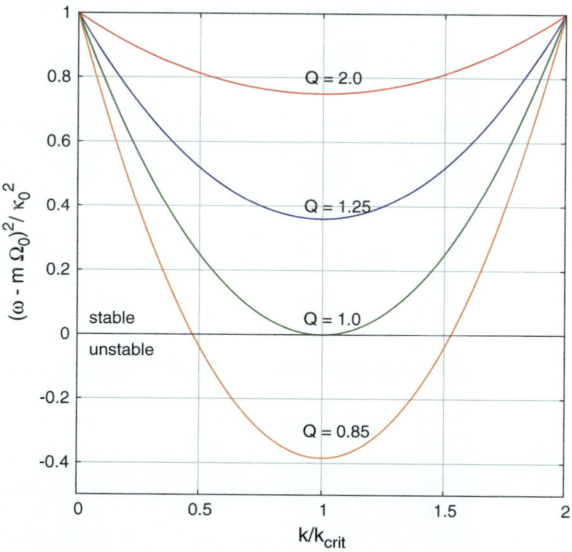

Fig. 2.15 The normalized dispersion relation (2.67) for perturbations in an infinitesimally thin disk. The critical wavenumber, k_{crit} is given by Eq. (2.64) and the Toomre number, Q by Eq. (2.65). For $Q = 0$ marginal stability is reached. For large k, i.e. small wavelengths, the disk is stabilized by pressure (sound waves) and for small k by a reduced density Σ_0. Increasing rotation κ_0 stabilizes as Q rises

2.5.3 Fragmentation Conditions

Many non-linear hydrodynamic simulations have been performed to study the fate of unstable disks. In such simulations, typically a young protostar with a mass in the range of $(0.5-1.0)\,M_\odot$ is surrounded by a disk having a mass of a few tenth of the stellar mass. When approaching the stability limit, which is usually done by increasing the disk mass with respect to the stellar mass, the main outcome is the formation of spiral arms of low order azimuthal wavenumber, very similar to galactic disks [146, 184].

To determine those areas which are most susceptible to fragmentation let us consider an accretion disk with a constant mass flow through the disk, $\dot{M} = 3\pi\nu\Sigma$. In an accretion disk the effective viscosity is given by

$$\nu = \alpha c_{\mathrm{s}} H = \alpha c_{\mathrm{s}}^2/\Omega\,, \tag{2.69}$$

and we find for the Toomre-Q

$$Q \propto \frac{c_{\mathrm{s}}^3}{\dot{M}}\,. \tag{2.70}$$

Assuming that $\dot{M}$ does not vary too strongly with radius this implies that Q falls off with radius because the disks become cooler at larger distances from the star. Hence, the most unstable region lies in the outer parts of the disk. This trend is clearly seen in early 3D grid-based simulations by A. Boss [40] where he studied the evolution of isothermal and adiabatic disks with a mass of about 140 M_{Jup} within 10 AU around a Solar mass star. In both cases clumps formed near the outer boundary. Similar

Fig. 2.16 An example of an SPH simulation of a self-gravitating disk. During the evolution spiral arms are forming that are later fragmenting into a number of individual planet like objects. For details of the simulation see Sect. 2.5.4 below. *Courtesy* Farzana Meru

results were obtained with the Smooth Particle Hydrodynamics (SPH) method. Using over one million particles in isothermal simulations is was shown that fragments of planetary mass can form easily [180] and on very short timescales of only a few hundred years. A typical example on how such a simulation looks like is shown in Fig. 2.16. This strong reduction in assembly-time clearly shows the great interest in the GI-mechanism as a possible path to giant planet formation. On the other hand for adiabatic simulations it was found that the forming clumps were sheared out and dispersed [181].

These results clearly indicate that the outcome, whether or not fragments are forming, will depend not only on the present value of the temperature but on the disk thermodynamics, that determines how the matter reacts upon compression. The internal temperature of the disk is determined by the balance of heating and cooling processes. If the cooling is higher than the heating, the disk will be unstable, if it is lower, the disk will be stable. For the disk we have the following *heating mechanisms* operating:

- internal dissipation in shock waves, for example produced by spiral arms
- effective viscosity, produced by the turbulent motion within the disk. This is typically modeled via standard viscous dissipation for example in an α-disk model as in Eq. (2.69)
- heating by external sources such as the central star, cosmic rays or nearby stars. This will be more important in the outer parts of the disk

On the other hand the following cooling mechanisms can be considered

- equation of state (EOS)
  The EOS determines the behavior of the gas upon compression. A medium that can be strongly compressed without heating up will be more susceptible to instability than a medium that heats up strongly. The following cases are often considered in running disk models:
  EOS1—(locally) isothermal. Here the disk temperature is a given function of radius that cannot change. This is equivalent to strong cooling, i.e. the gas cannot heat up upon compression
  EOS2—locally isothermal for low gas densities, then adiabatic above some suitable ρ_{crit}. This type of EOS models the turnover from an optically thin gas of low density and a denser medium that heats up upon compression. This type frequently used in star formation simulations [164]
- simple cooling laws
  Quite generally the cooling time is defined by

$$t_{\mathrm{cool}} = \frac{e_{\mathrm{th}}}{de_{\mathrm{th}}/dt}, \tag{2.71}$$

  where e_{th} is the thermal energy of the disk per surface area. Simple approximations are often used in the context of planet formation
  $t_{\mathrm{cool}}\Omega = const.$ (t_{cool} is a fixed fraction of the local rotational period)
  $t_{\mathrm{cool}} = const.$ (t_{cool} is fixed throughout)
- radiative cooling (from disk surfaces)
  The Rosseland mean opacity is proportional to $\kappa_{\mathrm{R}} \propto Z\,T^{\varepsilon}$, where the magnitude of κ_{R} is given by the amount of dust particles embedded in the disk, with the metal abundance Z. If one assumes that the energy is locally radiated away from the two disk surfaces with the flux, $F_{\mathrm{eff}} = \sigma_{\mathrm{B}} T_{\mathrm{eff}}$, then

$$t_{\mathrm{cool}} \simeq \frac{e_{\mathrm{th}}}{2\sigma_{\mathrm{B}} T_{\mathrm{eff}}^4} \propto T/T_{\mathrm{eff}}^4 \propto T^{-3+\varepsilon}\,Z \tag{2.72}$$

  where we have assumed an optically thick case where the midplane temperature is related to the surface temperature via

$$T_{\mathrm{eff}}^4 = T_{\mathrm{mid}}^4/\tau \tag{2.73}$$

  using the mean vertical optical depth, $\tau \sim \Sigma\kappa_{\mathrm{R}}$. Typically: $-3 < \varepsilon < 3$, such that t_{cool} grows with lower temperature

The last option, radiative cooling in combination with a realistic EOS is certainly the most realistic approximation one can make for flat two-dimensional disks but often the simple cooling laws are used.

As pointed out, the likelihood for fragmentation depends on the cooling timescale of the gas. For rapid cooling the system will be more likely to fragment than systems where the cooling rate is longer. Often, a simple β-*cooling* law is applied in numerical simulations with

$$t_{\text{cool}} = \beta\, \Omega^{-1}\,, \tag{2.74}$$

and a constant value of β. In a linear analysis for the local shearing sheet it was shown [90] that the instability is then determined directly by the value of β, in particular

$$t_{\text{cool}} \leq 3\Omega^{-1} \quad \Rightarrow \quad \text{fragmentation} \tag{2.75}$$

$$t_{\text{cool}} \geq 3\Omega^{-1} \quad \Rightarrow \quad \text{no fragmentation}\,. \tag{2.76}$$

This implies that $\beta_{\text{crit}} = 3$ is the critical value [90]. A simple estimate for β in accretion disks can be obtained from thermodynamic equilibrium where it is assumed that the internally produced heat is radiated away locally leading to the following cooling behaviour [227]

$$\frac{e_{\text{th}}}{t_{\text{cool}}} = \Sigma\nu \left(\frac{d\Omega}{dr}\right)^2\,. \tag{2.77}$$

Here, it was assumed that the heat generation, produced for example by gravo-turbulence for marginally stable self-gravitating disks, can be written as an effective viscous dissipation with kinematic viscosity ν. For an ideal gas with the thermal energy $e_{\text{th}} = c_{\text{v}}\Sigma T$, where c_{v} denotes the specific heat at constant volume, an α-viscosity as in Eq. (2.69), and Keplerian rotation, one finds for the equilibrium state

$$\Rightarrow \qquad t_{\text{cool}} \simeq \frac{4}{9}\frac{1}{\gamma(\gamma-1)\alpha}\,\Omega^{-1}\,. \tag{2.78}$$

As to be expected, the cooling ability of the gas directly determines the level of viscosity in the disk. locally. For $\alpha \sim 10^{-2}$ and $\gamma = 1.4$ we find with $e = \Sigma c_{\text{v}} T$ a cooling time of $t_{\text{cool}} \sim 12$ periods. This is roughly the timescale for changes of the thermal structure of an accretion disk.

One should keep in mind however that Eq. (2.78) describes the equilibrium situation for the disk and local variations are to be expected in realistic cases. Additionally, the simple β-cooling that went into the derivation of (2.78) is not a realistic cooling, as it does not depend on the density of the gas. When a gas clump is compressed it is to be expected that the cooling time rises and the clump will heat up preventing further collapse, i.e. for a realistic modeling more sophisticated cooling prescriptions, such as (2.72) will have to be applied. Applying the β-cooling prescription (2.74) simulations performed in the 2D shearing-sheet approximation using grid-based numerical models give results in rough agreement with the above fragmentation condition [90], as do corresponding global 3D disk models using the SPH method [240]. However, using very high resolution simulation in the 2D isothermal setup, it was shown that even for $\beta \approx 20$ fragments could form due to the stochastic nature of the turbulent flow [204].

As just mentioned, for fragmentation to occur in disks a cooling time shorter than $\approx 3\Omega^{-1}$ is required. From the definition of Q (2.65), using representative disk conditions with $H/r = 0.05$, one can obtain estimates on the cooling time. Assuming a disk close to possible instability by setting $Q = 1.5$, and using Eq. (2.73) for the optical depth one finds the following relation for the cooling time [6]

$$t_{\rm cool}\Omega \sim 10\,\tau \left(\frac{r}{5\,AU}\right)^{-1/2} . \tag{2.79}$$

Noticing that in optically thick disks the optical depth can easily reach 10^2 and more, this last relation seems to imply that there is no possibility for fragmentation at typical planet formation locations in the Solar System. This finding represents the fact that the fragmentation requirement, high mass (implying large Σ and high τ) and low cooling times (low τ) contradict each other. The presented arguments are based on the assumption of vertical energy transport by radiation (relation 2.73). Including vertical convective energy transport the disk can cool faster and the situation improves somewhat. For lower temperatures the opacities are lower and the cooling time is reduced again. Including additionally external perturbation, for example by a passing star that compresses the outer regions in the disk, a gravitational instability may be triggered. In summary one may conclude that, if at all, fragmentation can only be expected at very large radii from the star beyond 50 AU or so [230].

2.5.4 *Non-linear Simulations*

At the end of this section we would like to summarize briefly the results of a few recent numerical simulations concerning the fate of gravitational unstable disks. This is not an exhaustive review as there have been many simulations concerning this topic and the issue is not conclusively decided at present. A summary of the field was given in the Protostars and Planets V proceedings [79] where different numerical methods were directly compared using similar initial initial conditions. The codes used in the comparison included particle methods such as SPH (the public codes `GASOLINE` and `GADGET`) as well as the grid codes that used finite difference or finite volume methods either based on upwind schemes (`Indiana Code`) or Riemann-solvers (`FLASH`). It was noted that the outcome of the simulations depends crucially on numerical aspects such as spatial resolution (number of grid-points, or particles), regularization methods (smoothing length, artificial viscosity, flux-limiters), solver for self-gravity (smoothing criteria) and so on. Hence, it was concluded that in order to obtain reliable results it seems unavoidable to compare different methods on the same physical problem. Using appropriate criteria for the individual codes similar results could be obtained. One should keep in mind however, that the results obtained with different codes on this problem will never be identical because of the chaotic nature of the problem. Hence, the results can only be compared in a statistical sense.

2.5.4.1 The Convergence Issue

The sample study of Meru and Bate [185] illustrates one of main numerical problems very well. They considered the situation of a 0.1 $M_\odot$ disk around a 1 $M_\odot$ star spanning a factor of ten in radii, ranging from 0.25 to 25 (in dimensionless units). For the cooling they used the β-prescription (2.74) with different values for β within $2.0 \leq \beta \leq 18$. They performed SPH-simulations using a huge range of different particle numbers from $N = 31{,}250$ to $N = 16$ million particles. A typical result of such a simulation is displayed in Fig. 2.16. The results indicate that the value of β below which fragmentation occurs depends strongly on the particle number used in the simulations. As shown in Fig. 2.17 the critical value of β increases monotonically with N as given by the solid line. The hatched region indicates the fragmentation regime while above the line the disks are stable. Clearly, there is no real indication of conver-

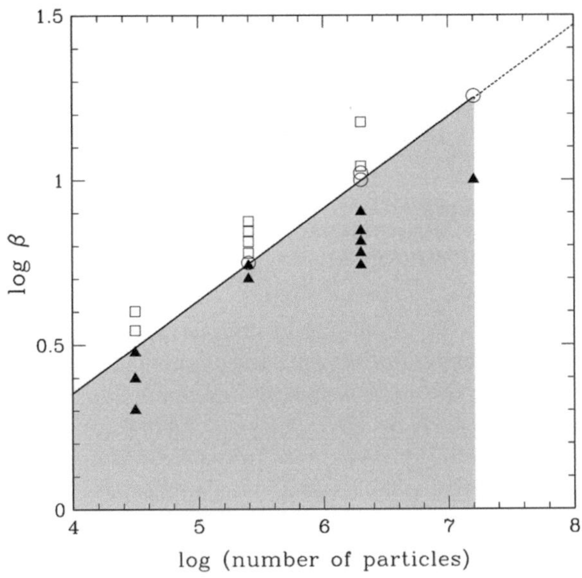

Fig. 2.17 The cooling rate β against numerical resolution (here the used particle number) for SPH-simulations of self-gravitating disks, here a 0.1 $M_\odot$ disk around a one solar mass star. The symbols denote the numerical outcome: non-fragmenting (open squares), fragmenting (solid triangles) and borderline (open circles) simulations. The borderline simulations are those that initially fragment but whose fragments are sheared apart. The solid black line separates fragmenting and non-fragmenting cases and the gray region marks the fragmentation regime. Taken from Fig. 3 in [185], © Royal Astronomical Society, by permission of Oxford University Press

gence in these simulations. Because the aforementioned SPH simulations are fully three-dimensional and hence require a very high particle number for numerical convergence, an alternative option that has been used frequently is the two-dimensional grid code FARGO [174] that has been empowered with a self-gravity module in 2D [19]. Comparing both codes, SPH and FARGO, Meru and Bate [186] find that the question of convergence hinges at the treatment of numerical viscosity, for both codes. They argue that convergence can be reached yielding critical β-values of about 20–30, making fragmentation easier. A detailed analyses and additional SPH simulations using an improved cooling prescription as well improved artificial viscosity (and consequently artificial dissipation) found numerical convergence and fragmentation occurring for cooling times between $\beta = 6$ and $\beta = 8$ for an adiabatic index $\gamma = 5/3$ [242]. They also find that fragmentation can occur only at very large distances beyond at least 50 AU, in agreement with the above expectations [230], and as seen in other studies [140, 268]. These cases demonstrate again the numerical problems that are still present in such self-gravitating disk simulations, and indicate the necessity to compare different codes and methods.

2.5.4.2 The Gravitational Potential

Care has to be taken when treating the disk in a flat 2D geometry only. Even though the disk may be vertically thin, the gravitational potential has to be modified to account correctly for the vertical extent of the disk [197] as illustrated in Fig. 2.18.

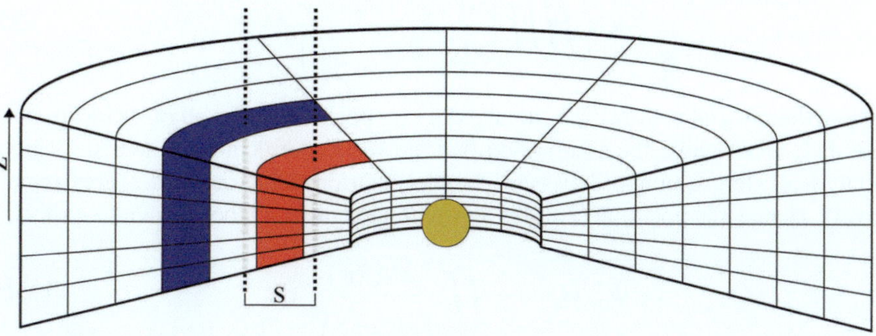

Fig. 2.18 Geometry of a protoplanetary disk around a central star to illustrate the requirements for the calculation of the disk's self-gravity in 2D simulations. The goal is to calculate the gravitational force exerted by a vertical slice of the disk (blue) on another vertical slice of the disk (red), that are separated by the projected distance s. As seen in the drawing, two vertical integrations have to be performed along the dashed lines that go through the cell centers assuming cylindrical coordinates. To obtain the total force between the two segments, this value has to be multiplied by the corresponding areas, see also Eq. (2.82). Taken from Fig. 1 in [197], © ESO, reproduced with permission

The potential at a point $\mathbf{r}$ generated by the remaining part of a self-gravitating disk is given by

$$\Psi_{\rm sg}(\mathbf{r}) = -\int_{\rm Disk} \frac{G\rho(\mathbf{r}')}{|\mathbf{r}-\mathbf{r}'|}\, d\mathbf{r}' . \tag{2.80}$$

Here, in contrast to the earlier expression (2.53) the full vertical stratification of the disk has been taken into account. As the standard 2D hydrodynamic equations (2.49)–(2.51) are obtained by an averaging process over the vertical direction, this has to be done for the gravitational potential as well. This is typically approximated by a suitably chosen smoothing parameter. For this purpose it is convenient to analyze the force between to individual elements (segments) of the disk. The potential at the location $\mathbf{r}$ generated by a disk element located at $\mathbf{r}'$ which is a projected distance s away is given by

$$\Psi_{\rm sg}(\mathbf{r}) = -\iiint \frac{G\rho(r',\varphi',z')}{\left(s^2+(z-z')^2\right)^{\frac{1}{2}}}\, dz'\, dA' . \tag{2.81}$$

Here dA' is the surface element of the disk in the disk's midplane (the $z = 0$ plane) that is located at the point $\mathbf{r}'$, separated by the projected distance s. The force at the position $\mathbf{r}$ due to the rest of disk is calculated from the gradient of the potential (2.81), and vertical averaging leads to the following force density [197]

$$\begin{aligned} F_{\rm sg}(s) &= -\int \rho(r,\varphi,z)\frac{\partial \Psi_{\rm sg}}{\partial s}\, dz \\ &= -Gs \iiiint \frac{\rho(r',\varphi',z')\,\rho(r,\varphi,z)}{\left(s^2+(z-z')^2\right)^{\frac{3}{2}}}\, dz'\, dA'\, dz , \end{aligned} \tag{2.82}$$

where the integral is over the vertical extent (at position $\mathbf{r}$) and all other locations of the disk. Even under the assumption that the vertical structure of the disk is given analytically (e.g. a simple Gaussian), the computation of the force (2.82) is extremely costly. Hence, the potential (2.81) is approximated often by the so-called ε-potential

$$\Psi_{\rm sg}^{\rm 2D}(s) = -\iint \frac{G\Sigma(\mathbf{r}')}{\left(s^2+\varepsilon_{\rm sg}^2\right)^{\frac{1}{2}}}\, dA' , \tag{2.83}$$

where $\varepsilon_{\rm sg}$ is the smoothing length for self-gravitating disks that takes into account the unresolved vertical extent of the disk. The force acting on each disk element is then calculated from the gradient of $\Psi_{\rm sg}^{\rm 2D}$. This formulation extends Eq. (2.53) for numerical simulations and the smoothing length does not only ensure that the numerical evaluation remains finite but is physically necessary due to the finite disk thickness. By comparing the results obtained with (2.83) to the exact formulation (2.81) (using a Gaussian vertical density profile) one obtains a good match with $\varepsilon_{\rm sg} \approx H$ [197].

2.5.4.3 Fragmentation Outcome and Longterm Evolution

As a massive protoplanetary disk may still be embedded somewhat in their envelope it is important to model the disk evolution in combination with infall onto the disk. Such studies have been performed through 2D numerical simulations of viscous disks that include radiative cooling, stellar irradiation, and mass infall [298]. The results show that fragmentation is possible when mass infall is included. However, two major problems have been identified: (a) Mass challenge: To be able to fragment the disk must have a high mass ($M_{\rm disk} > 0.3\, M_\odot$) and requires a high infall rate. The fragments then grow very fast and typically end up as Brown Dwarfs (BD) with masses well above the planetary regime. (b) Migration challenge: Once formed the fragments migrate inward rapidly on a timescale $<$1000 yrs. Only large masses survive, and small ones get tidally disrupted. The authors conclude that GI will lead to massive BDs or binaries. The migration problem will be discussed also in the following section. Similar findings have also been found in full 3D simulations of fragmenting disks [264].

Important further aspects concern the subsequent evolution of the clumps after they have formed. As just mentioned they migrate inward very rapidly and can be tidally disrupted upon reaching the inner region of the accretion disk close to the central star. This can lead to accretion events such as FU Ori outbursts that occur in the early phase of star formation when the disk is still very massive [269]. As an alternative, the *tidal downsizing scenario* considers another fate of the clumps. Upon inward migration the Roche radius of the massive clumps is shrinking continuously such that they might lose their envelopes by tidal effects through mass overflow across the L_1 Lagrange point, see Fig. 2.7. The final outcome could be objects with much smaller mass, sometimes even rocky planets [199]. In this scenario the above mentioned mass challenge would be resolved as the clumps loose the majority of their mass, and the gas giants could be assembled on a short timescale.

After having discussed two different pathways (by CA or GI) in forming giant planets, it is interesting to ask if one can possibly distinguish them observationally. The answer lies in the thermal evolution of the young planets. Planets formed by gravitational instability are thought to have higher entropy, larger radii, and possibly higher effective temperatures than core-accretion objects, because of the stored heat in the accumulated gas that did not have time to cool during the rapid accretion process. Following the longterm thermal evolution it has been shown that for planets with mass less than about 5 $M_{\rm Jup}$ differences in the effective temperature persist for a few million years only, while for planets of higher mass it can take up to about 100 million years before equilibration [255]. However, the feasibility of this distinction hinges on details of the models. For example, the accretion of additional solid material during the subsequent evolution will enhance the cooling of the planets formed by GI and reduce the differences to the CA planets, implying that it might be difficult to separate the two options observationally.

As a final remark, concerning possible difficulties of producing gaseous planets directly through the GI process we would like to point out that even the formation of planetesimals may be enhanced in the early phases of very massive disks. As shown in the previous section, small bodies experience gas drag and will drift towards pressure maxima occurring in the disk. As it turns out, in marginally stable disks the spiral arms may be sufficiently long lived to support the collection of particles in them such that the concentrated dust particles can collapse to their own self-gravity to form directly planetesimals in the early phase of planet formation [241].

Forming planets by direct gravitational instability of the disk has long been considered as an alternative option to the core instability pathway. An advantage would be the associated short time scale of the formation process. As stated by the well-known Toomre criterion, for the mechanism to work the disk must be sufficiently massive and at low temperature. In order for the collapse to proceed the gas must be able to cool sufficiently fast to get rid of the thermal energy. The required cooling times are only a few (3–5) dynamical timescales. These conditions, high disk mass and short cooling times can only be fulfilled in the very outer parts of a protoplanetary disk, probably only beyond a distance of about 50 AU. Hence, it represents a possible formation scenario for the distant planets detected by the direct imaging method.

2.6 Planet-Disk Interaction

In the preceding section we have implicitly assumed that the planets remain on their orbits during the growth phase. This in-situ formation process is based on the situation in the Solar System in which the large planets reside further out where they could accrete large amounts of gas, while the terrestrial planets are located further in where, due to the higher disk temperatures, only small amounts of gas were available to be accreted onto the rocky cores. Comparing this situation with the observed semi-major axis distribution of the extrasolar planets (as shown in Fig. 2.2) it was clear from the very beginning, in fact already after the discovery of the very first object, 51 Peg, by Mayor and Queloz [182], that the presence of close-in planets is difficult to explain by an in-situ formation process. This is because close to the star the amount of solids available for core formation is not sufficient and the disk temperature is too high to condense the gas. For an alternative point of view see for example [37]. This difficulty led to the immediate suggestion that 51 Peg had formed in the standard way at larger distances, similar to Jupiter in the Solar System, and subsequently drifted inward driven by planet-disk interaction [155]. Such a migration process is a natural consequence of the planet formation process because the gravitational interactions between the planet and the ambient disk will necessarily lead to forces acting on the planet causing it to change its orbital elements. In this section we will describe the physical processes of disk-planet interaction in more detail. Recent reviews of

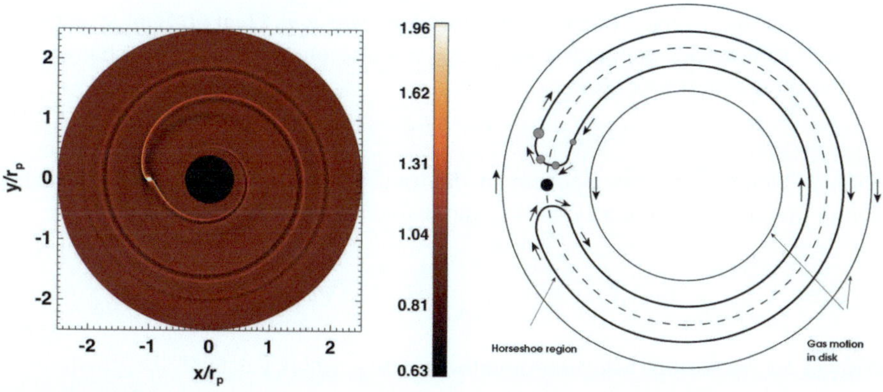

Fig. 2.19 The influence of a planet that is embedded in a protoplanetary disk. The planet moves counterclockwise around the central star which is located in the center of the two plots. The left panel shows the surface density distribution (in dimensionless units) of a 10 M_{Earth} planet on a circular orbit that is embedded in a disk of constant density. Shown is the configuration at about 20 orbits after insertion of the planet. The planet is a unit distance away from the star at the location $x = -1$, $y = 0$. The right panel shows a schematic plot of the motion of gas particles (smaller gray dots) on orbits close to the planet. The motion is depicted using arrows in a reference frame corotating with the planet. The horseshoe region that is corotating with the planet is contained within the region enclosed by the thick solid line, while the inner and outer circulating orbits are shown in the lighter color. Based on Fig. 1 in [133]

the topic are presented in the Protostars and Planets (V and VI) proceedings [15, 213] or in [20, 133].

2.6.1 *Basic Concepts*

Embedding a planet into the disk leads to characteristic disturbances in the density of the disk. Due to the shearing motion in the disk the sound waves emanating from the planet's location are sheared out and spiral waves are forming. As shown in Fig. 2.19 on the left panel, the inner spiral is leading and the outer one trailing, where the planet moves counterclockwise around the star. These spiral waves represent over densities in the gas distribution and generate a net force acting on the planet that changes its orbit. The second contribution to the forces comes from the corotation region of the planet which is depicted in the right panel of Fig. 2.19. For single particle trajectories the dynamics on both sides of the planet would be identical. However, for fluid motion there exists a small asymmetry between the U-turns of the gas in front and behind of the planet which generates again a net force on the planet. For a planet of mass m_{p} on a *circular orbit* its angular momentum is given by

$$J_{\mathrm{p}} = m_{\mathrm{p}} r_{\mathrm{p}}^2 \Omega_{\mathrm{p}}\,, \tag{2.84}$$

where r_p is the distance of the planet from the star and Ω_p its orbital speed. The density asymmetries generate a torque, Γ_p, acting on the planet that changes the angular momentum according to

$$\dot{J}_p = \Gamma_p\,. \tag{2.85}$$

Here, Γ_p denotes the z-component of the total torque that needs to be calculated adding the individual contributions of the whole disk

$$\Gamma_p = -\int_{\rm disk} \Sigma(\mathbf{r}_p \times \mathbf{F})_z\, df \;=\; \int_{\rm disk} \Sigma(\mathbf{r}_p \times \nabla\psi_p)_z\, df \;=\; \int_{\rm disk} \Sigma \frac{\partial \psi_p}{\partial \varphi} df\,. \tag{2.86}$$

In Eq. (2.86) we assume a flat two-dimensional disk, where **F** denotes the gravitational force per unit mass between the planet (located at $\mathbf{r}_p$) and a small surface element, df, of the disk which has the surface density, $\Sigma(r, \varphi)$. ψ_p denotes the gravitational potential of the planet. From Eq. (2.86) we can infer that it is exactly the asymmetry in the azimuthal direction that changes the angular momentum of the planet. From Eqs. (2.84, 2.85) and using Keplerian rotation

$$\Omega_K(r) = \sqrt{\frac{GM_*}{r^3}}\,, \tag{2.87}$$

we can calculate the change in the planet's distance as

$$\frac{1}{r_P}\frac{dr_P}{dt} \equiv \frac{1}{\tau_M} = 2\frac{\Gamma_P}{J_P}\,, \tag{2.88}$$

where we introduced the *migration timescale* τ_M. To infer the importance of the planetary migration, this timescale needs to be compared to the overall evolution time of disk.

Obviously, the analysis of the migration process comes down to the calculation of the torques acting on a planet. As we shall see below in Sect. 2.6.5.1 this statement refers to planets on circular orbits only, otherwise the power acting on the planet has to be considered as well. In principle, according to Eq. (2.86), this appears to be an easy evaluation. The problem lies in the fact that the total torque is always the difference between a positive and negative contribution that have nearly the same magnitude. As a consequence the total torque depends delicately on physical aspects of the disk such as viscosity, or heat diffusion. As we shall see, this does not only influence the speed of the migration but can change even the overall direction.

The calculation of the total torque can be performed by different approaches. For lower mass planets that perturb the density of the disk only mildly one can perform a *linearization* of the hydrodynamic equations around the unperturbed disk without the planet, similar to the methods shown in the previous section. Here, the basic state is given by an axisymmetric disk that is in pure rotation around the star with $\Omega(r)$ and has a given $\Sigma(r)$ profile and possibly temperature gradients. The planet is then added as a small disturbance via an additional potential that rotates with the speed Ω_p. Linearizing the equations yields a new perturbed disk structure, whose back reaction on the planets gives the torque, and hence the migration rate. In the azimuthal direction the potential can be expanded in a Fourier-series due to the periodicity. Using this procedure, the strongest perturbations in the disk are excited at particular resonances where the pattern speed of individual Fourier-components of the perturbed potential, $\omega = m(\Omega(r) - \Omega_{\rm p})$, matches a natural frequency of oscillation in the disk. The first resonance is the corotation resonance that occurs for material at the location of the planet which has $\Omega = \Omega_{\rm p}$. These are related closely to the horse-shoe drag that we will discuss below. Other resonances are excited at the those locations where the pattern speed ω matches the epicyclic frequency κ, see Eq. (2.60). These are the *Lindblad resonances* that may occur inside of the planetary orbit or outside.

In the second approach the torques are obtained from full non-linear hydrodynamic simulations of embedded planets in disks. Here, the disk is initialized with an equilibrium profile of density and temperature and then the planet is added as a perturbing object to the flow. In one type of models the initial planetary orbit is fixed in time and only the disk structure changes due to the action of the embedded planet. The simulations are then evolved in time until the system has reached a quasistationary equilibrium. The torque on the fixed planet is continuously monitored and the theoretical migration speed can be calculated by Eq. (2.88). In other models the planet is allowed to change its orbital elements according to the forces acting on it, and the change in orbital elements is directly calculated. Will not go into details of these methods but just quote the main results.

2.6.2 Type I Migration

In this section we deal with *lower mass planets*, up to about a few Earth masses that do not perturb the disk structure significantly. In particular, these planets do not open gaps in the disk. The torques can be then analysed through linear, as well as non-linear approaches. This migration regime where the disk is only weakly disturbed is termed *type I migration*. In this case both contributions are important, spiral arms (Lindblad torques) as well as the corotation region.

As seen from the left panel in Fig. 2.19 the spiral arms are the most prominent features visible in the perturbed density distribution in disk as generated by the planet. Clearly, from a purely mechanical point of view, and loosely speaking, the inner spiral arm pulls the planet forward which leads to an acceleration of the planet and a gain of angular momentum. On the other hand, the outer spiral pulls the planet back which leads to a loss in angular momentum. This feature is reflected in the mass weighted radial torque density distribution that is displayed in the right panel of Fig. 2.20. The part inside of the planet is positive while the part outside is negative. The total torque is given by the integral over this distribution

$$\Gamma_{\rm tot} = 2\pi \int \frac{d\Gamma}{dm}(r)\, \Sigma(r)\, r dr \tag{2.89}$$

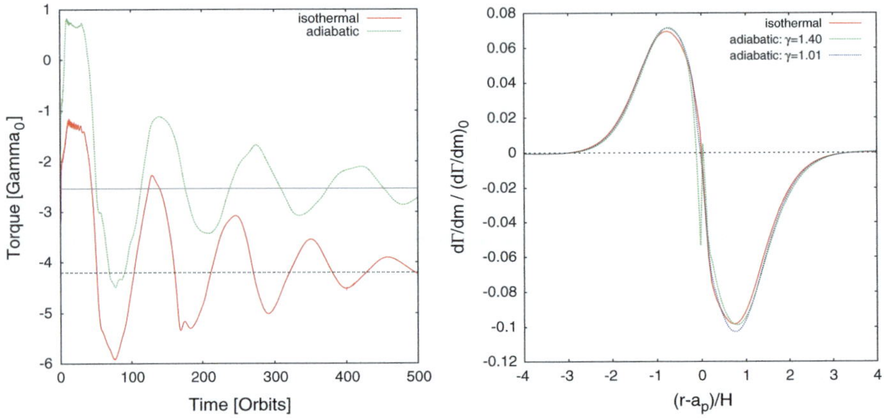

Fig. 2.20 The torque acting on a low-mass planet on a fixed circular orbit that is embedded in a disk that has a constant surface density and very low viscosity. Shown are results from 2D hydrodynamic simulations for an adiabatic and a locally isothermal disk. The **left panel** shows the time evolution of the total torque acting on planet. At the start of the simulations, when the gas performs its first U-turn near the planet, the torques reach a maximum (fully unsaturated values). Then they drop to reach the final saturated values through a series of oscillations. The horizontal lines refer to the 2D isothermal and adiabatic Lindblad torques [205]. Based on Fig. 2a in [133]. The **right panel** shows the radial torque density in the final state, again for an isothermal and two adiabatic simulations for $\gamma = 1.01$ and $\gamma = 1.4$ (adapted from Fig. 20 in [132]). The units Γ_0 and $(d\Gamma/dm)_0$ are stated in the main text, see Eqs. (2.90) and (2.94)

and, as already hinted above, it is a balance between positive inner and negative outer torques. The torque density, $d\Gamma(r)/dm$, scales with the planet to star mass ratio squared and as $(H/r)^{-4}$, and we rescale our results in units of

$$\left(\frac{d\Gamma}{dm}\right)_0 = \Omega_{\rm p}^2 r_{\rm p}^2 \left(\frac{m_{\rm p}}{M_*}\right)^2 \left(\frac{H}{r_{\rm p}}\right)^{-4}, \tag{2.90}$$

where quantities with index p are evaluated at the planet's position, e.g. $\Omega_{\rm p} = \Omega(r_{\rm p})$. We note that in evaluating the pressure scale height one has to choose either the isothermal or adiabatic sound speed such that $H_{\rm adi} = \sqrt{\gamma} H_{\rm iso}$. Using this scaling the torque densities for all simulations, isothermal and adiabatic, are identical (right panel in Fig. 2.20).

From a wave point of view the spiral arms can be treated as the superposition of the sounds waves emanating from the planet. The torques generated can then be analyzed by the mentioned linearization procedure which leads to the Lindblad resonances where $m(\Omega(r) - \Omega_{\rm p}) = \pm\kappa(r)$, with the epicyclic frequency κ. The plus sign, i.e., $\Omega(r) = \Omega_{\rm p} + \kappa(r)/m$, refers to inner Lindblad resonances (interior to $r_{\rm p}$) where the disk rotates faster than the planet. The minus sign refers to the outer Lindblad resonances, i.e., $\Omega(r) = \Omega_{\rm p} - \kappa(r)/m$ (exterior to $r_{\rm p}$ and the corotation region in Fig. 2.19). For a Keplerian disk the epicyclic frequency matches exactly the

Keplerian speed, i.e. $\kappa = \Omega_K$ (2.87), and the locations of the Lindblad resonances are given by

$$r_L = \left(\frac{m}{m \pm 1}\right)^{-2/3} r_p \,. \tag{2.91}$$

These resonances are named after Bertil Lindblad, who studied the dynamics of galaxies. In case of a radial pressure gradient in the disk the rotation rate is not exactly Keplerian and the resonances are shifted slightly such as to avoid a divergence at the planetary radius for large m [7, 96]. This causes the torque density to peak at a certain distance, typically a pressure scale length, H, away from the planet, as shown in the right panel of Fig. 2.19.

Detailed results for the Lindblad and corotation torques in locally isothermal disks with a fixed $T(r)$ stratification have been obtained for lower mass planets, of a few $M_\oplus$, through full 3D analyses. These were obtained either analytically by linear analysis [260] or through fully non-linear simulations [61]. The results for the total torque can be summarized as [61]

$$\Gamma_{\rm tot} = -(1.36 + 0.62\beta_\Sigma + 0.43\beta_T)\,\Gamma_0 \,, \tag{2.92}$$

where β_Σ and β_T denote the gradients of the density and temperature in the disk that are defined through the following power laws

$$\Sigma(r) = \Sigma_0\, r^{-\beta_\Sigma} \qquad \text{and} \qquad T(r) = T_0\, r^{-\beta_T} \,. \tag{2.93}$$

The fit formula (2.92) has been obtained from 3D local isothermal simulations with varying gradients in Σ and T, and for a viscosity of $\alpha = 0.004$ [61]. For the definition of the α-viscosity see Eq. (2.69). The torque normalization is given by

$$\Gamma_0 = \left(\frac{m_p}{M_*}\right)^2 \left(\frac{H}{r_p}\right)^{-2} \left(\Sigma_p r_p^2\right) r_p^2 \Omega_p^2 \,, \tag{2.94}$$

which reflects the scaling of the torque with the physical properties of the disk. Here, Σ_p denotes the *unperturbed* surface density at the location of the planet. As seen, the torque scales with square of the planet mass, m_p. This is due to the fact that the perturbations in the disk induced by the planet scale directly with m_p, and the back reaction on the planets adds the second factor. The opening angle of the spiral arms is given by the temperature in the disk, here through the scale height, $T \propto H^2$. Lower temperatures lead to tighter spirals and consequently larger torques because the torque maxima are closer to the planet (see Fig. 2.20). The direct scaling with the disk's density follows immediately from the definition of the torque in Eq. (2.86). The migration rate follows then from Eq. (2.88). The torque formula, Eq. (2.92) shows that for the typical gradients in the disk (β_Σ and β_T larger than zero) the migration will be directed inwards. While this is encouraging in explaining the close-in observed exoplanets, the short timescale of the migration causes problems, however. Using the slope values for the minimum mass solar nebula (MMSN), $\beta_\Sigma = 1.5$ and $\beta_T = 0.5$, we obtain $\Gamma_{\rm tot} = 2.5\Gamma_0$ and then from Eqs. (2.84), (2.88) and (2.92) we find for the migration time

$$\tau_{\rm mig} \approx \frac{1}{5} \left(\frac{m_{\rm p}}{M_*}\right)^{-1} \left(\frac{m_{\rm d}}{M_*}\right)^{-1} \left(\frac{H}{r_{\rm p}}\right)^2 \Omega_{\rm p}^{-1}, \tag{2.95}$$

where we used $m_{\rm d} = \Sigma_{\rm p} r_{\rm p}^2$ as a proxy for the disk mass. Looking at the speed of migration we find for a 1 $M_{\rm Earth}$ planet at 5 AU a migration timescale of a few 10^5 yrs. This is shorter than typical lifetimes of protoplanetary disks, which is a general result for type I migration in locally isothermal disks. Consequently, to prevent this too rapid inward radial drift, during the past years much effort has been put into finding mechanisms to slow it down.

The key to the solution lies in the corotation, or horseshoe region. Looking at the left panel of Fig. 2.20 one notices that directly after insertion of the planet the torques acting on the planet are much smaller than in the final equilibrium case. This is true for the isothermal, as well as the adiabatic model. In the latter, the torques are even reversed (positive) initially which would imply an outward migration. This torque reduction or reversal is caused by the flow dynamics in the corotation region. Directly after the insertion of the planet the flow begins to perform U-turns in the vicinity of the planet. As indicated in Fig. 2.19 by the black dots, material from the inside moves outwards and by doing so has to gain a certain amount of angular momentum from the planet. On the other side the process is reversed and the material gives angular momentum to the planet. Clearly, if there is an asymmetry in the U-turn motion in front and behind the planet there will be a net transfer of angular momentum to the planet. Because in this U-turn motion only material within the horseshoe region is involved, the asymmetry is driven by radial gradients of particular disk properties across the horseshoe region, and the resulting torque is called the *horseshoe drag* [275]. The oscillatory settling of the torque is due to the motion of the material along their horseshoe orbits. The oscillation period is given approximately by the libration time of matter in the horseshoe because after each libration period the matter reaches again the planet and receives a new kick. It is the mixing of the matter in the horseshoe region that tends to wipe out the initial gradients of the flow, and eventually the corotation torques saturate [275]. From linear analyses [95], it is known that the corotation torque is proportional to the radial gradient in the specific vorticity (sometimes also called *vortensity*)

$$\tilde{\omega} = \frac{\omega_z}{\Sigma} = \frac{(\nabla \times \mathbf{u})|_z}{\Sigma}, \tag{2.96}$$

and proportional to the radial gradient in the entropy, S, see [18, 209]. The quantity ω_z in Eq. (2.96) is the z-component of the vorticity of the disk flow. For inviscid and barotropic ($p = p(\rho)$) flows one can show that $\tilde{\omega}$ and S are constant along streamlines, i.e.

$$\frac{d}{dt}\left(\frac{\omega_z}{\Sigma}\right) = 0 \quad \text{and} \quad \frac{dS}{dt} = 0\,. \tag{2.97}$$

As a consequence any initial gradients (in S and $\tilde{\omega}$) are wiped out in the corotation region in the course of the simulations and the horseshoe drag vanishes [12] and only the (negative) Lindblad torques remain, exactly as seen in the nearly inviscid simulation presented in Fig. 2.20, see also [175]. Hence, the necessary physical mechanisms to maintain those gradients are viscosity and radiative diffusion, or radiative cooling [18, 209].

Indeed, through a series of 2D and 3D hydrodynamical simulations it has been shown in more realistic disks models which include radiative transport that the migration of planets can be slowed down and even directed outwards, see Paardekooper and Mellema [208] for the very first simulation of this kind. Subsequently, model sequences using varying planet masses in 2D [128] and 3D [127] for active, viscous disks with radiative transport have shown that there exists is a typical mass range in which the total torque is positive and the migration can be directed outwards. For maximum efficiency of the torque desaturation the viscous and radiative diffusion timescales should lie within the libration and U-turn timescales of the gas in the horseshoe region. Hence, it depends on the mass of the planet, and disk properties such as viscosity and opacity. An example of such a study for the 2D case with embedded planets located at 5 AU is shown in Fig. 2.21. For this disk regime the mass range with the fastest outward migration lies between 25 and 40 $M_{\rm Earth}$. To study the overall evolution of planets in disks it is useful to apply formulae that provide reliable estimates of the torque and can be computed fast and easily. These were provided in [176, 206] and confirmed by full 3D radiative simulations [34]. The torque formulae indicate that for small planets of a few Earth masses the migration should be directed inward, a feature hardly visible in Fig. 2.21. Indeed, the inward migration of planets below a mass of about 3 $M_{\rm Earth}$ has been verified in 3D high resolution simulations [149]. Very recently, the migration of low mass planets in this mass range has been addressed by including the effect of accretional heating from

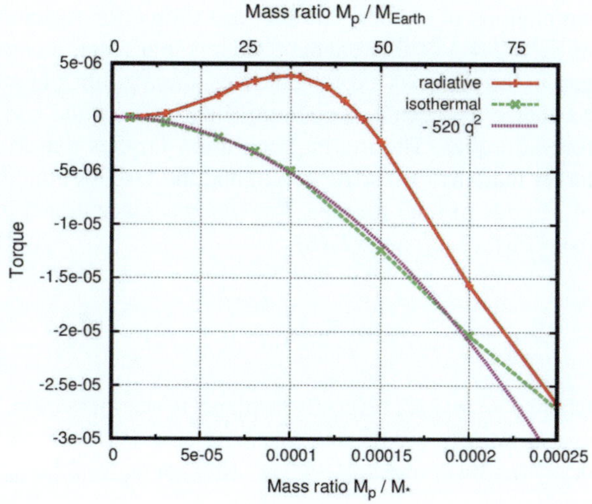

Fig. 2.21 The torque acting on embedded planets in 2D, viscous disks using $\alpha = 0.004$. Results are shown for a locally isothermal disk (green) and radiative models (red) that include viscous heating, radiative transport in the midplane, as well as radiative cooling from the disk surfaces. The purple line shows a parabola that scales with the square of the planet mass. Adapted from [128]

the (solid) material added to the planet in the simulations. Taking this additional heat source into account, it has been found that the migration can be directed outwards for very low mass planets [24]. This *heating torque* effect might play an important role in preventing the too rapid inward migration of planetary cores towards their host stars. These studies indicate that the migration direction and rate of an embedded planet depends on the details of the disk flow in the close vicinity of the planet. This requires more high resolution simulations similar to the nested-grid simulations presented in [60], or the more recent example by [88].

As mentioned above, the torques acting on a planet depend on the physical disk properties (viscosity and opacity). During the evolution of the disk these will change in space and time. To account for this, sequences of 2D axisymmetric disk models that include radiative cooling and viscous and stellar heating have been computed for different evolutionary states of the host star [31] as a function of the disk's metallicity, accretion rate, and lifetime. The parameterized disk structures in conjunction with the torque formulae [206] form an ideal setup for studying the longterm evolution of planets in disks.

2.6.3 Type II Migration

The analysis just presented above refers to the situation of embedded low-mass planets that do not change the disk's original structure significantly. However, for larger planet masses, the interaction becomes increasingly nonlinear, and the density profile in the disk will be modified. In the following, we describe the consequences of this process in more detail.

Before evaluating the impact on migration we will first look at the process of gap formation. An example on how the presence of a planet impacts the surface density profile of the disk is presented in Fig. 2.22, where results of 2D hydrodynamical simulations of isothermal disks are shown for various planet masses. As explained in Sect. 2.4.3 of this chapter, planets that reach a certain mass will tend to open a gap in the disk. As explained there, the equilibrium gap structure is determined by a balance between gap opening (tidal forces) and gap closing (pressure, viscosity) forces/torques. The thermal condition implies that the radius of the Hill sphere is larger than the disk scale height and the viscous condition follows from the balance of viscous to tidal torques. For the necessary planet masses to open a gap we find from Eqs. (2.45) and (2.46)

$$q > q_{\rm therm} \equiv 3 \left(\frac{H}{r}\right)^3_{\rm p} = 3h_{\rm p}^3 \quad \text{and} \quad q > q_{\rm visc} \equiv 30\pi\alpha h^2 \,, \qquad (2.98)$$

where $q = m_{\rm p}/M_*$ refers to the planet to star mass ratio. For a typical sample disk with aspect ratio $h = 0.05$ and viscosity $\alpha = 10^{-2}$ one obtains $q_{\rm therm} \approx 1.25 \cdot 10^{-4}$, or $m_{\rm p} \geq 0.13 m_{\rm Jup}$ and $q_{\rm visc} \approx 2.4 \cdot 10^{-3}$, or $m_{\rm p} \geq 2.5 m_{\rm Jup}$. Hence, for these parameters

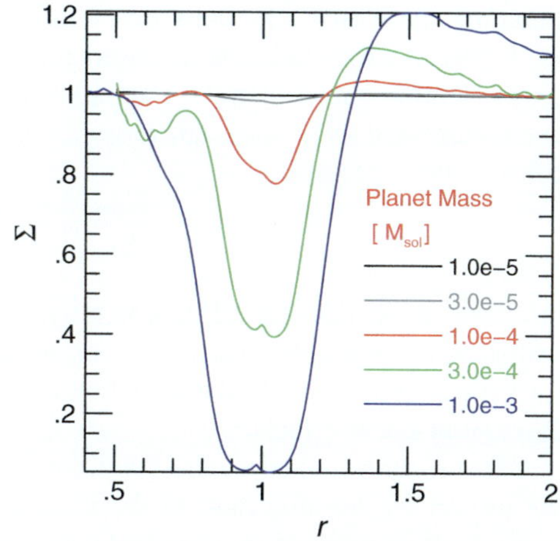

Fig. 2.22 The radial surface density profile of embedded planets with different masses in 2D, viscous disks using $\alpha = 0.004$. Results are shown for a locally isothermal disk with an aspect ratio $H/r = 0.05$. The radius is normalized to the position of the planet and the density to the unperturbed value

the criteria imply that the gap opening will be driven initially by the thermal criterion rather than the viscosity. However, from Eq. (2.98) we also infer that low mass planets of a few Earth masses will also open up gaps for very low disk viscosity [229]. This has been seen in numerical simulations [153] where a strongly reduced migration rate was noticed as well. As also mentioned above, the two criteria (2.98) have been combined to a single one [58] in Eq. (2.47). While Eq. (2.98) gives some estimate under what conditions gaps are to be expected it does not say anything about the depth of a gap that is carved out by the planet. Recently, some progress has been made in this direction and fits to numerical simulations have shown [75] that the determining parameter is given by

$$K \equiv q^2 \, h^{-5} \, \alpha^{-1} \,. \tag{2.99}$$

The depth of a gap is typically defined as the ratio of the surface density at the location of the gap, $\Sigma_{\rm gap}$, to the unperturbed density, Σ_0, of the disk without the planet. The simulations show that $\Sigma_{\rm gap}/\Sigma_0$ scales as K^{-1} [75]. This can be understood from a simple analytical argument [89], that we summarize here. Assuming that the Lindblad torques are generated within the gap, they are proportional to Γ_0 (see Eq. 2.94) with $\Sigma_{\rm p}$ replaced by $\Sigma_{\rm gap}$. This replacement can be justified because in linear theory the prime contribution to the torque comes from a radial region separated by one scaleheight H from the planet as shown in Fig. 2.20. Hence,

$$\Gamma_{\rm L} \sim q^2 h^{-3} \Sigma_{\rm gap} \, \Omega_{\rm p}^2 r_{\rm p}^4 \,. \tag{2.100}$$

This torque has to be balanced essentially by the viscous torque which scales as the derivative of $\Sigma \nu (d\Omega/dr)$. Substituting now $d\Sigma/dr$ with Σ_0/r and $\Omega_{\rm K}$ for Ω the viscous torque scales as

$$\Gamma_{\rm visc} \sim \Sigma_0 \nu \Omega_{\rm K} r^2 \,. \tag{2.101}$$

Equating these two equations one obtains with $\nu = \alpha H^2 \Omega_K$ exactly $\Sigma_{gap}/\Sigma_0 \propto K^{-1}$ [89]. As this estimate is based on linear theory the scaling agrees well for up to Neptune sized planets. For larger planets slightly different exponents have been suggested [89]. Through a more careful analysis of the angular momentum conservation the scaling

$$\frac{\Sigma_{gap}}{\Sigma_0} = \frac{1}{1 + 0.04K} \tag{2.102}$$

has been suggested [72, 123], which agrees well with the simpler scaling for large values of K. While Eq. (2.102) gives good agreement with numerical simulations for values up to $K \approx 10$, the deviations are larger in the range $K \sim 10$–1000 [72], due to non-linear effects.

After having looked at the conditions for gap formation we will now analyze the impact on the migration of the planet. For gap opening planets the corotation region will be reduced in mass leading to a lowering of the corotation torques. For larger planet masses the gap widens such that the Lindblad torques will be reduced as well, and the migration rate will be slowed down considerably. Given by the strong lowering in the density (Fig. 2.22) the perturbations are now nonlinear and the corresponding regime of migration was coined *type II migration* [276], in order to distinguish it from the previous linear type I regime. The main idea of type II migration assumes that after having opened an annular gap in the disk, the planet will remain always in the middle of this gap during the migration, i.e. the planet has to move with the same speed as the gap. Because the disk (with its gap) evolves on a viscous timescale it was expected that the planet is coupled entirely to the viscous evolution of the disk [157] and has to move with the same timescale

$$\tau_{mig,II} = \tau_{visc} = \frac{r_p^2}{\nu} = \frac{r_p^2}{\alpha c_s H} = \frac{1}{\alpha h^2 \Omega_p} \,. \tag{2.103}$$

In this type II migration regime the rate is independent of the planet mass, and depends only on disk and stellar parameter. This can only be valid if the disk mass, $m_d \approx \pi \Sigma_0 r_p^2$, is larger than the planet mass m_p. Otherwise the planetary inertia will require a reduction factor such that $\tau_{mig,II} = m_d/m_p \tau_{visc}$ [117, 258]. Numerical simulations of a migrating Jupiter type planet over long timescales found a good overall qualitative agreement with these expectations [202].

Recently however, the assumption (2.103) has been brought into question and it has been argued that the migration of gap-opening planets is not locked to the viscous evolution of the disk [74]. By performing sequences of numerical simulations of massive planets in the disk it was found that the migration rate is not given by Eq. (2.103) but can be faster as well as slower than this rate, a feature that was noted already earlier [82]. The reason for this behaviour lies in the fact that a gap does not separate the inner and outer disk completely as is typically assumed. During the migration of the planet, mass is transferred from one side of the planet to the other side across the planetary orbit and hence the gap will always be dynamically recreated. If the planet migrates faster than the viscous speed then mass is transferred

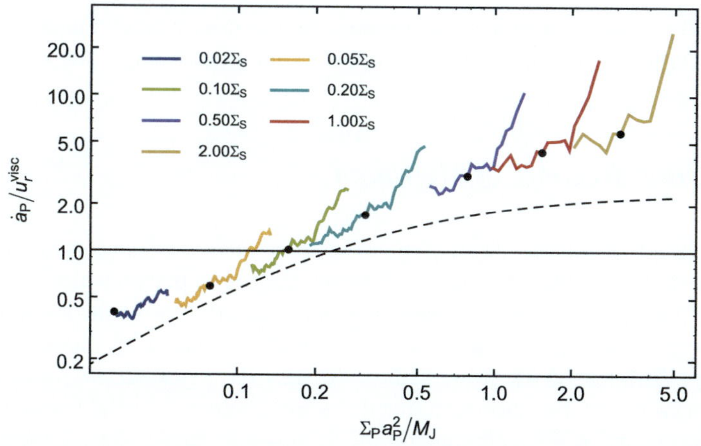

Fig. 2.23 The migration speed, $\dot{a}_\mathrm{P}$, of a massive Jupiter mass planet through locally isothermal disks with different surface densities. The speed is normalized to the viscous inflow velocity of the disk, $u_r^{visc} = 3/2\,\nu/r$, where $\nu = \alpha c_\mathrm{s} H$. Each of the models, as denoted by the different colors has a specific, constant mass accretion rate where the unit surface density, Σ_S, belongs to $\dot{M} = 10^{-7}\,M_\odot$/yr. All planets start at unit distance ($r = 1$) at the top right end of the individual curves. The black dots refer to the location $r = 0.7$ The dashed line refers to the results by [74] where a globally constant viscosity ν was used. Taken from Fig. 15 in [80], © ESO, reproduced with permission

from the inside to the outside and the other way around for migration speeds that are slower than the viscous speed. More recently, models of migrating planets in accreting disks have been constructed [80] and very similar results have been found. As shown in Fig. 2.23 the typical migration rate of a massive planet can be faster or slower as the viscous rate. From a physical perspective this result was expected because the planet can only be moved through the disk by torques acting on it, such that the actual migration speed is always given by Eq. (2.88).

The total torque acting on a migrating planet can be written as the product of a stationary torque and a dynamical correction factor [80]. The first contribution, the stationary torque $\Gamma_\mathrm{still} = f_\mathrm{still}\Gamma_0$, refers to the situation of a planet that does not move through the disk but remains at its actual distance from the planet. Here, Γ_0 is the normalization from above, see Eq. (2.94), and f_still is a correction factor that depends for a given disk parameter (viscosity, thickness) on the mass of the planet. For smaller mass planets that can be treated in the linear regime, f_still is a constant, given for example by Eq. (2.92). However, larger planets clear out a gap such that f_still becomes lower [276]. Simulations of embedded planets in accreting disks with non-zero $\dot{M}$ have been performed that first keep the planet fixed on its orbit in order to calculate f_still [80]. Then the planets were released and the subsequent evolution followed. Interestingly, for locally isothermal disks it was found that the normalized torque Γ/Γ_0 was constant *during* each migration process, which implies that one can write, $\Gamma = f_\mathrm{mig}\, f_\mathrm{still}\, \Gamma_0$, with a constant f_mig. This means that the factor f_mig does not depend on the local disk mass but only on the planet mass and disk viscosity [80]. It is typically smaller than one, only for rapid type III migration it is larger.

2.6.4 *Other Regimes of Migration*

Having studied the standard form of migration of planets in viscous, laminar disks we will now briefly describe other forms of planet migration that consider more aspects of disk physics that have hitherto been neglected. In addition to the mentioned type I and II migration regimes there exists a regime dominated by rapid, or runaway migration [178], for which subsequently the new term *type III migration* [9, 277] has been established. Because accretion disks are driven by internal turbulence either hydrodynamically or via a combination with magnetic fields, the gas motion is not laminar but rather shows random fluctuations. The interaction of these variable flow field with embedded planets leads to an additional type of migration, sometimes termed *stochastic migration*. Finally, in more massive disks the effects of disk self-gravity cannot be neglected anymore and will influence the density structure and subsequently the planet migration process.

2.6.4.1 Type III Migration

In contrast to most of the previous calculations (with the exception of the type II discussion) now the torque acting on a moving planet is considered. Let us focus on a planet undergoing inward radial migration on a circular orbit. Material located in the vicinity of the separatrix, between inner disk material that is on circulating streamlines and material on librating horseshoe orbits, will undergo a single U-turn upon a close encounter with the planet and will cross the entire horseshoe region. In doing so, it will directly move from the inner to the outer disk and not be trapped in the horseshoe region. This transfer of disk material from one side of the planet to the other implies an exchange of angular momentum with the planet. The inner material that crosses horseshoe gains angular momentum that the planet has to loose. The exerted torque can be written as [133]

$$\Gamma_{\text{flow}} = 2\pi \, \Sigma_s \, \dot{r}_p \Omega_p r_p^3 x_s \,, \qquad (2.104)$$

where x_s is the width of the horseshoe region and Σ_s the surface density at the inner separatrix. This "flow-through corotation torque" depends on the migration speed, $\dot{r}_p$, of the planet, and there is a positive feedback between the migration rate and the torque, i.e. the torque increases with the migration speed. Hence, in principle this can lead to a runaway migration [8, 178]. More detailed analysis [178] shows that the efficiency of the migration will depend on the mass contained within the horseshoe region. This is because the planet has to carry this material with it upon migrating inward which acts as additional inertia that will tend to counteract the runaway process. On the other if the horseshoe is too empty, i.e. the planet too massive, there will be too little material left over (Σ_s too small) to drive fast migration. The results show that the highest efficiency requires a partially emptied horseshoe region, where the difference between the mass that would be contained in the horseshoe region if the disk was unperturbed by the planet and the mass that is actually contained in this region (sometimes called coorbital mass-deficit, δm) equals the planet mass. For the MMSN the most promising planet mass for type III migration lies in the Saturn mass range beyond a distance of about 10 AU. An increase in disk mass will enhance the runaway migration such that disks that are prone to type III migration are always close to being at the limit of gravitational instability [178]. Due to the symmetry implied by Eq. (2.104) the direction of type III migration can be inward or outward, depending on the initial perturbation to the planet's motion. This feature has been demonstrated explicitly by numerical simulations [216, 217], where is was also shown that even massive planets of a few Jupiter masses can enter the rapid type III migration regime. The fast migration terminates when the planet moves into disk conditions that do no longer favor type III migration, i.e. the coorbital mass deficit, δm, no longer matches approximately the planet mass. During type III migration not only the planet itself has to be moved very fast through the disk but all of the material that resides bound to the planet within its Hill sphere. For massive planets this material is stored in a circumplanetary disk orbiting the planet and hence the details of type III efficiency will depend on the total mass within this disk [216, 217]. Type III migration may have played an important role during the early evolution of the Solar System, as a potential driving mechanism for motions of Saturn and Jupiter with the Grand Tack scenario [274]. The fact that type III migration is more efficient for massive disks takes us directly to a discussion of migration in disks where *self-gravity* can play a role.

2.6.4.2 Migration in Disks with Self-gravity

In situations where the total mass of the disk, $m_{\rm disk}$, is no longer negligible with respect to the star, M_*, it is not anymore sufficient to consider only the gravitational potential of the star but one has to add a contribution from the disk, i.e. consider an additional term from the disk's self-gravity $\Psi = \Psi_* + \Psi_{\rm d}$, where the latter is given by an integral over the whole disk (2.53). This change in the potential leads to a (slightly) different angular velocity of the disk that is now no longer Keplerian. In turn this leads to a radial shift in the location of the Lindblad resonances between the Fourier components of the planet and the disk material, see Eq. (2.91). As the torques on an embedded planet can be calculated in linear theory by summing over different resonances one may expect a change in the migration rate for embedded planets [19, 221]. Typical planet migration simulations with a 'life planet' that is allowed to move freely according to the gravitational forces of the disk assume that only the planet feels the disk's gravity but not the disk itself, i.e. neglect the disk's self-gravity. As pointed out [19], this is obviously an inconsistent situation, because planet and disk move in different gravitational fields. Through direct 2D-hydrodynamical simulations that include the disk's self-gravity they showed that for a disk only 3 times the MMSN this inconsistency may lead to a difference in the torque and hence migration rate by a factor of 2. They point out that a self-consistent simulation can be constructed easily by omitting the axisymmetric contribution of the density in calculating the force on the planet. By including the effects of disk self-gravity in the linear calculations it was found that for disks a few times the MMSN the change in the torque is only about 10% of that in a non-self-gravitating disk [19, 221]. This difference can be accounted for by the radial shift in the Lindblad resonances, leading to a migration rate is slightly enhanced over the standard value.

In the past years hydrodynamical simulations following the evolution of massive planets embedded in more massive disks have been presented in 2D [21] and 3D [188]. These simulations assume that a planet might have formed by gravitational instability at larger distances in the disk and then migrated to its present location. In the simulations presented in [21] the disk setup is chosen such that they are marginally stable, i.e. have a Toomre number of about $Q \sim 2$ in the main part. Under these conditions the disk does not fragment into individual blobs but generates stochastic spiral disturbances that can produce an effective viscosity and a heating of the disk. This heat is assumed to be radiated away from the disk surfaces, a process modeled by a radially varying cooling function, where the cooling time is given by $\tau_{\rm cool} = \beta \Omega^{-1}$, so called β-cooling, see Eq. (2.74). Hence, a statistically stationary disk setup can be produced where the effective viscosity, as generated by the gravo-turbulent fluctuations of the disk, is given by

$$\alpha_{\rm sg} = \frac{4}{9} \frac{1}{\gamma(\gamma-1)\beta} . \tag{2.105}$$

This relation can quite generally be derived [90] just by assuming that the turbulent heat generated acts like a viscous dissipation $\propto \Sigma \nu (d\Omega/dr)^2$ which is balanced by the local cooling $de_{\rm th}/dt \propto e_{\rm th}/\tau_{\rm cool}$, with the thermal energy density $e_{\rm th}$, see Sect. 2.5.3. And this is independent of the source of the turbulence. For a value of $\beta = 20$ [21] found in their simulations an $\alpha_{\rm sg} \approx (2\text{–}3) \cdot 10^{-2}$ in agreement with the expectation (2.105). Then massive planets are embedded in these disks at large distances (about 100 AU) and their subsequent evolution is followed through 2D hydrodynamical calculations. The simulations show that the planets experience a very rapid inward migration. For example, a Jupiter mass planet embedded into a disk with aspect ratio $h = 0.1$ ($\beta = 20$) migrated from 100 AU all the way to only 20 AU within less than 10^4 yrs. This fast migration is somewhat reminiscent to the type III migration that was discussed earlier. However, the planets do not open up a clear gap such that the coorbital mass deficit equals the planet mass, because the gap opening times are longer than the migration timescale. Hence, the migration is basically of type I with stochastic episodes superimposed, that allow even for brief phases of outward migration. These results have been supported by full 3D simulations of migrating Jupiter type planets [188] where a Jupiter planet embedded at 25 AU into a $0.14\,M_\odot$ disk plunges in only 10^3 yrs to 6 AU. These short migration timescales of clumps formed in more massive disks represent a serious challenge to planet formation via gravitational instability scenario [298], as shown in the previous section.

2.6.4.3 Stochastic Migration

As pointed out already just above, planets embedded in a disk that exhibits turbulent fluctuations will display not a monotonous smooth migration but will experience stochastic fluctuations in their actual migration rate. To study this process under realistic conditions, numerical simulations of turbulent disks with embedded planets have been performed using full magneto-hydrodynamical (MHD) computations in 3D. Magnetized disks are susceptible to the magneto-rotational instability (MRI) which is believed to be responsible for responsible for driving accretion at least in sufficiently ionized disks [11, 109]. Several direct simulations of planets embedded in turbulent discs have been performed for the case of MRI unstable discs [200, 201, 266]. In MHD simulations with embedded low mass planets (up to 30 M_{Earth}) is was found that the planets experience a random walk process where on average the migration resembles standard type I inward migration, however, with possible long term ($\sim$100 orbits) phases of outward migration [201]. The fluctuations can also lead to a non-zero eccentricity of the embedded planets of the order of a few percent [200]. For a Jupiter mass planet it has been found that the gap is somewhat wider than in the corresponding laminar case using the same effective α-value for the disk [133, 266]. Here, the (relative) turbulent fluctuations become weaker and the migration resembles that of standard laminar disks. For intermediate planet masses, phases of sustained outward migration have been reported [266]. In simulations with very low planet masses (a few M_{Earth}) is was found [16] that the structure of the (averaged) flow in the corotation region in full MHD simulations is comparable to the laminar results, indicated the presence of corotation torques also in turbulent magnetic disks. To circumvent the expensive full 3D magnetohydrodynamical simulations, approximate treatments have been developed where the random features of the MHD turbulence is modeled by stochastic forces that can be added in standard hydrodynamical simulations [17, 147]. We will no go into more details at this point but rather refer to other review articles [15, 133].

2.6.5 Eccentricity and Inclination

In addition to a change in the distance from the central star, embedded planets will suffer a change in the other orbital elements as well. Most important here are the eccentricity and inclination evolution. The circular and coplanar orbits in the Solar System were taken already early on, for example by Kant and Laplace [124, 145], as evidence that the planets form in a flat disk-like configuration. The growing sample of extrasolar planets with their huge variety in orbital elements has shown that a large fraction of planets display significant orbital eccentricity and some systems show a large tilt with respect to the stellar rotation axis, that is expected to be aligned approximately with the rotation axis of the protoplanetary nebula. The question arises whether these orbital characteristics of the extrasolar planets can be a result of disk-planet interaction or have another dynamical origin.

2.6.5.1 Eccentricity

In calculating the change of the eccentricity of the planet, we have to start from the angular momentum and energy of a planet orbiting the star. These are given by

$$J_{\rm p} = m_{\rm p} \sqrt{GM_* a} \sqrt{1 - e^2} \quad \text{and} \quad E_{\rm p} = -\frac{1}{2} \frac{GM_* m_{\rm p}}{a} \tag{2.106}$$

where we dropped the subscript and wrote for simplicity just a and e for the planetary semi-major axis and eccentricity, respectively. Classical mechanics tells us that the angular momentum is changed by the torque and the energy by the power acting on the planet by the disk. These are given by

$$\Gamma_{\rm disk} = -\int_{\rm disk} \Sigma \; (\mathbf{r}_{\rm p} \times \mathbf{F})\big|_z \, df \quad \text{and} \quad P_{\rm disk} = \int_{\rm disk} \Sigma \, \dot{\mathbf{r}}_{\rm p} \cdot \mathbf{F} \, df \,, \tag{2.107}$$

where $\Gamma_{\rm disk}$ was defined already before in Eq. (2.86). The changes in $E_{\rm p}$ and $J_{\rm p}$ are now given as

$$\frac{\dot{J}_{\rm p}}{J_{\rm p}} = \frac{1}{2}\frac{\dot{a}}{a} - \frac{e^2}{1-e^2}\frac{\dot{e}}{e} = \frac{\Gamma_{\rm disk}}{J_{\rm p}} \tag{2.108}$$

and

$$\frac{\dot{E}_{\rm p}}{E_{\rm p}} = -\frac{\dot{a}}{a} = \frac{P_{\rm disk}}{E_{\rm p}} \,. \tag{2.109}$$

Having calculated $\Gamma_{\rm disk}$ and $P_{\rm disk}$ from the simulations, one can directly evaluate the changes in the orbital elements of a and e. Obviously, the change in the semi-major axis is given solely by $P_{\rm disk}$ and not by the torque as assumed above in Eq. (2.88). The reason for this apparent discrepancy lies in the fact that we assumed circular orbits in the discussion in Sect. 2.6.1. Setting $e = 0$ we find from Eqs. (2.108) and (2.109) that $P_{\rm disk} = \Omega_{\rm K} \, \Gamma_{\rm disk}$ such that for circular orbits one can use indeed the torque to estimate the migration rate.

For low mass planets that do not open a significant gap within the disk, linear perturbation theory can be applied to analyze the eccentricity evolution. In contrast to the circular case an eccentric planet generates additional resonances in the disk [96]. Similar to a planet on a circular orbit, Lindblad resonances occur at disk locations where the perturbation frequency experienced by a fluid element equals the local epicyclic frequency, and corotation resonances occur wherever the pattern speed equals the local angular velocity of the disk. External Lindblad resonances act to increase e, while corotation and co-orbital Lindblad resonances act to damp it [94]. The outcome depends on the relative contribution of damping and exciting contributions and, similar to the migration, and the net result is the (small) difference of larger individual contributions. Linear analyses shows that (for not too high initial values) the eccentricity is damped exponentially on a timescale much shorter than the migration timescale, where typically

$$\tau_{\rm ecc} \sim \left(\frac{H}{r}\right)^2 \tau_{\rm mig} \,. \tag{2.110}$$

Equation (2.110) applies for initial eccentricities smaller than the relative scale height, $e \lesssim H/r$ [261]. This rapid decline in e is caused primarily by the damping action of the corotation resonances. For larger initial eccentricities with $e \gtrsim H/r$ the increased speed of the planet with respect to the disk material makes the interaction nonlinear and the damping rate changes to [211]

$$\frac{de_{\rm p}}{dt} \propto -e_{\rm p}^{-2} \,. \tag{2.111}$$

These results on the evolution of $e_{\rm p}$ (here we added the subscript 'p' for clarity), including the turnover from the exponential to the quadratic damping in Eq. (2.111), have been confirmed through full non-linear multi-dimensional hydrodynamical simulations of embedded low-mass planets up to 30 $M_{\rm Earth}$ in 2D disks [56] and up to 200 $M_{\rm Earth}$ in 3D radiative disks [32]. In Fig. 2.24 we display the time evolution of the planetary eccentricity for different initial eccentricities (left panel) and different

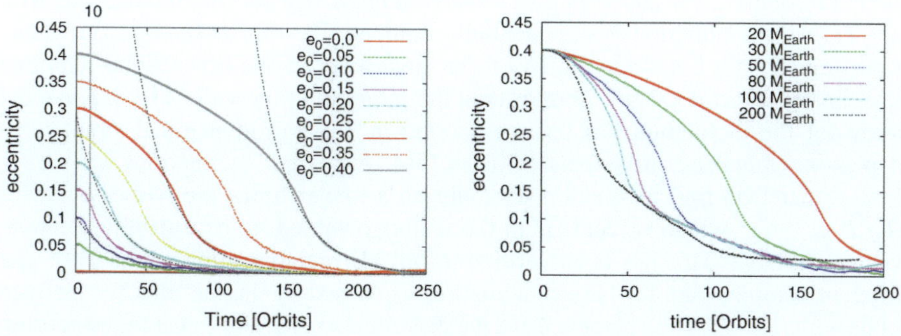

Fig. 2.24 The change in eccentricity of an embedded planet due to planet-disk interaction. The disk models have been calculated using full 3D radiation hydrodynamical simulations. **Left panel** The change in eccentricity of a $m_{\rm p} = 20 M_{\rm Earth}$ planet for various initial eccentricities. The planet mass is switched on gradually during the first 10 orbits (vertical dotted line). The dashed lines indicate exponential decay laws. **Right panel** The change in eccentricity of planets with different masses all starting from the same initial eccentricity, $e_0 = 0.4$. Taken from Figs. 15 (bottom panel) and 23 (bottom panel) in [32], © ESO, reproduced with permission

planet masses (right panel). From the time evolution it is clear that the decay rate is indeed different for large and small e_p, it changes from that of Eq. (2.111) to the exponential rate at the turnover eccentricity of $e_p \approx 0.1$ which is equivalent to is $e_p \approx 2H/r$. The same behaviour was found in the 2D simulations as well [56]. The right panel of Fig. 2.24 shows that the damping rate increases with increasing mass of the planet, at least for large initial e_p. This behaviour is in conflict with theoretical expectations based on linear theory that there should be a regime where eccentricity excitation due to planet disk interaction is possible [94]. The authors argue that the damping effect of the corotation region diminishes when the planet becomes massive enough to open a gap. Hence, they expect a possible eccentricity excitation for planet masses around Saturn or Jupiter. However, the results displayed in Fig. 2.24 from [32] and more recent 3D simulations [77] indicate otherwise. Only for very large planets with masses beyond 10 $M_{\rm Jup}$ planet disk interactions will lead to a significant rise in the e_p [62, 214]. This increase in the planet's eccentricity is caused in that case by the back-reaction of an eccentric disk on the embedded planet where the disk's eccentricity has been generated by the presence of the planet [129]. However, very recent simulations have hinted at possible eccentricity growth for massive planets in disks with low viscosity and small pressure [73], demonstrating that more studies are required to settle this issue. Nevertheless, the numerical results indicate that the observed high eccentricities in the extrasolar planets cannot in general be the outcome of planet-disk interaction but are rather a result of multi-body gravitational interactions.

2.6.5.2 Inclination

Similar to the eccentricity changes, the disk's impact on the inclination has been studied through linear analyses [261] which indicate that for low inclinations with $i_p \lesssim H/r$ the inclination is exponentially damped, i.e. $di_p/dt \propto -i_p$. The decay occurs again on a timescale that is of the same order as the eccentricity damping time, i.e. by a factor $(H/r)^2$ shorter than the migration timescale. The exponential decay for the inclination has been supported in 3D hydrodynamical simulations of planets embedded in isothermal disks [56]. For larger inclinations $i_p \gtrsim H/r$, these researchers find inclination damping on a longer timescale with a behavior $di_p/dt \propto -i_p^{-2}$ which is identical to the scaling obtained for eccentricity damping when e_p is large. This has been verified in full 3D radiative simulations [33]. The different damping rate for larger inclinations is caused by the fact that for inclined orbits with $i_p \gtrsim H/r$ the planets leave the disk region vertically twice for each orbit. The damping action of the disk is massively reduced and can only operate during the short time intervals while the planet is crossing the disk. This process can then be treated by a dynamical friction approach [238]. Additionally, it has been shown that the inclination is damped by disk-planet interaction for all planet masses up to about 1 $M_{\rm Jup}$ [33, 172]. In recent 3D simulations of planet-disk interaction for massive, inclined planets it has been shown that the inclination seems typically damped, however with a damping timescale that increases for larger initial inclinations [30, 238, 295].

In summary, due to the much shorter timescales of eccentricity and inclination damping in contrast to the migration time, it is to be expected that for isolated planets embedded in disks these quantities should be very small, at least in laminar disks; i.e., even if planets formed in disks with non-zero e_p and i_p, they would be driven rapidly to a state with $e_p = 0$ and $i_p = 0$.

Growing planets that are still embedded in the disk experience gravitational forces due to the interaction with the disk that change the orbital parameter of the planet. Most important is the radial migration of the planet as this implies that the locations where planets are observed today do not necessarily coincide with their formation locations. Initial results (for isothermal disks) indicated very rapid inward migration such that too many planets would have been lost into the star. Recent results have shown that migration can in fact be directed inward or outward, depending on the physical parameters of the disk. On the other hand, eccentricity and inclination are typically damped on much shorter timescales. Hence, the large eccentricities and inclinations of the observed exoplanets are most likely not caused by planet-disk interaction but have their origin in multi-body dynamical interactions.

2.7 Multi-body Systems

So far we have studied the growth and evolution of single planets in the disk. However, the observations indicate that planets tend to be found in systems with several objects rather than being lonesome travelers. In fact, through the Kepler space-mission hundreds of transiting multi-planet systems were discovered [41]. Out of the 1200 planetary systems observed by now (mid 2015), about 500 are multiple planet systems (source: www.exoplanet.eu). The occurrence and architecture of planetary systems has nicely been reviewed in [290].

The presence of additional objects in the system induces gravitational disturbances that can modify the evolution of the individual planets as well the overall evolution of the whole system. This can lead to planets with extreme orbital elements such as high eccentricity or inclination and to the formation of resonant planetary systems. Furthermore, planets are found in binary star systems where their formation pathway is made more challenging through the additional strong gravitational disturbance by the companion star.

In this part we will discuss the dynamics of multi-planet systems, with some focus on the formation of resonant, or near resonant systems, as well as planets in binary star systems.

2.7.1 Resonances

When talking about resonances, in the context of these lecture notes we will restrict ourselves to mean motion resonances (MMRs) which are defined by an integer ratio of the mean orbital motions of two planets,

$$n_1/n_2 = p/q \qquad \text{with} \qquad q, p \in \mathbb{N}_{>0}\,, \tag{2.112}$$

where n_i denotes the mean orbital velocity of the i-th planet. We will denote the inner planet by $i = 1$ and the outer one by $i = 2$, hence $p > q$. The *order* m of the resonance is given by

$$p = q + m\,. \tag{2.113}$$

Among the 8 planets in the Solar System there is no MMR. Taking into account the dwarf planets, one finds that Neptune and Pluto are in fact engaged in a 3:2 MMR, where the orbital period of Neptune is 165 yrs and that of Pluto 248 yrs, and their semi-major axes are 29.7 AU and 35.5 AU, respectively. This means that for 3 orbits of Neptune, Pluto performs exactly 2 orbits around the Sun. The high eccentricity of Pluto's orbit ($e_\mathrm{P} = 0.25$) implies that Pluto's perihelion lies inside the orbit of Neptune such that the two orbits are actually crossing each other. In general, planetary orbits are only stable if the two planets never approach each other to a distance closer than their mutual Hill radius. For crossing orbits, this condition can only be satisfied if the orbits are resonant. Hence, Pluto's orbit is protected from being unstable by the special 3:2 resonance that avoids close encounters and maximizes Pluto's separation from Neptune. For the Pluto-Neptune system additional stabilization arises from the fact that their orbits are mutually inclined by about 17°. The fact that the orbit of Pluto is shared by many other plutinos has led eventually to the degradation of Pluto from full planet to dwarf planet status, as decided by the International Astronomical Union in 2006. Similarly, in an exoplanetary system resonances are a way to protect the system from dynamical instability.

To specify a particular resonant configuration of a system, it is useful to define the resonant angles Φ_k by

$$\Phi_k = p\lambda_2 - q\lambda_1 - p\varpi_1 + q\varpi_2 - k(\varpi_2 - \varpi_1)\,, \tag{2.114}$$

with $p > q$. In Eq. (2.114) the λ_i denote the mean longitudes and ϖ_i the longitudes of periapse of two planets. The integers k satisfy $q \le k \le p$ and can take $p - q + 1$ possible values. Of the $p - q + 1$ resonant angles, at most two are linearly independent, and a resonance condition is given if at least one angle will librate, i.e. its values do not extend over the whole range of 360° but is limited to a smaller range. Let us take as an example a system with a 2:1 resonance. Here, $p = 2$ and $q = 1$, such that we have

$$\Phi_1 = 2\lambda_2 - \lambda_1 - \varpi_1, \quad \Phi_2 = 2\lambda_2 - \lambda_1 - \varpi_2 \qquad (\Delta\varpi = \Phi_2 - \Phi_1)\,, \tag{2.115}$$

where we defined as an additional variable the difference $\Delta\varpi$ (sometimes also $\Phi_1 - \Phi_2$) between the longitudes of periapse which is often used to replace one of the resonant angles Φ_1 or Φ_2. The resonant angles Φ_k follow from an expansion of the perturbation potential between the two planets that would move otherwise on unperturbed Kepler-orbits. Upon expanding the perturbation in a Fourier-series, the resonant angles are the stationary phases in the expansion. If an angle Φ_k is in libration (i.e. covers a range smaller than $[0, 2\pi]$), then the system is said to be in a MMR. For more detailed information of resonances see the excellent book of *Solar System Dynamics* by Murray and Dermott [198]. For a 2:1 resonance, if both angles Φ_k librate (i.e. $\Delta\varpi$ as well), then the system is said to be in a state of apsidal corotation resonance (ACR), where the two lines of periapse rotate with the same speed (on average) [23]. For Keplerian orbits the resonant condition (2.112) for two planets implies a ratio of their semimajor axes of $a_2/a_1 = (n_1/n_2)^{2/3}$, which follow from Kepler's third law.

One of the most famous examples of a planetary system in a first order 2:1 resonant configuration is GJ 876. The initial discovery [167] noted two massive planets (GJ 876 b and c) with minimum masses of 0.56 and 1.89 $M_{\rm Jup}$ that orbit a M-type star of only 0.3 $M_{\odot}$. These two planets have periods of 30 and 60 days and their orbits are fully aligned, i.e. $\Delta\varpi$ is in libration with $\Delta\varpi_{\rm max} \approx 30°$ [167], hence GJ 876 is in an ACR state. Another example is the system HD 128311 where two massive planets are again in a 2:1 resonance but here only one resonant angle is librating, while the second one is circulating, hence the system is not in an ACR state [267]. A system that is believed to be in a second order 3:1 resonance is HD 60532 where two massive planets orbit an F-type star [65].

Later two additional planets have been discovered in GJ 876, one (d) very close-in on a 2 day orbit around the host star, and another fourth Uranus mass planet (e) in a 124 day orbit beyond planets b and c. This places the three outer planets of GJ 876 in a Laplace-type 4:2:1 configuration [244]. This first known *resonant chain* amongst extrasolar planets bears some similarity to the Laplace resonance of the Galilean satellites Io-Europa-Ganymede orbiting Jupiter but it is dynamically very different as it exhibits a longterm chaotic dynamics [171, 244]. Another famous example of a planetary system that is believed to be in a 3-body Laplace resonance is the directly imaged system HR 8799 where three massive planets are orbiting a young A-star [98]. During the last years more resonant chains have been discovered using *Kepler*-data. In Kepler-60 three planets with masses around $\sim$4 $M_{\oplus}$ are engaged in a 5:4:3 Laplace MMR [99]. The four Neptune mass planets in Kepler-223 have periods in a ratio close to 3:4:6:8 [189]. This four planet resonant chain has been taken as another clear indication that migration of planets through the disk must have taken place.

2.7.1.1 Formation of Resonant Planetary Systems

A resonant configuration between two planets is a special dynamical state and it is not very likely that the two planets formed directly in situ at their observed locations, or were captured into such a state. Instead, the formation of resonant planetary systems is a natural outcome of a differential migration process. Here, it is assumed that several planets form in a disk and migrate according to the torques acting on them. Typically, the radial drift speeds are not identical and a differential migration process ensues. In case the inner planet migrates slower than the outer one, the radial distance between the two planets becomes smaller and they approach each other. Whenever the location of a resonance is crossed, i.e. the current orbital periods of the planets have an integer ratio, the mutual interaction becomes stronger due to the periodic perturbations. As a consequence the planets can be trapped in a resonance that is maintained during the subsequent migration process. A typical situation of such a process is shown in Fig. 2.25, which shows the surface density distribution of a disk with two massive embedded planets, obtained from a numerical model that simulates the formation of the system HD 73526 [246]. Because the planets are relatively close, they orbit within a joint deep gap where the outer planet feels only the torque by the outer disk (which is negative) and the inner planet feels only the torque of the inner disk (which is positive). Consequently, the two planets approach each other and a convergent migration process sets in, such that eventually the planets are captured in a resonance, here 2:1. In the subsequent migration process the planets maintain the resonance and move jointly together towards the central star. The resonant capture is

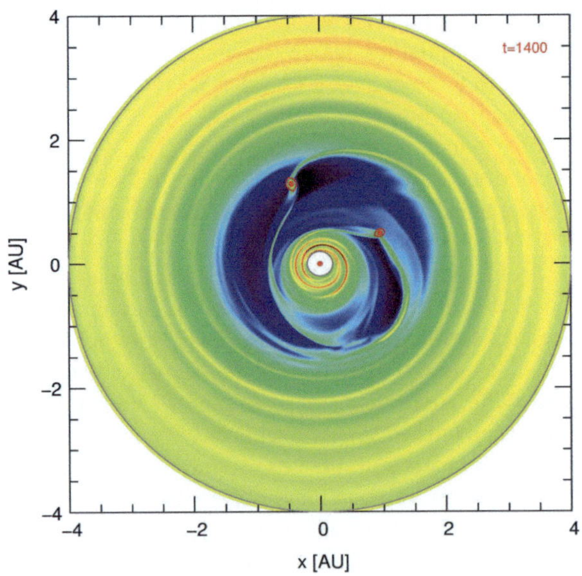

Fig. 2.25 Outcome of a two-dimensional hydrodynamical simulation of two embedded planets in a protoplanetary disk, for parameters of the system HD 73526 where two planets of about 2.5 $M_{\rm Jup}$ each orbit a central star. Shown is the gas surface density where yellow denotes higher density and blue lower values. The central red dot denotes the position of the star while the locations of the planets are given by the two red dots and the red lines indicate their Roche lobes. The scale of the x, y axes is in AU. Shown is the configuration 1400 yrs into the simulations after capture in a 2:1 resonance has occurred, see Fig. 2.26. Adapted from [246]

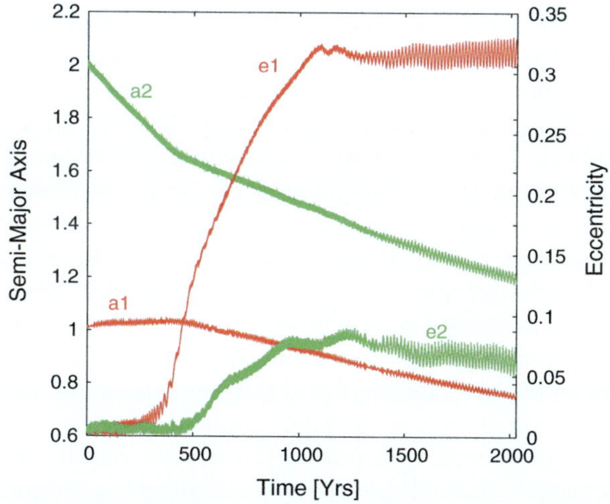

Fig. 2.26 The evolution of the semi-major axes and eccentricities of two embedded planets in a disk (those shown in Fig. 2.25) for a full hydrodynamical model. The semimajor axis and eccentricity of the inner planet (a_1, e_1) are indicated by the red curves and of the outer planet (a_2, e_2) by the green curves. Starting from their initial positions ($a_1 = 1.0$ and $a_2 = 2.0$ AU) the outer planet migrates inward (driven by the outer disk) and the inner one very slowly outward (due to the inner disk). After about 400 yrs into the simulations they are captured in a 2:1 mean motion resonance and they remain coupled during their inward migration. The eccentricities increase rapidly after resonant capture and settle to equilibrium values for longer time. Taken from Fig. 7 in [246], © ESO, reproduced with permission

accompanied by an increase in the orbital eccentricities of the planets. The evolution of the semimajor axis and eccentricity for such a resonant capture process is displayed in Fig. 2.26.

The modeling of the resonant capture process is very time consuming because multi-dimensional hydrodynamic simulations have to be performed, and a full evolution takes typically several weeks to run due to the long evolutionary times to be simulated. Hence, a simple model for this type of process, using only 3-body simulations consisting of one star and two planets has been suggested where the action of the disk is taken care of by adding additional damping forces using a parameterization of the migration and eccentricity damping. This type of a modeling takes only a few minutes. Specifically, the following prescription for the time evolution of the semi-major axis, a, and eccentricity, e, of embedded planets has been used in simulations performed to explain the observed state of GJ 876 [148]

$$\frac{\dot{a}}{a} = \frac{1}{\tau_{\rm a}}, \quad \frac{\dot{e}}{e} = \frac{1}{\tau_{\rm e}} \equiv K \, \frac{\dot{a}}{a} \,. \tag{2.116}$$

Here $\tau_{\rm a}$ and $\tau_{\rm e}$ denote the damping time scales for the semi-major axis and eccentricity of the planets, respectively, and K is a constant that describes the ratio between these two. This prescription for $\dot{a}$ and $\dot{e}$ can be used in principle for both planets because they are both in contact with the disk, but often it is applied only to the outer one assuming that the inner disk has been accreted already onto the star. The parameter K specifies the ratio of migration and eccentricity damping. In a study to model GJ 876 Lee and Peale [148] have applied exactly this model (damping of the outer planet only) and showed that a configuration similar to the observed one can be obtained by choosing a value $K = 100$ for the eccentricity damping. Smaller values for K typically result in too small damping and instability of the system. In follow-up studies, using full hydrodynamic simulations it was found that the disk damping for these systems with massive planets will produce values of $K \sim 10$ only [134]. Later it was shown that by taking the inner disk into account, as displayed in Fig. 2.25, it is possible to obtain full hydrodynamic results in agreement with the observed state of GJ 876 [59]. The system ends up in apsidal corotation, with correct eccentricities.

This type of differential migration process is a very natural outcome of the dynamical evolution of planets embedded in protoplanetary disks, and very many resonant configurations should be expected, the only question remaining, in which resonance are the planets captured. To become captured in a resonance the migration has to be sufficiently slow such that the excitation mechanism can operate long enough for them to be captured. If the migration is too fast, the planets cross the resonance and the system becomes more unstable. In addition the interaction between the two planets has to be strong enough for capture. As a consequence, massive planets in the Jupiter mass range and larger are typically captured into a 3:1 or 2:1 resonance [134]. In fact observationally there is indeed a clustering of massive planets within the 2:1 resonance [290]. On the other hand, for smaller mass planets, where the outer planet has up to a Saturn mass are typically captured into a 3:2 resonance [223].

Given that embedded planets have to migrate through the disk due to the disk torques and given that resonant capture is a very probable outcome during the evolution, then the small number of extrasolar planet pairs actually being in resonance is quite surprising. Indeed, the observed distribution of period ratios between two adjacent extrasolar planets is relatively smooth and does show only a mild enhancement of planet pairs near the mostly expected 3:2 and 2:1 resonances. Considering the turbulent nature of the protoplanetary disk the migration process is not smooth anymore but has a random component due to the gravitational disturbances of individual turbulent eddies. Through a large set of N-body simulations taking disk turbulence into account it was shown that the overall distribution of the period ratios (see Fig. 2.27 for the observational data) seems to be in rough agreement with standard migration including stochastic forces [237]. This is in agreement with results showing that stochastic perturbations can break the resonances between two planets [3]. This breaking of resonances allows for the possibility of bringing two planets very close together and producing systems of closely spaced low-mass planets. For the system Kepler-36 it was shown that this mechanism may be responsible to explain the unusually close configuration near the high degree first order 7:6 mean motion resonance [210].

Another feature apparent in Fig. 2.27 is the fact that the over abundances of period ratios near the 2:1 and 3:2 resonances do not occur at exact resonance position but lie

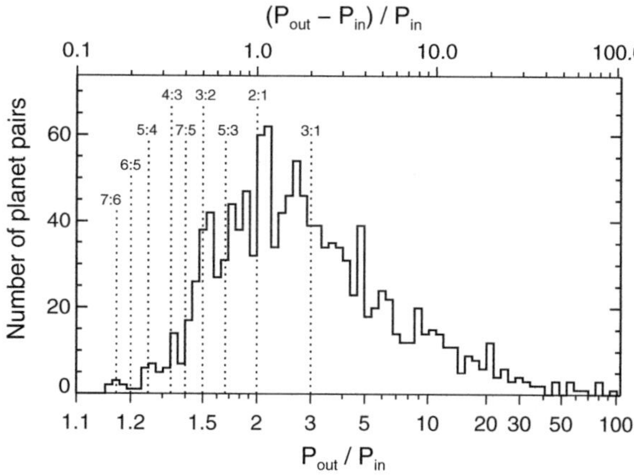

Fig. 2.27 Histogram of the number of two adjacent planets with a given period ratio for multi-planet system. The overall distribution is smooth but enhancements are visible near the 3:2, 2:1 and 3:1 resonances. Courtesy: D. Fabrycky. Based on Fig. 6 in [290]

just outside of it, a fact noticed earlier [22]. It has been argued that this small shift of period ratios just outside of the nominal values may be the result of resonant repulsion or dissipative divergence in the presence of tidal dissipation within the inner planet [22, 161]. Recently, it has been argued however, that tides alone cannot be held responsible for this enhancement near the 2:1 resonance, because this would imply too large initial eccentricities [254]. Another recent suggestion is the interaction of the planets with the remaining planetesimal disk. Here, it has been found that for the case of a sufficiently massive planetesimal disk the resonances can be disrupted and planet system remains just outside of the nominal values during the subsequent evolution [49].

2.7.2 Dynamics

Already after the first detections of extrasolar planets it was noted that they display a large variation of orbital eccentricities similar to spectroscopic binaries [265]. Both e-distributions cover the whole range from $0 \le e \le 1$ and have high mean values (with $e \sim 0.3$) in obvious contrast to the Solar System where the majority of the planets (despite Mercury and Mars) have a rather small eccentricity. For the overall e-distribution the following form has been suggested [252]

$$\frac{dN}{de} \propto \left(\frac{1}{(1+e)^b} - \frac{e}{2^b} \right) , \tag{2.117}$$

where dN is the number of planets in an eccentricity interval $[e, e + de]$. According to [252] $b = 4$ gives a good match to the observed distribution which gives a smooth distribution which peaks at $e = 0$. There is a tendency for lower mass planets to have a lower eccentricity, in particular for $m_\mathrm{p} < 50\, M_\mathrm{Earth}$ the maximum eccentricity is around $e \approx 0.4$ while larger mass planets reach over $e > 0.9$. This finding is in agreement with the fact that lower mass planets are preferentially found in multi-planet systems that have on average lower eccentricities. For an overview of the observational properties and more references see [290].

In the previous Sect. 2.6 we showed that planet-disk interaction will typically lead to eccentricity and inclination damping on a timescale shorter than the migration time. Consequently, planet-disk interaction cannot be responsible for the observed eccentricity distribution and two other processes (planet-planet scattering or the Kozai mechanism) have been suggested that both require additional objects. In the first scenario several planets have formed in the disk and undergo initially a convergent migration process which leads to a compact configuration either in resonance or close to it. Upon disk dissipation its stabilizing effect disappears and the planetary system my become dynamically unstable. This will increase the eccentricity and inclination of the objects leading eventually to a scattering processes [2]. Some bodies will be scattered towards the star and some outwards, or be thrown out altogether. As shown by several simulations using many realizations of such models, these type of planet-planet scattering processes can lead to eccentricity distributions very similar to the one observed for the ensemble of extrasolar planets [86, 122].

2.7.2.1 The Kozai Mechanism

In the past 20 years or so a dynamical process operating in a hierarchical 3-body system has gained much of attention to explain certain properties of extrasolar planetary systems. The mechanism was originally proposed to explain the dynamics of main-belt asteroids due to the perturbations of Jupiter by Y. Kozai [139]. The geometry of a hierarchical system is sketched in Fig. 2.28. A central 2-body system with

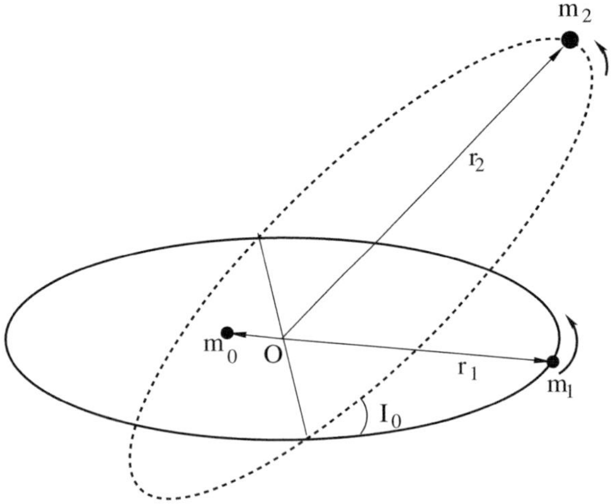

Fig. 2.28 The geometry of a hierarchical 3-body system susceptible to the Kozai mechanism. A binary system with masses m_0 and m_1 is orbited by a third object m_2 that is on an inclined orbit. The inclination of the two orbital planes is initially I_0. Please note that the distance (semimajor axis) of the third object (m_2) is much larger than that of the secondary object (m_1) so the real orbits do not intersect

masses m_0 and m_1 is orbited by a third object (m_2) at a larger distance, $r_2 > r_1$. The two orbital planes are inclined by a certain angle I with each other. Such a configuration describes for example the asteroid case in the Solar System, where: m_0 = Sun, m_1 = asteroid and m_2 = Jupiter, or a protoplanetary case where: m_0 = host star, m_1 = planet and m_2 = secondary star. The longterm evolution of such systems can result in a periodic exchange of angular momentum between the inner binary and the third, distant object, given that the initial inclination exceeds a certain critical value. As Kozai showed, for the satellite (the object m_1), the orbit averaged equations of motion have a conserved quantity, the z-component of the angular momentum $J_z = \cos(I)\sqrt{1-e^2}$, where z is the direction perpendicular to the orbit of the binary, m_2. If the initial inclination of the body's orbital plane with that of the perturber, I_0, is higher than a critical value $I_{\rm crit}$ then the orbit of the secondary (m_1) will become oscillatory with phases of high eccentricity (with a reduced inclination I) and low eccentricity (with high inclination), always maintaining a constant J_z. The value of $I_{\rm crit}$ depends on the separation of m_2 and is about 39° for very distant objects with $r_2 \gg r_1$. Even though presented by Kozai in 1962, for the first 35 years his work drew little attention with about 60 citations. However, after the discovery of the first highly eccentric exoplanet orbiting 16 Cyg B [54] in 1997, immediately the importance of the mechanism in the context of exoplanet research was noted [115], and the number of citations increased significantly to about 15 per year. Then, in 2003 it was noticed by Wu and Murray [292] that this mechanism could be used for a new type of migration process, namely eccentric or tidally driven migration. After that, the recognition of Kozai's work has increased even further to now over 70 citations per year (according to the ADS). Sometimes the mechanism is also referred to as *Kozai-Lidov* mechanism attributing the work by M. Lidov in the same year of 1962 [139].

We will now briefly explain how this mechanism operates in shrinking the orbit of the planet, i.e. how it reduces the semimajor axis, and produces planets on highly inclined orbits. If a planetary system consisting of a star and a planet is accompanied by a distant companion star with an orbit that is inclined with respect to the planet's orbit, then the perturbations will give rise to Kozai oscillations with periodic phases of very high planet eccentricity, reaching nearly $e \approx 1$. In this case, the planet comes very close to the star and energy is dissipated inside the central star which is taken from the planetary orbit. Hence, at each pericenter passage the planetary orbits shrinks by a small amount. Over the course of possibly several Gyrs the orbit can shrink significantly producing a large scale migration of the planet. Because at the same time, the inclination oscillates between large and small values this mechanism can produce compact systems with close-in planets on possibly highly eccentric orbits. Wu and Murray [292] and later Fabrycky and Tremaine [83] applied this model to the planetary system HD 80606 which today has the following observed orbital parameter: planet mass $m_{\rm p} = 3.94\, M_{\rm Jup}$, $a_{\rm p} = 0.449$ AU, $e_{\rm p} = 0.933$. Up to now this is the highest planet eccentricity observed among the sample of extrasolar planets, and as a consequence the distance from the planet to the stars varies between 0.03 and 0.88 AU. The central star of about $m_0 = 0.9\, M_\odot$ is orbited by a lower mass companion star at a distance of about 1200 AU. To model the longterm evolution of this system Wu and Murray used as starting parameter for the planet a mass of $m_p = 3.94\,M_{Jup}$, a semi-major axis $a_p = 5$ AU and an eccentricity $e_p = 0.1$. They assumed an initial inclination of the planet and binary star orbit of $I_0 = 85.6°$. Using these initial parameter and a tidal Q-value for the star of about 10^6 they find an initial period of about 0.022 Gyr for the Kozai oscillations and maximum eccentricities close too unity. Continuing the simulations until several Gyrs they show that indeed the system approaches today's parameter at an age of about 3 Gyrs, ending up with the observed eccentricity and an inclination of $I = 50°$.

This working example of tidally driven migration was used to suggest that a large fraction of the close-in Jupiter has migrated not by standard planet-disk interaction but rather by this process [83]. It was shown that the circularization timescale of the planet is comparable to the migration time which implies that aligned planets are most likely formed coplanar [14]. Noting that the sky projected spin-orbit angle (stellar obliquity, β) is higher for hotter central stars (with $T_{\rm eff} > 6250$ K) than for cooler stars, it has been suggested that tidally driven migration may even be the dominant mode of migration [289]. Assuming that most of the planets start with large obliquities lower mass stars with lower $T_{\rm eff}$ will tend to damp this misalignment more rapidly due to an active convection zone while more massive stars have smaller damping and the initial misalignment is maintained. However, using a sample of 61 transiting hot Jupiters with measured projected spin-orbit angle β it was shown that the observed distribution of the angles β is compatible with the assumption that most hot Jupiters were transported by smooth migration inside a proto-planetary disk [57].

2.7.3 Multi-planet Systems

The large sample of planets discovered by the Kepler space mission has shown that the main group of extrasolar planets are smaller planets within a radius range between 1.25 and about 5 $R_{\rm Earth}$ that we call here altogether Super-Earths [41]. Amongst those, true planetary systems with 2 and more planets are the rule as Kepler has discovered hundreds of transiting multi-planet systems, showing that statistically about half of all solar-type stars host at least one planet of this type where additionally for the multi-planet systems the inclinations between the individual planetary orbits are very small [290]. Mostly, the radial spacing between the planets is very small and they are close to their host stars. As a consequence these compact systems would fit well into the orbit of Mercury, i.e. they resemble a Solar System scaled down by a factor of about 10 with somewhat larger planets, though. Typically, the systems are not in resonance but show separations similar to the Solar System planets, in terms of their Hill radii [233]. The ubiquity of these compact systems of lower mass planets is probably related to their longterm stability because closely packed systems of larger mass planets would be dynamically unstable with scattering processes and possible ejections (as shown above), while lower mass planets might collide and stabilize (see also [86, 290]). The flatness of the complete system is again an indication for a disk-driven evolution.

Concerning the formation of these compact, tightly packed planetary systems different scenarios have been considered such as (1) in situ formation in a massive disk, (2) accretion during inward type I migration, (3) shepherding by interior mean-motion resonances of inwardly migrating massive planets, (4) shepherding by interior secular resonances of inwardly migrating massive planets, (5) circularization of high-eccentric planets by tidal interactions with the star, and (6) photo-evaporation of close-in giant planets. In [233] it is argued that the possibilities (3)–(6) are in conflict with some observational facts because (3) and (4) require a giant planet to push the planets inward and an additional disk to damp orbital eccentricities. In both scenarios a giant planet should be present just outside of the outermost planet which is not observed. For scenario (5) circularization is in principle possible (as shown above for tidal migration scenario) but it requires large initial eccentricities that would result in scattering events and a remaining single-planet system, which is in conflict with the observations. For (6) evaporation is possible only for very close-in planets and it requires several Gyrs to operate. The strong dependence on distance from the central star would only allow the innermost planet to be evaporated. Hence, it is concluded that only scenarios (1) (in situ) and (2) (inward type I migration) are leading contestors [106, 233].

The in-situ formation scenario requires a lot of material ($\approx$20–40 $M_\oplus$) within a fraction of an AU in the inner protoplanetary disk, unless radial transport of material is considered [108], but this might also lead to an enhancement of the mass of the outermost planet. Within the MMSN context a very steep surface density power law, $\Sigma = \Sigma_0 (r/AU)^{-x}$, with $x \approx 1.6-1.7$ and a 10 times higher normalization Σ_0 in comparison to the Solar System is required to form the planets within the observed mass range. This is problematic because the high mass pushes the disk towards instability with respect to fragmentation. Additionally, the observations indicate shallower profiles with $x \approx 0.5-1.0$ [286]. Despite these problems, some simulations of in-situ formation show that orbital properties (eccentricities, inclinations, separations) of the planets seem to match the observations [51, 233]. However, the absence of a distinct slope in the radial mass profile of extrasolar planetary systems is taken as an indication that a radial rearrangement of solids must have taken place for the majority of systems [232]. For scenario (2), simulations of migrating Super-Earths indicate the formation of resonant chains with ongoing destabilization and subsequent collision and accretion events, forming systems similar to the observed ones [55]. A possible distinction between models might be the occurrence of naked high-density rocks for scenario (1) and possibly lower density material containing ice for scenario (2) but planetary atmospheres could possibly hide the effect [233].

2.7.4 *Circumbinary Planets*

One of the recent findings of the Kepler space mission has been the discovery of circumbinary planets that orbit around two stars very similar to the Tatooine system in the *Star Wars* series. The first of such systems has been Kepler 16 where a Saturn mass planet orbits a central binary (with orbital period $P_{\rm bin} = 41$ days) with a period of about $P_{\rm p} = 229$ days [69]. By now about 10 systems are known and in most cases the planets orbit the central binary star very close to the mechanical stability limit, i.e. they have the closest possible distance to the binary star. A tighter orbit would lead to dynamical instability with a possible subsequent loss of the planet through a scattering event with the central binary star. This stability limit can be determined by performing sequences of 3-body integrations consisting of a low mass object (test particle) that orbit around a binary star. It is determined by the mass ratio $q_{\rm bin} = m_1/m_2$, and the eccentricity $e_{\rm bin}$ of the binary star and lies in the range of 2–5 times the semimajor axis of the binary, $a_{\rm bin}$ [81]. Through extensive 3-body simulations Holman and Wiegert [116] show that the dependence on $q_{\rm bin}$ is weak and they give the following approximate formula for the critical distance

$$a_{\rm crit} = (2.28 + 3.82 e_{\rm bin} - 1.71 e_{\rm bin}^2)\, a_{\rm bin} \,. \tag{2.118}$$

Beyond $a_{\rm crit}$ the orbits are in principle stable but newer simulations show in addition that for planetary semi-major axes that are near the mean motion resonances 4:1, 5:1 and so on with the central binary the orbits are very unstable [50], such that planets at those locations are easily scattered out of the system. Consequently, the observed locations of the planets around binary stars are typically between these resonances. In two systems (Kepler 16, Kepler 38) the planets are located between the 5:1 and 6:1 resonances and in 4 others (Kepler 35, 47, 64 and 413) they are between the 6:1 and 7:1 resonances. These cases comprise over half of the known systems and the closeness of the planets to the edge of instability raises the question as to their formation. Table 2.1 gives an overview of the orbital parameter of some well known systems, and shows the closeness of the planets to the boundary of stability, $a_{\rm p}/a_{\rm crit}$.

Typical formation scenarios for planets in general require the growth from smaller particles through a sequence of collisions as outlined in detail in Sects. 2.2 and 2.3. If the relative speed between two objects is too large the growth is seriously handicapped. As mentioned above, even under the conditions prevalent in the early Solar System growth is very difficult to achieve. Not surprisingly, the additional perturbations generated by the binary star in the center lead to increased particles velocities that make growth even more difficult, if not impossible [187, 207, 249]. Hence, it is typically assumed that the planets have formed further away from the binary star in environments that are dynamically much calmer, and then migrated inward to their present location through a planet-disk interaction process [173, 222]. This migration of the planet through the circumbinary disk would also naturally explain the flatness of these systems if one assumes that the disk is completely aligned with the orbital plane of the binary. In the following we shall concentrate on this migration scenario.

As mentioned above, the presence of a binary star inside a disk creates a large inner cavity cleared from the gas as shown in Fig. 2.29. This clearing is due to the transfer of angular momentum from the binary to the disk through tidal forces. The extent of the inner gap depends on the binary and disk parameter such as the binary's mass ratio, eccentricity and disk viscosity and temperature [10]. Numerical

Table 2.1 Parameters of selected circumbinary planets as listed in [290]. The upper table shows the parameter of the central binary star, and the bottom table the properties of the circumbinary planet. The last row contains the ratio of the observed semi-major axis to the critical value a_{crit} (Eq. 2.118) and indicates clearly the closeness to instability of the circumbinary planets

	Kepler-16	Kepler-38	Kepler-34	Kepler-35	Kepler-64	Kepler-413
Binary parameter						
M_1 ($M_\odot$)	0.62	0.95	1.05	0.89	1.38	0.82
M_2 ($M_\odot$)	0.20	0.25	1.02	0.81	0.38	0.54
P_{bin} (days)	41	18.6	28	21	20	10
a_{bin} (AU)	0.22	0.15	0.23	0.18	0.17	0.11
e_{bin}	0.16	0.10	0.52	0.14	0.22	0.04
Planet parameter						
m_p (M_{Jup})	0.33	0.36	0.21	0.13	< 0.1	0.21
P_p (days)	228	105.6	288	131	138	66
a_p (AU)	0.70	0.46	1.09	0.6	0.63	0.36
e_p	0.07	0.03	0.18	0.04	0.05	0.12
a_p/a_{crit}	1.09	1.13	1.28	1.13	1.15	1.36

Fig. 2.29 The structure of a circumbinary disk around a central binary star (indicated by the two red spheres in the middle). Color coded is the surface density of the disk and the vertical extension indicates the thickness (temperature) of the disk. *Courtesy* Richard Günther, University of Tübingen, based on simulations in [103]

simulations of hydrodynamic disks around binary stars show that the inner disk becomes eccentric and slowly precesses around the binary star where the precession rate is typically several 100 binary orbits, P_{bin} [78, 131]. However the full dynamical evolution of circumbinary disks around their central binary stars is complex and not fully understood yet [224].

Coming back to the evolution of planets embedded in disks we note that Saturn mass planets open a partial gap in the disk. Hence, the corotation torques are reduced and they are driven primarily by the Lindblad torques that will induce inward migration. This inward drift can be stopped at special locations, the so-called planet traps [177], where the total torque acting on the planet vanishes, such that the inward migration is terminated. Regions with a positive density slope near the inner edge of the disk which will act as a natural location for a planet trap. For single stars a positive density gradient can be generated for example by the magnetosphere of the young star while for binary stars the inner disk cavity is generated naturally by angular momentum transferal from the binary to the disk (see Fig. 2.29). The extent of the hole, i.e. the location of the disk's inner edge will then determine the final location of the planet. For typical disk parameter with $\alpha \approx 0.01$, $H/r \approx 0.05$ and binaries with moderate eccentricity, $e_{\rm bin} \approx 0.1$, such as Kepler-16 and Kepler-38 (see Table 2.1) it was shown that in the simulations the final parking position of planets is indeed near the the observed locations [130, 224]. In some simulations for Kepler 38 it was found that the planet was captured in the 5:1 resonance and remained at this location. This capture into resonance was accompanied by an increase of the eccentricity of the planetary orbit to about $e_{\rm bin} \approx 0.2$ [130]. As explained above, planets in such resonant orbits will typically be unstable in the longterm evolution and they can only be stable in the presence of a disk that tends to damp planetary eccentricities as shown in the previous section. So, after disk dispersal such a planet is expected to become unstable and be scattered out of the system. It is possible that several planets that initially formed in circumbinary disks experienced such a fate and are now a member of the free-floating planet population. In this case we are just observing those planets that luckily happened to end up in a final stable location in between the unstable resonances. As was shown, different disk parameter with a lower density and viscosity can lead to final parking positions near the observed location for the Kepler-38 system [130]. In Fig. 2.30 the final parking position is indicated by the innermost orange dot. It is right at the inner edge of the disk.

While the final position of planets of planets around binaries with moderate eccentricities (Kepler-16, Kepler-35 and Kepler-38) can be understood in terms of a migra-

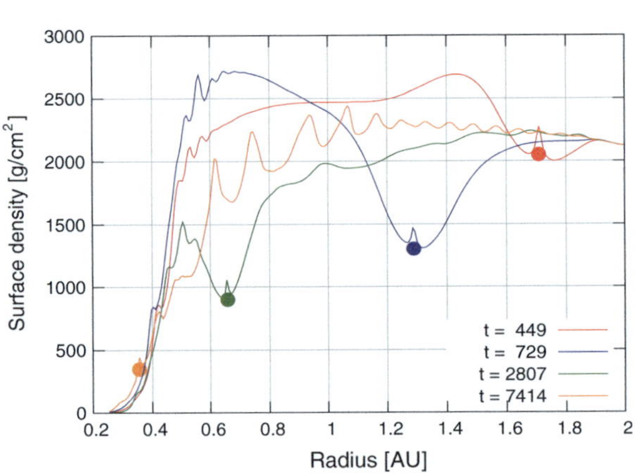

Fig. 2.30 The migration of a planet in the Kepler-38 system. Shown is the surface density at different times (in yrs) after insertion of the planet. The time levels are indicated by the labels. The dots indicate the radial position of the planet at those times, for illustration they have been moved to the corresponding surface density distribution. Clearly visible is the (partial) gap formed by the planet during its inward migration. The orange dot marks the final equilibrium position. Taken from Fig. 10 in [130], © ESO, reproduced with permission

tion process of a planet in a disk [130, 224], it is still difficult to explain the location of the observed planet in the Kepler-34 systems where the central binary has as large eccentricity of $e_{\rm bin} = 0.52$. This high eccentricity of the binary generates a large inner hole of the disk that is highly eccentric and precesses around the central binary with a period of about 120 yrs equivalent to nearly 2000 $P_{\rm bin}$. The final position of the planet is determined by the size of the inner hole. The planet settles in an elliptic orbit that precesses around the binary with exactly the same period as the disk, such that the pericenters of the disk and planet are always aligned. This final orbit has a semi-major axis that is substantially larger than the observed value indicated in Table 2.1. Hence, for the eccentric binary star Kepler-34, the observed position of the planet is still difficult to understand with a migration process but may be that better disk models with more realistic physics will lead to improved results.

In planetary systems with multiple objects interesting dynamics can take place. Disk-planet interaction will often lead to a convergence of orbits with capture in mean-motion resonances (MMR) between two adjacent planets. However, observationally there is only a slight overabundance of planets near the 3:2 and 2:1 MMRs, but stochastic and tidal processes can take planets out of exact resonance. The dynamical interaction between multiple massive planets can lead to eccentricity and inclination excitation. Another pathway to generate the high observed eccentricity and inclinations in some systems is through the Kozai-Lidov mechanism which requires the exchange of angular momentum with a third, distant object in the system. Concerning the observed abundance of planetary systems that contain several Super-Earths in a compact configuration, the most likely pathway appears to be the convergent inward migration process accompanied by stochastic forces with the disk. A very exciting discovery of the Kepler space mission has been the detection of circumbinary planets. The observed locations of the planets close to the dynamical stability region can be explained for most systems through a migration process of these planets through the circumplanetary disk.

Acknowledgements This text is based on a series of lectures on the topic *Planet formation and disk-planet interactions* given at the 45th "Saas-Fee Advanced Course" of the Swiss Society for Astrophysics and Astronomy (SSAA) held in Les Diablerets in March 2015. I acknowledge generous support from the SSAA, and would like to thank the organisers (Marc Audard, Yann Alibert and, Michael R. Meyer, Martine Logossou) for providing such a nice and stimulating atmosphere. I thank Giovanni Picogna for a reading of the manuscript.

References

1. Aarseth, S.J.: From NBODY1 to NBODY6: the growth of an industry. PASP **111**, 1333–1346 (1999). https://doi.org/10.1086/316455
2. Adams, F.C., Laughlin, G.: Migration and dynamical relaxation in crowded systems of giant planets. Icarus **163**, 290–306 (2003). https://doi.org/10.1016/S0019-1035(03)00081-2
3. Adams, F.C., Laughlin, G., Bloch, A.M.: Turbulence implies that mean motion resonances are rare. ApJ **683**, 1117–1128 (2008). https://doi.org/10.1086/589986
4. ALMA Partnership, Brogan, C.L., Pérez, L.M., Hunter, T.R., Dent, W.R.F., Hales, A.S., Hills, R.E., Corder, S., Fomalont, E.B., Vlahakis, C., Asaki, Y., Barkats, D., Hirota, A., Hodge, J.A., Impellizzeri, C.M.V., Kneissl, R., Liuzzo, E., Lucas, R., Marcelino, N., Matsushita, S., Nakanishi, K., Phillips, N., Richards, A.M.S., Toledo, I., Aladro, R., Broguiere, D., Cortes, J.R., Cortes, P.C., Espada, D., Galarza, F., Garcia-Appadoo, D., Guzman-Ramirez, L., Humphreys, E.M., Jung, T., Kameno, S., Laing, R.A., Leon, S., Marconi, G., Mignano, A., Nikolic, B., Nyman, L.A., Radiszcz, M., Remijan, A., Rodón, J.A., Sawada, T., Takahashi, S., Tilanus, R.P.J., Vila Vilaro, B., Watson, L.C., Wiklind, T., Akiyama, E., Chapillon, E., de Gregorio-Monsalvo, I., Di Francesco, J., Gueth, F., Kawamura, A., Lee, C.F., Nguyen Luong, Q., Mangum, J., Pietu, V., Sanhueza, P., Saigo, K., Takakuwa, S., Ubach, C., van Kempen, T., Wootten, A., Castro-Carrizo, A., Francke, H., Gallardo, J., Garcia, J., Gonzalez, S., Hill, T., Kaminski, T., Kurono, Y., Liu, H.Y., Lopez, C., Morales, F., Plarre, K., Schieven, G., Testi, L., Videla, L., Villard, E., Andreani, P., Hibbard, J.E., Tatematsu, K.: The 2014 ALMA Long Baseline Campaign: first results from high angular resolution observations toward the HL Tau region. ApJ **808**, L3 (2015). https://doi.org/10.1088/2041-8205/808/1/L3
5. Andrews, S.M., Williams, J.P.: High-resolution submillimeter constraints on circumstellar disk structure. ApJ **659**, 705–728 (2007). https://doi.org/10.1086/511741
6. Armitage, P.J.: Astrophysics of Planet Formation (2010)
7. Artymowicz, P.: On the wave excitation and a generalized torque formula for Lindblad resonances excited by external potential. ApJ **419**, 155 (1993). https://doi.org/10.1086/173469
8. Artymowicz, P.: Dynamics of gaseous disks with planets. In: Caroff, L., Moon, L.J., Backman, D., Praton, E. (eds.) Debris Disks and the Formation of Planets. Astronomical Society of the Pacific Conference Series, vol. 324, p. 39 (2004)
9. Artymowicz, P.: Migration Type III. In: KITP Conference: Planet Formation: Terrestrial and Extra Solar (2004)
10. Artymowicz, P., Lubow, S.H.: Dynamics of binary-disk interaction. 1: Resonances and disk gap sizes. ApJ **421**, 651–667 (1994). https://doi.org/10.1086/173679
11. Balbus, S.A., Hawley, J.F.: A powerful local shear instability in weakly magnetized disks. I—Linear analysis. II—Nonlinear evolution. ApJ **376**, 214–233 (1991). https://doi.org/10.1086/170270
12. Balmforth, N.J., Korycansky, D.G.: Non-linear dynamics of the corotation torque. MNRAS **326**, 833–851 (2001). https://doi.org/10.1046/j.1365-8711.2001.04619.x
13. Baraffe, I., Chabrier, G., Fortney, J., Sotin, C.: Planetary Internal Structures. Protostars and Planets VI, pp. 763–786 (2014). https://doi.org/10.2458/azu_uapress_9780816531240-ch033
14. Barker, A.J., Ogilvie, G.I.: On the tidal evolution of Hot Jupiters on inclined orbits. MNRAS **395**, 2268–2287 (2009). https://doi.org/10.1111/j.1365-2966.2009.14694.x
15. Baruteau, C., Crida, A., Paardekooper, S.J., Masset, F., Guilet, J., Bitsch, B., Nelson, R., Kley, W., Papaloizou, J.: Planet-Disk Interactions and Early Evolution of Planetary Systems. Protostars and Planets VI, pp. 667–689 (2014). https://doi.org/10.2458/azu_uapress_9780816531240-ch029
16. Baruteau, C., Fromang, S., Nelson, R.P., Masset, F.: Corotation torques experienced by planets embedded in weakly magnetized turbulent discs. A&A **533**, A84 (2011). https://doi.org/10.1051/0004-6361/201117227
17. Baruteau, C., Lin, D.N.C.: Protoplanetary migration in turbulent isothermal disks. ApJ **709**, 759–773 (2010). https://doi.org/10.1088/0004-637X/709/2/759

18. Baruteau, C., Masset, F.: On the corotation torque in a radiatively inefficient disk. ApJ **672**, 1054–1067 (2008). https://doi.org/10.1086/523667
19. Baruteau, C., Masset, F.: Type I planetary migration in a self-gravitating disk. ApJ **678**, 483–497 (2008). https://doi.org/10.1086/529487
20. Baruteau, C., Masset, F.: Recent developments in planet migration theory. In: Souchay, J., Mathis, S., Tokieda, T. (eds.) Lecture Notes in Physics, vol. 861, p. 201. Springer, Berlin (2013). https://doi.org/10.1007/978-3-642-32961-6_6
21. Baruteau, C., Meru, F., Paardekooper, S.J.: Rapid inward migration of planets formed by gravitational instability. MNRAS **416**, 1971–1982 (2011). https://doi.org/10.1111/j.1365-2966.2011.19172.x
22. Batygin, K., Morbidelli, A.: Dissipative divergence of resonant orbits. AJ **145**, 1 (2013). https://doi.org/10.1088/0004-6256/145/1/1
23. Beaugé, C., Ferraz-Mello, S., Michtchenko, T.A.: Extrasolar planets in mean-motion resonance: apses alignment and asymmetric stationary solutions. ApJ **593**, 1124–1133 (2003). https://doi.org/10.1086/376568
24. Benítez-Llambay, P., Masset, F., Koenigsberger, G., Szulágyi, J.: Planet heating prevents inward migration of planetary cores. Nature **520**, 63–65 (2015). https://doi.org/10.1038/nature14277
25. Benz, W., Anic, A., Horner, J., Whitby, J.A.: The origin of Mercury. Space Sci. Rev. **132**, 189–202 (2007). https://doi.org/10.1007/s11214-007-9284-1
26. Benz, W., Asphaug, E.: Impact simulations with fracture. I—Method and tests. Icarus **107**, 98 (1994). https://doi.org/10.1006/icar.1994.1009
27. Benz, W., Asphaug, E.: Catastrophic disruptions revisited. Icarus **142**, 5–20 (1999). https://doi.org/10.1006/icar.1999.6204
28. Benz, W., Slattery, W.L., Cameron, A.G.W.: The origin of the moon and the single-impact hypothesis. I. Icarus **66**, 515–535 (1986). https://doi.org/10.1016/0019-1035(86)90088-6
29. Binney, J., Tremaine, S.: Galactic dynamics (1987)
30. Bitsch, B., Crida, A., Libert, A.S., Lega, E.: Highly inclined and eccentric massive planets. I. Planet-disc interactions. A&A **555**, A124 (2013). https://doi.org/10.1051/0004-6361/201220310
31. Bitsch, B., Johansen, A., Lambrechts, M., Morbidelli, A.: The structure of protoplanetary discs around evolving young stars. A&A **575**, A28 (2015). https://doi.org/10.1051/0004-6361/201424964
32. Bitsch, B., Kley, W.: Orbital evolution of eccentric planets in radiative discs. A&A **523**, A30 (2010). https://doi.org/10.1051/0004-6361/201014414
33. Bitsch, B., Kley, W.: Evolution of inclined planets in three-dimensional radiative discs. A&A **530**, A41 (2011). https://doi.org/10.1051/0004-6361/201016179
34. Bitsch, B., Kley, W.: Range of outward migration and influence of the disc's mass on the migration of giant planet cores. A&A **536**, A77 (2011). https://doi.org/10.1051/0004-6361/201117202
35. Blum, J., Wurm, G.: The growth mechanisms of macroscopic bodies in protoplanetary disks. ARA&A **46**, 21–56 (2008). https://doi.org/10.1146/annurev.astro.46.060407.145152
36. Blum, J., Wurm, G., Kempf, S., Poppe, T., Klahr, H., Kozasa, T., Rott, M., Henning, T., Dorschner, J., Schräpler, R., Keller, H.U., Markiewicz, W.J., Mann, I., Gustafson, B.A., Giovane, F., Neuhaus, D., Fechtig, H., Grün, E., Feuerbacher, B., Kochan, H., Ratke, L., El Goresy, A., Morfill, G., Weidenschilling, S.J., Schwehm, G., Metzler, K., Ip, W.H.: Growth and form of planetary seedlings: results from a microgravity aggregation experiment. Phys. Rev. Lett. **85**, 2426 (2000). https://doi.org/10.1103/PhysRevLett.85.2426
37. Bodenheimer, P., Hubickyj, O., Lissauer, J.J.: Models of the in situ formation of detected extrasolar giant planets. Icarus **143**, 2–14 (2000). https://doi.org/10.1006/icar.1999.6246
38. Bodenheimer, P., Pollack, J.B.: Calculations of the accretion and evolution of giant planets: the effects of solid cores. Icarus **67**, 391–408 (1986). https://doi.org/10.1016/0019-1035(86)90122-3

39. Boley, A.C.: The two modes of gas giant planet formation. ApJ **695**, L53–L57 (2009). https://doi.org/10.1088/0004-637X/695/1/L53
40. Boss, A.P.: Giant planet formation by gravitational instability. Science **276**, 1836–1839 (1997). https://doi.org/10.1126/science.276.5320.1836
41. Burke, C.J., Bryson, S.T., Mullally, F., Rowe, J.F., Christiansen, J.L., Thompson, S.E., Coughlin, J.L., Haas, M.R., Batalha, N.M., Caldwell, D.A., Jenkins, J.M., Still, M., Barclay, T., Borucki, W.J., Chaplin, W.J., Ciardi, D.R., Clarke, B.D., Cochran, W.D., Demory, B.O., Esquerdo, G.A., Gautier III, T.N., Gilliland, R.L., Girouard, F.R., Havel, M., Henze, C.E., Howell, S.B., Huber, D., Latham, D.W., Li, J., Morehead, R.C., Morton, T.D., Pepper, J., Quintana, E., Ragozzine, D., Seader, S.E., Shah, Y., Shporer, A., Tenenbaum, P., Twicken, J.D., Wolfgang, A.: Planetary candidates observed by Kepler IV: planet sample from Q1–Q8 (22 months). ApJS **210**, 19 (2014). https://doi.org/10.1088/0067-0049/210/2/19
42. Cameron, A.G.W.: Physics of the primitive solar accretion disk. Moon Planets **18**, 5–40 (1978). https://doi.org/10.1007/BF00896696
43. Cassen, P.: Protostellar disks and planet formation. In: Queloz, D., Udry, S., Mayor, M., Benz, W., Cassen, P., Guillot, T., Quirrenbach, A. (eds.) Saas-Fee Advanced Course 31: Extrasolar planets, pp. 369–448 (2006). https://doi.org/10.1007/978-3-540-31470-7-3
44. Chabrier, G., Johansen, A., Janson, M., Rafikov, R.: Giant Planet and Brown Dwarf Formation. Protostars and Planets VI, pp. 619–642 (2014). https://doi.org/10.2458/azu_uapress_9780816531240-ch027
45. Chambers, J.E.: A hybrid symplectic integrator that permits close encounters between massive bodies. MNRAS **304**, 793–799 (1999). https://doi.org/10.1046/j.1365-8711.1999.02379.x
46. Chambers, J.E.: Making more terrestrial planets. Icarus **152**, 205–224 (2001). https://doi.org/10.1006/icar.2001.6639
47. Chambers, J.E.: Planetary accretion in the inner Solar System. Earth Planet. Sci. Lett. **223**, 241–252 (2004). https://doi.org/10.1016/j.epsl.2004.04.031
48. Chambers, J.E., Wetherill, G.W.: Making the terrestrial planets: N-body integrations of planetary embryos in three dimensions. Icarus **136**, 304–327 (1998). https://doi.org/10.1006/icar.1998.6007
49. Chatterjee, S., Ford, E.B.: Planetesimal interactions can explain the mysterious period ratios of small near-resonant planets. ApJ **803**, 33 (2015). https://doi.org/10.1088/0004-637X/803/1/33
50. Chavez, C.E., Georgakarakos, N., Prodan, S., Reyes-Ruiz, M., Aceves, H., Betancourt, F., Perez-Tijerina, E.: A dynamical stability study of Kepler circumbinary planetary systems with one planet. MNRAS **446**, 1283–1292 (2015). https://doi.org/10.1093/mnras/stu2142
51. Chiang, E., Laughlin, G.: The minimum-mass extrasolar nebula: in situ formation of close-in super-Earths. MNRAS **431**, 3444–3455 (2013). https://doi.org/10.1093/mnras/stt424
52. Chiang, E., Youdin, A.N.: Forming planetesimals in solar and extrasolar nebulae. Annu. Rev. Earth Planet. Sci. **38**, 493–522 (2010). https://doi.org/10.1146/annurev-earth-040809-152513
53. Chiang, E.I., Goldreich, P.: Spectral energy distributions of T Tauri stars with passive circumstellar disks. ApJ **490**, 368–376 (1997)
54. Cochran, W.D., Hatzes, A.P., Butler, R.P., Marcy, G.W.: The discovery of a planetary companion to 16 Cygni B. ApJ **483**, 457–463 (1997). https://doi.org/10.1086/304245
55. Cossou, C., Raymond, S.N., Hersant, F., Pierens, A.: Hot super-Earths and giant planet cores from different migration histories. A&A **569**, A56 (2014). https://doi.org/10.1051/0004-6361/201424157
56. Cresswell, P., Dirksen, G., Kley, W., Nelson, R.P.: On the evolution of eccentric and inclined protoplanets embedded in protoplanetary disks. A&A **473**, 329–342 (2007). https://doi.org/10.1051/0004-6361:20077666
57. Crida, A., Batygin, K.: Spin-orbit angle distribution and the origin of (mis)aligned hot Jupiters. A&A **567**, A42 (2014). https://doi.org/10.1051/0004-6361/201323292
58. Crida, A., Morbidelli, A., Masset, F.: On the width and shape of gaps in protoplanetary disks. Icarus **181**, 587–604 (2006). https://doi.org/10.1016/j.icarus.2005.10.007

59. Crida, A., Sándor, Z., Kley, W.: Influence of an inner disc on the orbital evolution of massive planets migrating in resonance. A&A **483**, 325–337 (2008). https://doi.org/10.1051/0004-6361:20079291
60. D'Angelo, G., Kley, W., Henning, T.: Orbital migration and mass accretion of protoplanets in three-dimensional global computations with nested grids. ApJ **586**, 540–561 (2003). https://doi.org/10.1086/367555
61. D'Angelo, G., Lubow, S.H.: Three-dimensional disk-planet torques in a locally isothermal disk. ApJ **724**, 730–747 (2010). https://doi.org/10.1088/0004-637X/724/1/730
62. D'Angelo, G., Lubow, S.H., Bate, M.R.: Evolution of giant planets in eccentric disks. ApJ **652**, 1698–1714 (2006). https://doi.org/10.1086/508451
63. Demory, B.O., Seager, S.: Lack of inflated radii for Kepler giant planet candidates receiving modest stellar irradiation. ApJS **197**, 12 (2011). https://doi.org/10.1088/0067-0049/197/1/12
64. Desch, S.J.: Mass distribution and planet formation in the solar nebula. ApJ **671**, 878–893 (2007). https://doi.org/10.1086/522825
65. Desort, M., Lagrange, A.M., Galland, F., Beust, H., Udry, S., Mayor, M., Lo Curto, G.: Extrasolar planets and brown dwarfs around A-F type stars. V. A planetary system found with HARPS around the F6IV-V star HD 60532. A&A **491**, 883–888 (2008). https://doi.org/10.1051/0004-6361:200810241
66. Dodson-Robinson, S.E., Veras, D., Ford, E.B., Beichman, C.A.: The formation mechanism of gas giants on wide orbits. ApJ **707**, 79–88 (2009). https://doi.org/10.1088/0004-637X/707/1/79
67. Dominik, C., Tielens, A.G.G.M.: Resistance to rolling in the adhesive contact of two elastic spheres. Philos. Mag. Part A **72**, 783–803 (1995). https://doi.org/10.1080/01418619508243800
68. Dominik, C., Tielens, A.G.G.M.: The physics of dust coagulation and the structure of dust aggregates in space. ApJ **480**, 647–673 (1997)
69. Doyle, L.R., Carter, J.A., Fabrycky, D.C., Slawson, R.W., Howell, S.B., Winn, J.N., Orosz, J.A., Prša, A., Welsh, W.F., Quinn, S.N., Latham, D., Torres, G., Buchhave, L.A., Marcy, G.W., Fortney, J.J., Shporer, A., Ford, E.B., Lissauer, J.J., Ragozzine, D., Rucker, M., Batalha, N., Jenkins, J.M., Borucki, W.J., Koch, D., Middour, C.K., Hall, J.R., McCauliff, S., Fanelli, M.N., Quintana, E.V., Holman, M.J., Caldwell, D.A., Still, M., Stefanik, R.P., Brown, W.R., Esquerdo, G.A., Tang, S., Furesz, G., Geary, J.C., Berlind, P., Calkins, M.L., Short, D.R., Steffen, J.H., Sasselov, D., Dunham, E.W., Cochran, W.D., Boss, A., Haas, M.R., Buzasi, D., Fischer, D.: Kepler-16: a transiting circumbinary planet. Science **333**, 1602 (2011). https://doi.org/10.1126/science.1210923
70. Draine, B.T.: Interstellar dust grains. ARA&A **41**, 241–289 (2003). https://doi.org/10.1146/annurev.astro.41.011802.094840
71. Dubrovinsky, L., Dubrovinskaia, N., Prakapenka, V.B., Abakumov, A.M.: Implementation of micro-ball nanodiamond anvils for high-pressure studies above 6 Mbar. Nat. Commun. **3**, 1163 (2012). https://doi.org/10.1038/ncomms2160
72. Duffell, P.C.: A simple analytical model for gaps in protoplanetary disks. ApJ **807**, L11 (2015). https://doi.org/10.1088/2041-8205/807/1/L11
73. Duffell, P.C., Chiang, E.: Eccentric Jupiters via disk-planet interactions. ApJ **812**, 94 (2015). https://doi.org/10.1088/0004-637X/812/2/94
74. Duffell, P.C., Haiman, Z., MacFadyen, A.I., D'Orazio, D.J., Farris, B.D.: The migration of gap-opening planets is not locked to viscous disk evolution. ApJ **792**, L10 (2014). https://doi.org/10.1088/2041-8205/792/1/L10
75. Duffell, P.C., MacFadyen, A.I.: Gap opening by extremely low-mass planets in a viscous disk. ApJ **769**, 41 (2013). https://doi.org/10.1088/0004-637X/769/1/41
76. Duncan, M.J., Levison, H.F., Lee, M.H.: A multiple time step symplectic algorithm for integrating close encounters. AJ **116**, 2067–2077 (1998). https://doi.org/10.1086/300541
77. Dunhill, A.C., Alexander, R.D., Armitage, P.J.: A limit on eccentricity growth from global 3D simulations of disc-planet interactions. MNRAS **428**, 3072–3082 (2013). https://doi.org/10.1093/mnras/sts254

78. Dunhill, A.C., Cuadra, J., Dougados, C.: Precession and accretion in circumbinary discs: the case of HD 104237. MNRAS **448**, 3545–3554 (2015). https://doi.org/10.1093/mnras/stv284
79. Durisen, R.H., Boss, A.P., Mayer, L., Nelson, A.F., Quinn, T., Rice, W.K.M.: Gravitational Instabilities in Gaseous Protoplanetary Disks and Implications for Giant Planet Formation. Protostars and Planets V, pp. 607–622 (2007)
80. Dürmann, C., Kley, W.: Migration of massive planets in accreting disks. A&A **574**, A52 (2015). https://doi.org/10.1051/0004-6361/201424837
81. Dvorak, R., Froeschle, C., Froeschle, C.: Stability of outer planetary orbits (P-types) in binaries. A&A **226**, 335–342 (1989)
82. Edgar, R.G.: Giant planet migration in viscous power-law disks. ApJ **663**, 1325–1334 (2007). https://doi.org/10.1086/518591
83. Fabrycky, D., Tremaine, S.: Shrinking binary and planetary orbits by Kozai cycles with tidal friction. ApJ **669**, 1298–1315 (2007). https://doi.org/10.1086/521702
84. Fabrycky, D.C., Murray-Clay, R.A.: Stability of the directly imaged multiplanet system HR 8799: resonance and masses. ApJ **710**, 1408–1421 (2010). https://doi.org/10.1088/0004-637X/710/2/1408
85. Fahr, H., Willerding, E.A.: Die Entstehung von Sonnensystemen. Spektrum Akademischer Verlag (1998)
86. Ford, E.B., Rasio, F.A.: Origins of eccentric extrasolar planets: testing the planet-planet scattering model. ApJ **686**, 621–636 (2008). https://doi.org/10.1086/590926
87. Fortney, J.J., Nettelmann, N.: The interior structure, composition, and evolution of giant planets. Space Sci. Rev. **152**, 423–447 (2010). https://doi.org/10.1007/s11214-009-9582-x
88. Fung, J., Artymowicz, P., Wu, Y.: The 3D flow field around an embedded planet. ApJ **811**, 101 (2015). https://doi.org/10.1088/0004-637X/811/2/101
89. Fung, J., Shi, J.M., Chiang, E.: How empty are disk gaps opened by giant planets? ApJ **782**, 88 (2014). https://doi.org/10.1088/0004-637X/782/2/88
90. Gammie, C.F.: Nonlinear outcome of gravitational instability in cooling, gaseous disks. ApJ **553**, 174–183 (2001). https://doi.org/10.1086/320631
91. Garaud, P., Meru, F., Galvagni, M., Olczak, C.: From dust to planetesimals: an improved model for collisional growth in protoplanetary disks. ApJ **764**, 146 (2013). https://doi.org/10.1088/0004-637X/764/2/146
92. Geretshauser, R.J., Meru, F., Speith, R., Kley, W.: The four-population model: a new classification scheme for pre-planetesimal collisions. A&A **531**, A166 (2011). https://doi.org/10.1051/0004-6361/201116901
93. Goldreich, P., Lithwick, Y., Sari, R.: Planet formation by coagulation: a focus on Uranus and Neptune. ARA&A **42**, 549–601 (2004). https://doi.org/10.1146/annurev.astro.42.053102.134004
94. Goldreich, P., Sari, R.: Eccentricity evolution for planets in gaseous disks. ApJ **585**, 1024–1037 (2003). https://doi.org/10.1086/346202
95. Goldreich, P., Tremaine, S.: The excitation of density waves at the Lindblad and corotation resonances by an external potential. ApJ **233**, 857–871 (1979). https://doi.org/10.1086/157448
96. Goldreich, P., Tremaine, S.: Disk-satellite interactions. ApJ **241**, 425–441 (1980). https://doi.org/10.1086/158356
97. Goldreich, P., Ward, W.R.: The formation of planetesimals. ApJ **183**, 1051–1062 (1973). https://doi.org/10.1086/152291
98. Goździewski, K., Migaszewski, C.: Multiple mean motion resonances in the HR 8799 planetary system. MNRAS **440**, 3140–3171 (2014). https://doi.org/10.1093/mnras/stu455
99. Goździewski, K., Migaszewski, C., Panichi, F., Szuszkiewicz, E.: The Laplace resonance in the Kepler-60 planetary system. MNRAS **455**, L104–L108 (2016). https://doi.org/10.1093/mnrasl/slv156
100. Greenzweig, Y., Lissauer, J.J.: Accretion rates of protoplanets. Icarus **87**, 40–77 (1990). https://doi.org/10.1016/0019-1035(90)90021-Z

101. Guillot, T.: The interiors of giant planets: models and outstanding questions. Annu. Rev. Earth Planet. Sci. **33**, 493–530 (2005). https://doi.org/10.1146/annurev.earth.32.101802.120325
102. Guillot, T., Gautier, D.: Giant Planets. ArXiv e-prints (2014)
103. Günther, R., Schäfer, C., Kley, W.: Evolution of irradiated circumbinary disks. A&A **423**, 559–566 (2004). https://doi.org/10.1051/0004-6361:20040223
104. Güttler, C., Blum, J., Zsom, A., Ormel, C.W., Dullemond, C.P.: The outcome of protoplanetary dust growth: pebbles, boulders, or planetesimals?. I. Mapping the zoo of laboratory collision experiments. A&A **513**, A56 (2010). https://doi.org/10.1051/0004-6361/200912852
105. Güttler, C., Krause, M., Geretshauser, R.J., Speith, R., Blum, J.: The physics of protoplanetesimal dust agglomerates. IV. Toward a dynamical collision model. ApJ **701**, 130–141 (2009). https://doi.org/10.1088/0004-637X/701/1/130
106. Haghighipour, N.: The formation and dynamics of super-earth planets. Annu. Rev. Earth Planet. Sci. **41**, 469–495 (2013). https://doi.org/10.1146/annurev-earth-042711-105340
107. Hamel, J.: Geschichte der Astronomie. Von den Anfängen bis zur Gegenwart (1998)
108. Hansen, B.M.S., Murray, N.: Migration then assembly: formation of Neptune-mass planets inside 1 AU. ApJ **751**, 158 (2012). https://doi.org/10.1088/0004-637X/751/2/158
109. Hawley, J.F., Balbus, S.A.: A powerful local shear instability in weakly magnetized disks. II. Nonlinear evolution. ApJ **376**, 223 (1991). https://doi.org/10.1086/170271
110. Hayashi, C.: Structure of the solar nebula, growth and decay of magnetic fields and effects of magnetic and turbulent viscosities on the nebula. Prog. Theor. Phys. Suppl. **70**, 35–53 (1981). https://doi.org/10.1143/PTPS.70.35
111. Hayes, W., Tremaine, S.: Fitting selected random planetary systems to Titius-Bode laws. Icarus **135**, 549–557 (1998). https://doi.org/10.1006/icar.1998.5999
112. Heißelmann, D., Blum, J., Fraser, H.J., Wolling, K.: Microgravity experiments on the collisional behavior of saturnian ring particles. Icarus **206**, 424–430 (2010). https://doi.org/10.1016/j.icarus.2009.08.009
113. Helled, R., Bodenheimer, P., Podolak, M., Boley, A., Meru, F., Nayakshin, S., Fortney, J.J., Mayer, L., Alibert, Y., Boss, A.P.: Giant Planet Formation, Evolution, and Internal Structure. Protostars and Planets VI, pp. 643–665 (2014). https://doi.org/10.2458/azu_uapress_9780816531240-ch028
114. Hertz, H.: über die Berührung fester elastischer Körper. J. reine und angewandte Mathematik **94**, 156–171 (1882)
115. Holman, M., Touma, J., Tremaine, S.: Chaotic variations in the eccentricity of the planet orbiting 16 Cygni B. Nature **386**, 254–256 (1997). https://doi.org/10.1038/386254a0
116. Holman, M.J., Wiegert, P.A.: Long-term stability of planets in binary systems. AJ **117**, 621–628 (1999). https://doi.org/10.1086/300695
117. Ivanov, P.B., Papaloizou, J.C.B., Polnarev, A.G.: The evolution of a supermassive binary caused by an accretion disc. MNRAS **307**, 79–90 (1999). https://doi.org/10.1046/j.1365-8711.1999.02623.x
118. Jankowski, T., Wurm, G., Kelling, T., Teiser, J., Sabolo, W., Gutiérrez, P.J., Bertini, I.: Crossing barriers in planetesimal formation: the growth of mm-dust aggregates with large constituent grains. A&A **542**, A80 (2012). https://doi.org/10.1051/0004-6361/201218984
119. Johansen, A., Blum, J., Tanaka, H., Ormel, C., Bizzarro, M., Rickman, H.: The Multifaceted Planetesimal Formation Process. Protostars and Planets VI, pp. 547–570 (2014). https://doi.org/10.2458/azu_uapress_9780816531240-ch024
120. Johansen, A., Oishi, J.S., Mac Low, M.M., Klahr, H., Henning, T., Youdin, A.: Rapid planetesimal formation in turbulent circumstellar disks. Nature **448**, 1022–1025 (2007). https://doi.org/10.1038/nature06086
121. Johnson, K.L., Kendall, K., Roberts, A.D.: Surface energy and the contact of elastic solids. R. Soc. Lond. Proc. Ser. A **324**, 301–313 (1971). https://doi.org/10.1098/rspa.1971.0141
122. Jurić, M., Tremaine, S.: Dynamical origin of extrasolar planet eccentricity distribution. ApJ **686**, 603–620 (2008). https://doi.org/10.1086/590047
123. Kanagawa, K.D., Muto, T., Tanaka, H., Tanigawa, T., Takeuchi, T., Tsukagoshi, T., Momose, M.: Mass estimates of a giant planet in a protoplanetary disk from the gap structures. ApJ **806**, L15 (2015). https://doi.org/10.1088/2041-8205/806/1/L15

124. Kant, I.: Allgemeine Naturgeschichte und Theorie des Himmels (1755)
125. Kippenhahn, R., Weigert, A.: Stellar Structure and Evolution (1990)
126. Kley, W.: Mass flow and accretion through gaps in accretion discs. MNRAS **303**, 696–710 (1999). https://doi.org/10.1046/j.1365-8711.1999.02198.x
127. Kley, W., Bitsch, B., Klahr, H.: Planet migration in three-dimensional radiative discs. A&A **506**, 971–987 (2009). https://doi.org/10.1051/0004-6361/200912072
128. Kley, W., Crida, A.: Migration of protoplanets in radiative discs. A&A **487**, L9–L12 (2008). https://doi.org/10.1051/0004-6361:200810033
129. Kley, W., Dirksen, G.: Disk eccentricity and embedded planets. A&A **447**, 369–377 (2006). https://doi.org/10.1051/0004-6361:20053914
130. Kley, W., Haghighipour, N.: Modeling circumbinary planets: the case of Kepler-38. A&A **564**, A72 (2014). https://doi.org/10.1051/0004-6361/201323235
131. Kley, W., Haghighipour, N.: Evolution of circumbinary planets around eccentric binaries: the case of Kepler-34. A&A **581**, A20 (2015). https://doi.org/10.1051/0004-6361/201526648
132. Kley, W., Müller, T.W.A., Kolb, S.M., Benítez-Llambay, P., Masset, F.: Low-mass planets in nearly inviscid disks: numerical treatment. A&A **546**, A99 (2012). https://doi.org/10.1051/0004-6361/201219719
133. Kley, W., Nelson, R.P.: Planet-disk interaction and orbital evolution. ARA&A **50**, 211–249 (2012). https://doi.org/10.1146/annurev-astro-081811-125523
134. Kley, W., Peitz, J., Bryden, G.: Evolution of planetary systems in resonance. A&A **414**, 735–747 (2004). https://doi.org/10.1051/0004-6361:20031589
135. Kokubo, E.: Planetary accretion: from planitesimals to protoplanets. In: Schielicke R.E. (ed.) Reviews in Modern Astronomy, vol. 14, p. 117 (2001)
136. Kokubo, E., Ida, S.: Oligarchic growth of protoplanets. Icarus **131**, 171–178 (1998). https://doi.org/10.1006/icar.1997.5840
137. Kokubo, E., Ida, S.: Formation of protoplanet systems and diversity of planetary systems. ApJ **581**, 666–680 (2002). https://doi.org/10.1086/344105
138. Kominami, J., Ida, S.: The effect of tidal interaction with a gas disk on formation of terrestrial planets. Icarus **157**, 43–56 (2002). https://doi.org/10.1006/icar.2001.6811
139. Kozai, Y.: Secular perturbations of asteroids with high inclination and eccentricity. AJ **67**, 591 (1962). https://doi.org/10.1086/108790
140. Kratter, K.M., Murray-Clay, R.A., Youdin, A.N.: The runts of the litter: why planets formed through gravitational instability can only be failed binary stars. ApJ **710**, 1375–1386 (2010). https://doi.org/10.1088/0004-637X/710/2/1375
141. Krause, M., Blum, J.: Growth and form of planetary seedlings: results from a sounding rocket microgravity aggregation experiment. Phys. Rev. Lett. **93**(2), 021103 (2004). https://doi.org/10.1103/PhysRevLett.93.021103
142. Kretke, K.A., Lin, D.N.C.: Grain retention and formation of planetesimals near the snow line in MRI-driven turbulent protoplanetary disks. ApJ **664**, L55–L58 (2007). https://doi.org/10.1086/520718
143. Kuiper, G.P.: On the origin of the solar system. Proc. Natl. Acad. Sci. **37**, 1–14 (1951). https://doi.org/10.1073/pnas.37.1.1
144. Lambrechts, M., Johansen, A.: Rapid growth of gas-giant cores by pebble accretion. A&A **544**, A32 (2012). https://doi.org/10.1051/0004-6361/201219127
145. Laplace, P.S.: Exposition du système du monde (1776)
146. Laughlin, G., Korchagin, V., Adams, F.C.: The dynamics of heavy gaseous disks. ApJ **504**, 945–966 (1998). https://doi.org/10.1086/306117
147. Laughlin, G., Steinacker, A., Adams, F.C.: Type I planetary migration with MHD turbulence. ApJ **608**, 489–496 (2004). https://doi.org/10.1086/386316
148. Lee, M.H., Peale, S.J.: Dynamics and origin of the 2:1 orbital resonances of the GJ 876 planets. ApJ **567**, 596–609 (2002). https://doi.org/10.1086/338504
149. Lega, E., Crida, A., Bitsch, B., Morbidelli, A.: Migration of earth-sized planets in 3D radiative discs. MNRAS **440**, 683–695 (2014). https://doi.org/10.1093/mnras/stu304

150. Leinhardt, Z.M., Richardson, D.C.: N-body simulations of planetesimal evolution: effect of varying impactor mass ratio. Icarus **159**, 306–313 (2002). https://doi.org/10.1006/icar.2002.6909
151. Leinhardt, Z.M., Richardson, D.C., Quinn, T.: Direct N-body simulations of rubble pile collisions. Icarus **146**, 133–151 (2000). https://doi.org/10.1006/icar.2000.6370
152. Leinhardt, Z.M., Stewart, S.T.: Full numerical simulations of catastrophic small body collisions. Icarus **199**, 542–559 (2009). https://doi.org/10.1016/j.icarus.2008.09.013
153. Li, H., Lubow, S.H., Li, S., Lin, D.N.C.: Type I planet migration in nearly laminar disks. ApJ **690**, L52–L55 (2009). https://doi.org/10.1088/0004-637X/690/1/L52
154. Lin, C.C., Shu, F.H.: On the spiral structure of disk galaxies. ApJ **140**, 646 (1964). https://doi.org/10.1086/147955
155. Lin, D.N.C., Bodenheimer, P., Richardson, D.C.: Orbital migration of the planetary companion of 51 Pegasi to its present location. Nature **380**, 606–607 (1996). https://doi.org/10.1038/380606a0
156. Lin, D.N.C., Papaloizou, J.: Tidal torques on accretion discs in binary systems with extreme mass ratios. MNRAS **186**, 799–812 (1979)
157. Lin, D.N.C., Papaloizou, J.: On the tidal interaction between protoplanets and the protoplanetary disk. III—Orbital migration of protoplanets. ApJ **309**, 846–857 (1986). https://doi.org/10.1086/164653
158. Lissauer, J.J.: Timescales for planetary accretion and the structure of the protoplanetary disk. Icarus **69**, 249–265 (1987). https://doi.org/10.1016/0019-1035(87)90104-7
159. Lissauer, J.J.: Planet formation. ARA&A **31**, 129–174 (1993). https://doi.org/10.1146/annurev.aa.31.090193.001021
160. Lissauer, J.J., Stewart, G.R.: Growth of planets from planetesimals. In: Levy, E.H., Lunine, J.I. (eds.) Protostars and Planets III, pp. 1061–1088 (1993)
161. Lithwick, Y., Wu, Y.: Resonant repulsion of Kepler planet pairs. ApJ **756**, L11 (2012). https://doi.org/10.1088/2041-8205/756/1/L11
162. Lubow, S.H., Seibert, M., Artymowicz, P.: Disk accretion onto high-mass planets. ApJ **526**, 1001–1012 (1999). https://doi.org/10.1086/308045
163. Lynden-Bell, D., Pringle, J.E.: The evolution of viscous discs and the origin of the nebular variables. MNRAS **168**, 603–637 (1974)
164. Mac Low, M.M., Klessen, R.S.: Control of star formation by supersonic turbulence. Rev. Mod. Phys. **76**, 125–194 (2004). https://doi.org/10.1103/RevModPhys.76.125
165. Madhusudhan, N., Lee, K.K.M., Mousis, O.: A possible carbon-rich interior in super-Earth 55 Cancri e. ApJ **759**, L40 (2012). https://doi.org/10.1088/2041-8205/759/2/L40
166. Marboeuf, U., Thiabaud, A., Alibert, Y., Cabral, N., Benz, W.: From planetesimals to planets: volatile molecules. A&A **570**, A36 (2014). https://doi.org/10.1051/0004-6361/201423431
167. Marcy, G.W., Butler, R.P., Fischer, D., Vogt, S.S., Lissauer, J.J., Rivera, E.J.: A pair of resonant planets orbiting GJ 876. ApJ **556**, 296–301 (2001). https://doi.org/10.1086/321552
168. Marcy, G.W., Isaacson, H., Howard, A.W., Rowe, J.F., Jenkins, J.M., Bryson, S.T., Latham, D.W., Howell, S.B., Gautier III, T.N., Batalha, N.M., Rogers, L., Ciardi, D., Fischer, D.A., Gilliland, R.L., Kjeldsen, H., Christensen-Dalsgaard, J., Huber, D., Chaplin, W.J., Basu, S., Buchhave, L.A., Quinn, S.N., Borucki, W.J., Koch, D.G., Hunter, R., Caldwell, D.A., Van Cleve, J., Kolbl, R., Weiss, L.M., Petigura, E., Seager, S., Morton, T., Johnson, J.A., Ballard, S., Burke, C., Cochran, W.D., Endl, M., MacQueen, P., Everett, M.E., Lissauer, J.J., Ford, E.B., Torres, G., Fressin, F., Brown, T.M., Steffen, J.H., Charbonneau, D., Basri, G.S., Sasselov, D.D., Winn, J., Sanchis-Ojeda, R., Christiansen, J., Adams, E., Henze, C., Dupree, A., Fabrycky, D.C., Fortney, J.J., Tarter, J., Holman, M.J., Tenenbaum, P., Shporer, A., Lucas, P.W., Welsh, W.F., Orosz, J.A., Bedding, T.R., Campante, T.L., Davies, G.R., Elsworth, Y., Handberg, R., Hekker, S., Karoff, C., Kawaler, S.D., Lund, M.N., Lundkvist, M., Metcalfe, T.S., Miglio, A., Silva Aguirre, V., Stello, D., White, T.R., Boss, A., Devore, E., Gould, A., Prsa, A., Agol, E., Barclay, T., Coughlin, J., Brugamyer, E., Mullally, F., Quintana, E.V., Still, M., Thompson, S.E., Morrison, D., Twicken, J.D., Désert, J.M., Carter, J., Crepp, J.R., Hébrard, G., Santerne, A., Moutou, C., Sobeck, C., Hudgins, D., Haas, M.R., Robertson, P.,

Lillo-Box, J., Barrado, D.: Masses, radii, and orbits of small Kepler planets: the transition from gaseous to rocky planets. ApJS **210**, 20 (2014). https://doi.org/10.1088/0067-0049/210/2/20
169. Marois, C., Macintosh, B., Barman, T., Zuckerman, B., Song, I., Patience, J., Lafrenière, D., Doyon, R.: Direct imaging of multiple planets orbiting the star HR 8799. Science **322**, 1348 (2008). https://doi.org/10.1126/science.1166585
170. Marois, C., Zuckerman, B., Konopacky, Q.M., Macintosh, B., Barman, T.: Images of a fourth planet orbiting HR 8799. Nature **468**, 1080–1083 (2010). https://doi.org/10.1038/nature09684
171. Martí, J.G., Giuppone, C.A., Beaugé, C.: Dynamical analysis of the Gliese-876 Laplace resonance. MNRAS **433**, 928–934 (2013). https://doi.org/10.1093/mnras/stt765
172. Marzari, F., Nelson, A.F.: Interaction of a giant planet in an inclined orbit with a circumstellar disk. ApJ **705**, 1575–1583 (2009). https://doi.org/10.1088/0004-637X/705/2/1575
173. Marzari, F., Thebault, P., Scholl, H., Picogna, G., Baruteau, C.: Influence of the circumbinary disk gravity on planetesimal accumulation in the Kepler-16 system. A&A **553**, A71 (2013). https://doi.org/10.1051/0004-6361/201220893
174. Masset, F.: FARGO: a fast eulerian transport algorithm for differentially rotating disks. A&AS **141**, 165–173 (2000). https://doi.org/10.1051/aas:2000116
175. Masset, F.S.: The co-orbital corotation torque in a viscous disk: numerical simulations. A&A **387**, 605–623 (2002). https://doi.org/10.1051/0004-6361:20020240
176. Masset, F.S., Casoli, J.: Saturated torque formula for planetary migration in viscous disks with thermal diffusion: recipe for protoplanet population synthesis. ApJ **723**, 1393–1417 (2010). https://doi.org/10.1088/0004-637X/723/2/1393
177. Masset, F.S., Morbidelli, A., Crida, A., Ferreira, J.: Disk surface density transitions as protoplanet traps. ApJ **642**, 478–487 (2006). https://doi.org/10.1086/500967
178. Masset, F.S., Papaloizou, J.C.B.: Runaway migration and the formation of Hot Jupiters. ApJ **588**, 494–508 (2003). https://doi.org/10.1086/373892
179. Mathis, J.S., Rumpl, W., Nordsieck, K.H.: The size distribution of interstellar grains. ApJ **217**, 425–433 (1977). https://doi.org/10.1086/155591
180. Mayer, L., Quinn, T., Wadsley, J., Stadel, J.: Formation of giant planets by fragmentation of protoplanetary disks. Science **298**, 1756–1759 (2002). https://doi.org/10.1126/science.1077635
181. Mayer, L., Quinn, T., Wadsley, J., Stadel, J.: The evolution of gravitationally unstable protoplanetary disks: fragmentation and possible giant planet formation. ApJ **609**, 1045–1064 (2004). https://doi.org/10.1086/421288
182. Mayor, M., Queloz, D.: A Jupiter-mass companion to a solar-type star. Nature **378**, 355–359 (1995). https://doi.org/10.1038/378355a0
183. McCaughrean, M.J., O'dell, C.R.: Direct imaging of circumstellar disks in the orion nebula. AJ **111**, 1977 (1996). https://doi.org/10.1086/117934
184. Mejía, A.C., Durisen, R.H., Pickett, M.K., Cai, K.: The thermal regulation of gravitational instabilities in protoplanetary disks. II. Extended simulations with varied cooling rates. ApJ **619**, 1098–1113 (2005). https://doi.org/10.1086/426707
185. Meru, F., Bate, M.R.: Non-convergence of the critical cooling time-scale for fragmentation of self-gravitating discs. MNRAS **411**, L1–L5 (2011). https://doi.org/10.1111/j.1745-3933.2010.00978.x
186. Meru, F., Bate, M.R.: On the convergence of the critical cooling time-scale for the fragmentation of self-gravitating discs. MNRAS **427**, 2022–2046 (2012). https://doi.org/10.1111/j.1365-2966.2012.22035.x
187. Meschiari, S.: Circumbinary planet formation in the Kepler-16 system. I. N-body simulations. ApJ **752**, 71 (2012). https://doi.org/10.1088/0004-637X/752/1/71
188. Michael, S., Durisen, R.H., Boley, A.C.: Migration of gas giant planets in gravitationally unstable disks. ApJ **737**, L42 (2011). https://doi.org/10.1088/2041-8205/737/2/L42
189. Mills, S.M., Fabrycky, D.C., Migaszewski, C., Ford, E.B., Petigura, E., Isaacson, H.: A resonant chain of four transiting, sub-Neptune planets. Nature **533**, 509–512 (2016). https://doi.org/10.1038/nature17445

190. Mizuno, H.: Formation of the Giant Planets. Prog. Theor. Phys. **64**, 544–557 (1980). https://doi.org/10.1143/PTP.64.544
191. Mizuno, H., Nakazawa, K., Hayashi, C.: Instability of a gaseous envelope surrounding a planetary core and formation of giant planets. Progr. Theor. Phys. **60**, 699–710 (1978). https://doi.org/10.1143/PTP.60.699
192. Monaghan, J.J.: Smoothed particle hydrodynamics. Rep. Prog. Phys. **68**, 1703–1759 (2005). https://doi.org/10.1088/0034-4885/68/8/R01
193. Montmerle, T., Augereau, J.C., Chaussidon, M., Gounelle, M., Marty, B., Morbidelli, A.: From Suns to life: a chronological approach to the history of life on Earth 3. Solar system formation and early evolution: the first 100 million years. Earth Moon Planets **98**, 39–95 (2006). https://doi.org/10.1007/s11038-006-9087-5
194. Morbidelli, A., Lunine, J.I., O'Brien, D.P., Raymond, S.N., Walsh, K.J.: Building terrestrial planets. Annu. Rev. Earth Planet. Sci. **40**, 251–275 (2012). https://doi.org/10.1146/annurev-earth-042711-105319
195. Mordasini, C., Alibert, Y., Klahr, H., Henning, T.: Characterization of exoplanets from their formation. I. Models of combined planet formation and evolution. A&A **547**, A111 (2012). https://doi.org/10.1051/0004-6361/201118457
196. Mordasini, C., Mollière, P., Dittkrist, K.M., Jin, S., Alibert, Y.: Global models of planet formation and evolution. Int. J. Astrobiol. **14**, 201–232 (2015). https://doi.org/10.1017/S1473550414000263
197. Müller, T.W.A., Kley, W., Meru, F.: Treating gravity in thin-disk simulations. A&A **541**, A123 (2012). https://doi.org/10.1051/0004-6361/201118737
198. Murray, C.D., Dermott, S.F.: Solar system dynamics (1999)
199. Nayakshin, S.: Formation of planets by tidal downsizing of giant planet embryos. MNRAS **408**, L36–L40 (2010). https://doi.org/10.1111/j.1745-3933.2010.00923.x
200. Nelson, R.P.: On the orbital evolution of low mass protoplanets in turbulent, magnetised disks. A&A **443**, 1067–1085 (2005). https://doi.org/10.1051/0004-6361:20042605
201. Nelson, R.P., Papaloizou, J.C.B.: The interaction of giant planets with a disc with MHD turbulence—IV. Migration rates of embedded protoplanets. MNRAS **350**, 849–864 (2004). https://doi.org/10.1111/j.1365-2966.2004.07406.x
202. Nelson, R.P., Papaloizou, J.C.B., Masset, F., Kley, W.: The migration and growth of protoplanets in protostellar discs. MNRAS **318**, 18–36 (2000). https://doi.org/10.1046/j.1365-8711.2000.03605.x
203. O'Brien, D.P., Morbidelli, A., Levison, H.F.: Terrestrial planet formation with strong dynamical friction. Icarus **184**, 39–58 (2006). https://doi.org/10.1016/j.icarus.2006.04.005
204. Paardekooper, S.J.: Numerical convergence in self-gravitating shearing sheet simulations and the stochastic nature of disc fragmentation. MNRAS **421**, 3286–3299 (2012). https://doi.org/10.1111/j.1365-2966.2012.20553.x
205. Paardekooper, S.J., Baruteau, C., Crida, A., Kley, W.: A torque formula for non-isothermal type I planetary migration—I. Unsaturated horseshoe drag. MNRAS **401**, 1950–1964 (2010). https://doi.org/10.1111/j.1365-2966.2009.15782.x
206. Paardekooper, S.J., Baruteau, C., Kley, W.: A torque formula for non-isothermal Type I planetary migration—II. Effects of diffusion. MNRAS **410**, 293–303 (2011). https://doi.org/10.1111/j.1365-2966.2010.17442.x
207. Paardekooper, S.J., Leinhardt, Z.M., Thébault, P., Baruteau, C.: How not to build tatooine: the difficulty of in situ formation of circumbinary planets Kepler 16b, Kepler 34b, and Kepler 35b. ApJ **754**, L16 (2012). https://doi.org/10.1088/2041-8205/754/1/L16
208. Paardekooper, S.J., Mellema, G.: Halting Type I planet migration in non-isothermal disks. A&A **459**, L17–L20 (2006). https://doi.org/10.1051/0004-6361:20066304
209. Paardekooper, S.J., Papaloizou, J.C.B.: On disc protoplanet interactions in a non-barotropic disc with thermal diffusion. A&A **485**, 877–895 (2008). https://doi.org/10.1051/0004-6361:20078702
210. Paardekooper, S.J., Rein, H., Kley, W.: The formation of systems with closely spaced low-mass planets and the application to Kepler-36. MNRAS **434**, 3018–3029 (2013). https://doi.org/10.1093/mnras/stt1224

211. Papaloizou, J.C.B., Larwood, J.D.: On the orbital evolution and growth of protoplanets embedded in a gaseous disc. MNRAS **315**, 823–833 (2000). https://doi.org/10.1046/j.1365-8711.2000.03466.x
212. Papaloizou, J.C.B., Nelson, R.P.: Models of accreting gas giant protoplanets in protostellar disks. A&A **433**, 247–265 (2005). https://doi.org/10.1051/0004-6361:20042029
213. Papaloizou, J.C.B., Nelson, R.P., Kley, W., Masset, F.S., Artymowicz, P.: Disk-Planet Interactions During Planet Formation. Protostars and Planets V, pp. 655–668 (2007)
214. Papaloizou, J.C.B., Nelson, R.P., Masset, F.: Orbital eccentricity growth through disc-companion tidal interaction. A&A **366**, 263–275 (2001). https://doi.org/10.1051/0004-6361:20000011
215. Paszun, D., Dominik, C.: Numerical determination of the material properties of porous dust cakes. A&A **484**, 859–868 (2008). https://doi.org/10.1051/0004-6361:20079262
216. Pepliński, A., Artymowicz, P., Mellema, G.: Numerical simulations of type III planetary migration—II. Inward migration of massive planets. MNRAS **386**, 179–198 (2008). https://doi.org/10.1111/j.1365-2966.2008.13046.x
217. Pepliński, A., Artymowicz, P., Mellema, G.: Numerical simulations of Type III planetary migration—III. Outward migration of massive planets. MNRAS **387**, 1063–1079 (2008). https://doi.org/10.1111/j.1365-2966.2008.13339.x
218. Perri, F., Cameron, A.G.W.: Hydrodynamic instability of the solar nebula in the presence of a planetary core. Icarus **22**, 416–425 (1974). https://doi.org/10.1016/0019-1035(74)90074-8
219. Perryman, M.: The Exoplanet Handbook (2011)
220. Petigura, E.A., Howard, A.W., Marcy, G.W.: Prevalence of Earth-size planets orbiting Sun-like stars. Proc. Natl. Acad. Sci. **110**, 19273–19278 (2013)
221. Pierens, A., Huré, J.M.: How does disk gravity really influence Type-I migration? A&A **433**, L37–L40 (2005). https://doi.org/10.1051/0004-6361:200500099
222. Pierens, A., Nelson, R.P.: On the migration of protoplanets embedded in circumbinary disks. A&A **472**, 993–1001 (2007). https://doi.org/10.1051/0004-6361:20077659
223. Pierens, A., Nelson, R.P.: Constraints on resonant-trapping for two planets embedded in a protoplanetary disc. A&A **482**, 333–340 (2008). https://doi.org/10.1051/0004-6361:20079062
224. Pierens, A., Nelson, R.P.: Migration and gas accretion scenarios for the Kepler 16, 34, and 35 circumbinary planets. A&A **556**, A134 (2013). https://doi.org/10.1051/0004-6361/201321777
225. Pollack, J.B., Hubickyj, O., Bodenheimer, P., Lissauer, J.J., Podolak, M., Greenzweig, Y.: Formation of the giant planets by concurrent accretion of solids and gas. Icarus **124**, 62–85 (1996). https://doi.org/10.1006/icar.1996.0190
226. Poppe, T., Blum, J., Henning, T.: Analogous experiments on the stickiness of micron-sized preplanetary dust. ApJ **533**, 454–471 (2000). https://doi.org/10.1086/308626
227. Pringle, J.E.: Accretion discs in astrophysics. ARA&A **19**, 137–162 (1981). https://doi.org/10.1146/annurev.aa.19.090181.001033
228. Quirrenbach, A.: Detection and characterization of extrasolar planets. In: Queloz, D., Udry, S., Mayor, M., Benz, W., Cassen, P., Guillot, T., Quirrenbach, A. (eds.) Saas-Fee Advanced Course 31: Extrasolar Planets, pp. 1–242 (2006). https://doi.org/10.1007/978-3-540-31470-7_1
229. Rafikov, R.R.: Planet migration and gap formation by tidally induced shocks. ApJ **572**, 566–579 (2002). https://doi.org/10.1086/340228
230. Rafikov, R.R.: Can giant planets form by direct gravitational instability? ApJ **621**, L69–L72 (2005). https://doi.org/10.1086/428899
231. Rauer, H., Catala, C., Aerts, C., Appourchaux, T., Benz, W., Brandeker, A., Christensen-Dalsgaard, J., Deleuil, M., Gizon, L., Goupil, M.J., Güdel, M., Janot-Pacheco, E., Mas-Hesse, M., Pagano, I., Piotto, G., Pollacco, D., Santos, C., Smith, A., Suárez, J.C., Szabó, R., Udry, S., Adibekyan, V., Alibert, Y., Almenara, J.M., Amaro-Seoane, P., Eiff, M.A.v., Asplund, M., Antonello, E., Barnes, S., Baudin, F., Belkacem, K., Bergemann, M., Bihain, G., Birch, A.C., Bonfils, X., Boisse, I., Bonomo, A.S., Borsa, F., Brandão, I.M., Brocato, E., Brun, S., Burleigh, M., Burston, R., Cabrera, J., Cassisi, S., Chaplin, W., Charpinet, S., Chiappini, C.,

Church, R.P., Csizmadia, S., Cunha, M., Damasso, M., Davies, M.B., Deeg, H.J., Díaz, R.F., Dreizler, S., Dreyer, C., Eggenberger, P., Ehrenreich, D., Eigmüller, P., Erikson, A., Farmer, R., Feltzing, S., de Oliveira Fialho, F., Figueira, P., Forveille, T., Fridlund, M., García, R.A., Giommi, P., Giuffrida, G., Godolt, M., Gomes da Silva, J., Granzer, T., Grenfell, J.L., Grotsch-Noels, A., Günther, E., Haswell, C.A., Hatzes, A.P., Hébrard, G., Hekker, S., Helled, R., Heng, K., Jenkins, J.M., Johansen, A., Khodachenko, M.L., Kislyakova, K.G., Kley, W., Kolb, U., Krivova, N., Kupka, F., Lammer, H., Lanza, A.F., Lebreton, Y., Magrin, D., Marcos-Arenal, P., Marrese, P.M., Marques, J.P., Martins, J., Mathis, S., Mathur, S., Messina, S., Miglio, A., Montalban, J., Montalto, M., Monteiro, M.J.P.F.G., Moradi, H., Moravveji, E., Mordasini, C., Morel, T., Mortier, A., Nascimbeni, V., Nelson, R.P., Nielsen, M.B., Noack, L., Norton, A.J., Ofir, A., Oshagh, M., Ouazzani, R.M., Pápics, P., Parro, V.C., Petit, P., Plez, B., Poretti, E., Quirrenbach, A., Ragazzoni, R., Raimondo, G., Rainer, M., Reese, D.R., Redmer, R., Reffert, S., Rojas-Ayala, B., Roxburgh, I.W., Salmon, S., Santerne, A., Schneider, J., Schou, J., Schuh, S., Schunker, H., Silva-Valio, A., Silvotti, R., Skillen, I., Snellen, I., Sohl, F., Sousa, S.G., Sozzetti, A., Stello, D., Strassmeier, K.G., Švanda, M., Szabó, G.M., Tkachenko, A., Valencia, D., Van Grootel, V., Vauclair, S.D., Ventura, P., Wagner, F.W., Walton, N.A., Weingrill, J., Werner, S.C., Wheatley, P.J., Zwintz, K.: The PLATO 2.0 mission. Exp. Astron. **38**, 249–330 (2014). https://doi.org/10.1007/s10686-014-9383-4
232. Raymond, S.N., Cossou, C.: No universal minimum-mass extrasolar nebula: evidence against in situ accretion of systems of hot super-Earths. MNRAS **440**, L11–L15 (2014). https://doi.org/10.1093/mnrasl/slu011
233. Raymond, S.N., Kokubo, E., Morbidelli, A., Morishima, R., Walsh, K.J.: Terrestrial Planet Formation at Home and Abroad. Protostars and Planets VI, pp. 595–618 (2014). https://doi.org/10.2458/azu_uapress_9780816531240-ch026
234. Raymond, S.N., O'Brien, D.P., Morbidelli, A., Kaib, N.A.: Building the terrestrial planets: constrained accretion in the inner Solar System. Icarus **203**, 644–662 (2009). https://doi.org/10.1016/j.icarus.2009.05.016
235. Raymond, S.N., Quinn, T., Lunine, J.I.: Making other earths: dynamical simulations of terrestrial planet formation and water delivery. Icarus **168**, 1–17 (2004). https://doi.org/10.1016/j.icarus.2003.11.019
236. Raymond, S.N., Quinn, T., Lunine, J.I.: High-resolution simulations of the final assembly of Earth-like planets I. Terrestrial accretion and dynamics. Icarus **183**, 265–282 (2006). https://doi.org/10.1016/j.icarus.2006.03.011
237. Rein, H.: Period ratios in multiplanetary systems discovered by Kepler are consistent with planet migration. MNRAS **427**, L21–L24 (2012). https://doi.org/10.1111/j.1745-3933.2012.01337.x
238. Rein, H.: Planet-disc interaction in highly inclined systems. MNRAS **422**, 3611–3616 (2012). https://doi.org/10.1111/j.1365-2966.2012.20869.x
239. Rein, H., Liu, S.F.: REBOUND: an open-source multi-purpose N-body code for collisional dynamics. A&A **537**, A128 (2012). https://doi.org/10.1051/0004-6361/201118085
240. Rice, W.K.M., Armitage, P.J., Bate, M.R., Bonnell, I.A.: The effect of cooling on the global stability of self-gravitating protoplanetary discs. MNRAS **339**, 1025–1030 (2003). https://doi.org/10.1046/j.1365-8711.2003.06253.x
241. Rice, W.K.M., Lodato, G., Pringle, J.E., Armitage, P.J., Bonnell, I.A.: Planetesimal formation via fragmentation in self-gravitating protoplanetary discs. MNRAS **372**, L9–L13 (2006). https://doi.org/10.1111/j.1745-3933.2006.00215.x
242. Rice, W.K.M., Paardekooper, S.J., Forgan, D.H., Armitage, P.J.: Convergence of simulations of self-gravitating accretion discs—II. Sensitivity to the implementation of radiative cooling and artificial viscosity. MNRAS **438**, 1593–1602 (2014). https://doi.org/10.1093/mnras/stt2297
243. Ricker, G.R., Winn, J.N., Vanderspek, R., Latham, D.W., Bakos, G.Á., Bean, J.L., Berta-Thompson, Z.K., Brown, T.M., Buchhave, L., Butler, N.R., Butler, R.P., Chaplin, W.J., Charbonneau, D., Christensen-Dalsgaard, J., Clampin, M., Deming, D., Doty, J., De Lee, N., Dressing, C., Dunham, E.W., Endl, M., Fressin, F., Ge, J., Henning, T., Holman, M.J., Howard, A.W., Ida, S., Jenkins, J.M., Jernigan, G., Johnson, J.A., Kaltenegger, L., Kawai, N., Kjeldsen,

H., Laughlin, G., Levine, A.M., Lin, D., Lissauer, J.J., MacQueen, P., Marcy, G., McCullough, P.R., Morton, T.D., Narita, N., Paegert, M., Palle, E., Pepe, F., Pepper, J., Quirrenbach, A., Rinehart, S.A., Sasselov, D., Sato, B., Seager, S., Sozzetti, A., Stassun, K.G., Sullivan, P., Szentgyorgyi, A., Torres, G., Udry, S., Villasenor, J.: Transiting exoplanet survey satellite (TESS). J. Astron. Telesc. Instrum. Syst. **1**(1), 014003 (2015). https://doi.org/10.1117/1.JATIS.1.1.014003
244. Rivera, E.J., Laughlin, G., Butler, R.P., Vogt, S.S., Haghighipour, N., Meschiari, S.: The Lick-Carnegie exoplanet survey: a Uranus-Mass fourth planet for GJ 876 in an extrasolar Laplace configuration. ApJ **719**, 890–899 (2010). https://doi.org/10.1088/0004-637X/719/1/890
245. Safronov, V.S.: Evolution of the protoplanetary cloud and formation of the earth and planets (1972)
246. Sándor, Z., Kley, W., Klagyivik, P.: Stability and formation of the resonant system HD 73526. A&A **472**, 981–992 (2007). https://doi.org/10.1051/0004-6361:20077345
247. Saumon, D., Guillot, T.: Shock compression of deuterium and the interiors of Jupiter and Saturn. ApJ **609**, 1170–1180 (2004). https://doi.org/10.1086/421257
248. Schlichting, H.E., Fuentes, C.I., Trilling, D.E.: Initial planetesimal sizes and the size distribution of small Kuiper belt objects. AJ **146**, 36 (2013). https://doi.org/10.1088/0004-6256/146/2/36
249. Scholl, H., Marzari, F., Thébault, P.: Relative velocities among accreting planetesimals in binary systems: the circumbinary case. MNRAS **380**, 1119–1126 (2007). https://doi.org/10.1111/j.1365-2966.2007.12145.x
250. Seizinger, A., Kley, W.: Bouncing behavior of microscopic dust aggregates. A&A **551**, A65 (2013). https://doi.org/10.1051/0004-6361/201220946
251. Seizinger, A., Speith, R., Kley, W.: Compression behavior of porous dust agglomerates. A&A **541**, A59 (2012). https://doi.org/10.1051/0004-6361/201218855
252. Shen, Y., Turner, E.L.: On the eccentricity distribution of exoplanets from radial velocity surveys. ApJ **685**, 553–559 (2008). https://doi.org/10.1086/590548
253. Shu, F.H.: The physics of astrophysics. Volume II: Gas dynamics (1992)
254. Silburt, A., Rein, H.: Tides alone cannot explain Kepler planets close to 2:1 MMR. MNRAS **453**, 4089–4096 (2015). https://doi.org/10.1093/mnras/stv1924
255. Spiegel, D.S., Burrows, A.: Spectral and photometric diagnostics of giant planet formation scenarios. ApJ **745**, 174 (2012). https://doi.org/10.1088/0004-637X/745/2/174
256. Stevenson, D.J.: Formation of the giant planets. Planet. Space Sci. **30**, 755–764 (1982). https://doi.org/10.1016/0032-0633(82)90108-8
257. Stewart, S.T., Leinhardt, Z.M.: Velocity-dependent catastrophic disruption criteria for planetesimals. ApJ **691**, L133–L137 (2009). https://doi.org/10.1088/0004-637X/691/2/L133
258. Syer, D., Clarke, C.J.: Satellites in discs: regulating the accretion luminosity. MNRAS **277**, 758–766 (1995). https://doi.org/10.1093/mnras/277.3.758
259. Szulágyi, J., Morbidelli, A., Crida, A., Masset, F.: Accretion of Jupiter-Mass planets in the limit of vanishing viscosity. ApJ **782**, 65 (2014). https://doi.org/10.1088/0004-637X/782/2/65
260. Tanaka, H., Takeuchi, T., Ward, W.R.: Three-dimensional interaction between a planet and an isothermal gaseous disk. I. Corotation and lindblad torques and planet migration. ApJ **565**, 1257–1274 (2002). https://doi.org/10.1086/324713
261. Tanaka, H., Ward, W.R.: Three-dimensional Interaction between a planet and an isothermal gaseous disk. II. Eccentricity waves and bending waves. ApJ **602**, 388–395 (2004). https://doi.org/10.1086/380992
262. Thommes, E.W., Duncan, M.J., Levison, H.F.: Oligarchic growth of giant planets. Icarus **161**, 431–455 (2003). https://doi.org/10.1016/S0019-1035(02)00043-X
263. Toomre, A.: On the gravitational stability of a disk of stars. ApJ **139**, 1217–1238 (1964). https://doi.org/10.1086/147861
264. Tsukamoto, Y., Machida, M.N., Inutsuka, S.i.: Formation, orbital and thermal evolution, and survival of planetary-mass clumps in the early phase of circumstellar disc evolution. MNRAS **436**, 1667–1673 (2013). https://doi.org/10.1093/mnras/stt1684

265. Udry, S., Santos, N.C.: Statistical properties of exoplanets. ARA&A **45**, 397–439 (2007). https://doi.org/10.1146/annurev.astro.45.051806.110529
266. Uribe, A.L., Klahr, H., Flock, M., Henning, T.: Three-dimensional magnetohydrodynamic simulations of planet migration in turbulent stratified disks. ApJ **736**, 85 (2011). https://doi.org/10.1088/0004-637X/736/2/85
267. Vogt, S.S., Butler, R.P., Marcy, G.W., Fischer, D.A., Henry, G.W., Laughlin, G., Wright, J.T., Johnson, J.A.: Five new multicomponent planetary systems. ApJ **632**, 638–658 (2005). https://doi.org/10.1086/432901
268. Vorobyov, E.I.: Formation of giant planets and brown dwarfs on wide orbits. A&A **552**, A129 (2013). https://doi.org/10.1051/0004-6361/201220601
269. Vorobyov, E.I., Basu, S.: The burst mode of protostellar accretion. ApJ **650**, 956–969 (2006). https://doi.org/10.1086/507320
270. Vorobyov, E.I., Basu, S.: Formation and survivability of giant planets on wide orbits. ApJ **714**, L133–L137 (2010). https://doi.org/10.1088/2041-8205/714/1/L133
271. Wada, K., Tanaka, H., Suyama, T., Kimura, H., Yamamoto, T.: Numerical simulation of dust aggregate collisions. I. Compression and disruption of two-dimensional aggregates. ApJ **661**, 320–333 (2007). https://doi.org/10.1086/514332
272. Wada, K., Tanaka, H., Suyama, T., Kimura, H., Yamamoto, T.: Collisional growth conditions for dust aggregates. ApJ **702**, 1490–1501 (2009). https://doi.org/10.1088/0004-637X/702/2/1490
273. Wada, K., Tanaka, H., Suyama, T., Kimura, H., Yamamoto, T.: The rebound condition of dust aggregates revealed by numerical simulation of their collisions. ApJ **737**, 36 (2011). https://doi.org/10.1088/0004-637X/737/1/36
274. Walsh, K.J., Morbidelli, A., Raymond, S.N., O'Brien, D.P., Mandell, A.M.: A low mass for Mars from Jupiter's early gas-driven migration. Nature **475**, 206–209 (2011). https://doi.org/10.1038/nature10201
275. Ward, W.R.: Horsehoe orbit drag. In: Lunar and Planetary Science Conference, vol. 22, p. 1463 (1991)
276. Ward, W.R.: Protoplanet migration by nebula tides. Icarus **126**, 261–281 (1997). https://doi.org/10.1006/icar.1996.5647
277. Ward, W.R.: On Type III Protoplanet Migration. AGU Fall Meeting Abstracts (2004)
278. Weidenschilling, S.J.: Aerodynamics of solid bodies in the solar nebula. MNRAS **180**, 57–70 (1977)
279. Weidenschilling, S.J.: The distribution of mass in the planetary system and solar nebula. Ap&SS **51**, 153–158 (1977). https://doi.org/10.1007/BF00642464
280. Weidenschilling, S.J., Cuzzi, J.N.: Formation of planetesimals in the solar nebula. In: Levy, E.H., Lunine J.I., (eds.) Protostars and Planets III, pp. 1031–1060 (1993)
281. Weidenschilling, S.J., Spaute, D., Davis, D.R., Marzari, F., Ohtsuki, K.: Accretional evolution of a planetesimal swarm. Icarus **128**, 429–455 (1997). https://doi.org/10.1006/icar.1997.5747
282. Weidling, R., Güttler, C., Blum, J.: Free collisions in a microgravity many-particle experiment. I. Dust aggregate sticking at low velocities. Icarus **218**, 688–700 (2012). https://doi.org/10.1016/j.icarus.2011.10.002
283. Weidling, R., Güttler, C., Blum, J., Brauer, F.: The physics of protoplanetesimal dust agglomerates. III. Compaction in multiple collisions. ApJ **696**, 2036–2043 (2009). https://doi.org/10.1088/0004-637X/696/2/2036
284. Weizsäcker, C.F.von: Über die Entstehung des Planetensystems. Mit 2 Abbildungen. ZAp **22**, 319 (1943)
285. Wetherill, G.W., Stewart, G.R.: Formation of planetary embryos—effects of fragmentation, low relative velocity, and independent variation of eccentricity and inclination. Icarus **106**, 190 (1993). https://doi.org/10.1006/icar.1993.1166
286. Williams, J.P., Cieza, L.A.: Protoplanetary disks and their evolution. ARA&A **49**, 67–117 (2011). https://doi.org/10.1146/annurev-astro-081710-102548
287. Windmark, F., Birnstiel, T., Güttler, C., Blum, J., Dullemond, C.P., Henning, T.: Planetesimal formation by sweep-up: how the bouncing barrier can be beneficial to growth. A&A **540**, A73 (2012). https://doi.org/10.1051/0004-6361/201118475

288. Windmark, F., Birnstiel, T., Ormel, C.W., Dullemond, C.P.: Breaking through: the effects of a velocity distribution on barriers to dust growth. A&A **544**, L16 (2012). https://doi.org/10.1051/0004-6361/201220004
289. Winn, J.N., Fabrycky, D., Albrecht, S., Johnson, J.A.: Hot stars with Hot Jupiters have high obliquities. ApJ **718**, L145–L149 (2010). https://doi.org/10.1088/2041-8205/718/2/L145
290. Winn, J.N., Fabrycky, D.C.: The occurrence and architecture of exoplanetary systems. ARA&A **53**, 409–447 (2015). https://doi.org/10.1146/annurev-astro-082214-122246
291. Woolfson, M.M.: The Origin and Evolution of the Solar System. The Institute of Physics Publishing (2000)
292. Wu, Y., Murray, N.: Planet migration and binary companions: the case of HD 80606b. ApJ **589**, 605–614 (2003). https://doi.org/10.1086/374598
293. Wurm, G., Blum, J., Colwell, J.E.: NOTE: a new mechanism relevant to the formation of planetesimals in the solar nebula. Icarus **151**, 318–321 (2001). https://doi.org/10.1006/icar.2001.6620
294. Wurm, G., Paraskov, G., Krauss, O.: Growth of planetesimals by impacts at 25 m/s. Icarus **178**, 253–263 (2005). https://doi.org/10.1016/j.icarus.2005.04.002
295. Xiang-Gruess, M., Papaloizou, J.C.B.: Interaction between massive planets on inclined orbits and circumstellar discs. MNRAS **431**, 1320–1336 (2013). https://doi.org/10.1093/mnras/stt254
296. Youdin, A.N.: From grains to planetesimals. In: Montmerle, T., Ehrenreich, D., Lagrange, A.M. (eds.) EAS Publications Series. EAS Publications Series, vol. 41, pp. 187–207 (2010). https://doi.org/10.1051/eas/1041016
297. Youdin, A.N., Goodman, J.: Streaming instabilities in protoplanetary disks. ApJ **620**, 459–469 (2005). https://doi.org/10.1086/426895
298. Zhu, Z., Hartmann, L., Nelson, R.P., Gammie, C.F.: Challenges in forming planets by gravitational instability: disk irradiation and clump migration, accretion, and tidal destruction. ApJ **746**, 110 (2012). https://doi.org/10.1088/0004-637X/746/1/110
299. Zsom, A., Ormel, C.W., Güttler, C., Blum, J., Dullemond, C.P.: The outcome of protoplanetary dust growth: pebbles, boulders, or planetesimals? II. Introducing the bouncing barrier. A&A **513**, A57 (2010). https://doi.org/10.1051/0004-6361/200912976

MIX
Papier aus verantwortungsvollen Quellen
Paper from responsible sources
FSC® C105338

FSC
www.fsc.org

If you have any concerns about our products,
you can contact us on
ProductSafety@springernature.com

In case Publisher is established outside the EU,
the EU authorized representative is:
Springer Nature Customer Service Center GmbH
Europaplatz 3, 69115 Heidelberg, Germany

Printed by Libri Plureos GmbH
in Hamburg, Germany